I0048567

Fundamentals of RF and Microwave Circuit Design

Second Edition

Fundamentals of RF and Microwave Circuit Design

Practical Analysis and Design Tools

Second Edition

Manou Ghanevati

Fundamentals of RF and Microwave Circuit Design

Second Edition

ISBN 978-0-578-57530-8

Copyright @ 2019 by Manou Ghanevati

Published in USA

First Published 2019

All rights reserved. Printed and bound in the United States of America. No part of this book may be reproduced, stored in a retrieval system, transmitted in any form or by any means without written permission from the author.

Contents

Acknowledgements

The author is grateful to Professor Behagi in co-authoring the first edition of *Fundamentals of RF and Microwave Circuit Design* and his support throughout the improvement of the original manuscript.

The author would like to thank Professor Mohammad Tofighi of Penn State University for reviewing parts of the manuscript and making very valuable comments and suggestions.

Many thanks also to Mr. Mike Engelhardt, Mr. John Hamburger, and the supporting team at Linear Technology Corporation for their encouragement and positive remarks about this work.

Finally, the author is thankful to the individuals who reviewed the book in detail, found, and reported many typos and mistakes for both editions of this book.

Preface to the Second Edition

The second edition of *Fundamentals of RF and Microwave Circuit Design* attempts at exploring more aspects of this field in an easy to understand manner that is combined with practical examples and applications with the aid of open source software. A new chapter is dedicated to sinewave oscillator circuit design, and time-domain analysis is introduced for the first time. New appendices have been added to the book to discuss in details transmission line parameters and most two-port networks. Modifications have been done throughout the book and new chapter sections have been added. The reader will find out that the combination of novel and unique use of collection of software tools together with simple presentation of important and yet complex topics can provide a powerful methodology in designing and analyzing many passive and active RF/Microwave circuits. Microwave Engineering covers extremely vast subject with topics ranging from semiconductor physics to electromagnetic theory. The applications of each field are, however, diverse and unique. Unlike many texts on the subject this book does not attempt to cover every aspect of RF and microwave engineering in a single volume. Almost all subject matter covered in the text is accompanied by examples.

The organization of the book is as follows: Chapter 1 presents a general explanation of RF and microwave concepts and components. Engineering students will be surprised to find out that resistors, inductors, and capacitors at high frequencies are no longer ideal elements but rather a network of circuit elements. Wideband transformers and Baluns are also discussed for the first time. In chapter 2 the transmission line theory is developed and several important parameters are defined. Transmission line theory and parameters are discussed in detail in newly added Appendix A. Popular types of transmission lines are introduced and their parameters are examined. In Chapter 3 network parameters and the application of Smith chart as a graphical tool in dealing with impedance behavior and reflection coefficient are discussed. Various network parameters with examples are discussed in newly added Appendix B.

Description of RF and microwave networks in terms of their scattering parameters, known as S- Parameters, is introduced. The subject of lumped and distributed resonant circuits and filters are discussed in Chapter 4. An introduction to the vast subject of filter synthesis is also treated in this chapter. In Chapter 5 the condition for maximum power transfer and the lumped element impedance matching are considered. The analytical equations for matching two complex impedances with lossless two-element networks are derived. Both analytical and graphical techniques are used to design narrowband and broadband matching networks. In Chapter 6 both narrowband and broadband distributed matching networks are analytically and graphically analyzed. In Chapter 7 single-stage amplifiers are designed by utilizing two different impedance matching objectives. The first amplifier is designed for maxim gain where the input and the output are conjugate matched to the source and load impedance; the second amplifier is designed as a low noise amplifier where the transistor is selectively mismatched to achieve a specific Noise Figure. In Chapter 8 sinewave oscillator circuits are discussed. Several important and widely used oscillator circuits including Colpitts, Clad, and Crystal oscillators are introduced through examples and many practical issues have been tackled in detail. This chapter also covers new trends in modern IC technology and oscillator circuits for microwave applications. Application of Monte Carlo in engineering design process and sensitivity is also covered in chapter 8.

Chapters 7 and 8 are focus of active RF and microwave circuit design. As such, a spice model has been utilized for design of multiple amplifiers. A DC analysis has been performed first and transistor DC-IV curves have been generated for proper selection of DC operating points. An AC analysis is then followed to generate S-parameters at desired DC biasing condition. From simulated two port parameters, RF parameters of interest including stability factors can be generated using LTspice equation editor. Furthermore, using internal capability of the software, a model has been developed to simulate and predict noise figure of the LNA circuit. Time-domain including transient analysis and Fast Fourier Transform (FFT) make design and analysis of oscillator circuits in both time domain and frequency domain possible.

Manou Ghanevati
January 2020

Preface to the First Edition

Microwave Engineering can be a fascinating and fulfilling career path. It is also an extremely vast subject with topics ranging from semiconductor physics to electromagnetic theory. Unlike many texts on the subject this book does not attempt to cover every aspect of RF and microwave engineering in a single volume. This text covers the subject from a computer aided design standpoint. This includes topics such as lumped element components, transmission lines, impedance matching, and basic linear amplifier design. Almost all subject matter covered in the text is accompanied by examples that are solved using the Linear Technology software. University students will find this a potent learning tool. Practicing engineers will find the book very useful as a reference guide to quickly setup designs using the LTspice software. The authors thoroughly cover the basics as well as introducing CAD techniques that may not be familiar to some engineers. This includes subjects such as the frequent use of the MATLAB scripting capability.

The organization of the book is as follows: Chapter 1 presents a general explanation of RF and microwave concepts and components. Engineering students will be surprised to find out that resistors, inductors, and capacitors at high frequencies are no longer ideal elements but rather a network of circuit elements. For example, a capacitor at one frequency may in fact behave as an inductor at another frequency. In chapter 2 the transmission line theory is developed and several important parameters are defined. It is shown how to simulate and measure these parameters using LTspice software. Popular types of transmission lines are introduced and their parameters are examined. In Chapter 3 network parameters and the application of Smith chart as a graphical tool in dealing with impedance behavior and reflection coefficient are discussed.

Description of RF and microwave networks in terms of their scattering parameters, known as S- Parameters, is introduced. The subject of lumped and distributed resonant circuits and filters are discussed in Chapter 4. An introduction to the vast subject of filter synthesis is also treated in this chapter. In

Chapter 5 the condition for maximum power transfer and the lumped element impedance matching are considered. The analytical equations for matching two complex impedances with lossless two-element networks are derived. Both analytical and graphical techniques are used to design narrowband and broadband matching networks. In Chapter 6 both narrowband and broadband distributed matching networks are analytically and graphically analyzed. In Chapter 7 single-stage amplifiers are designed by utilizing two different impedance matching objectives. The first amplifier is designed for maxim gain where the input and the output are conjugate matched to the source and load impedance; the second amplifier is designed as a low noise amplifier where the transistor is selectively mismatched to achieve a specific Noise Figure.

LTspice is capable of linear and non-linear circuit simulation. As such, a spice model has been utilized for design of above amplifiers. A DC analysis has been performed first and transistor DC-IV curves have been generated for proper selection of DC operating points. An AC analysis is then followed to generate S-parameters at desired DC biasing condition. From simulated two port parameters, RF parameters of interest including stability factors can be generated using LTspice equation editor. Furthermore, using internal capability of LTspice, a model has been developed to simulate and predict noise figure of the LNA circuit.

Ali A. Behagi and Manou Ghanevati
June 2017

Chapter 1

RF and Microwave Concepts and Components

1.1 Introduction

An electromagnetic wave is a propagating wave that consists of electric and magnetic fields. The electric field is produced by stationary electric charges while the magnetic field is produced by moving electric charges. A time-varying magnetic field produces an electric field and a time-varying electric field produces a magnetic field. The characteristics of electromagnetic waves are frequency, wavelength, phase, impedance, and power density. In free space, the relationship between the wavelength and frequency is given by Equation (1-1).

$$\lambda = \frac{c}{f} \qquad (1\text{-}1)$$

In the MKS system, λ is the wavelength of the signal in meters, c is the velocity of light approximately equal to 300,000 kilometers per second, and f is the frequency in cycles per second, or Hz.

The electromagnetic spectrum is the range of all possible frequencies of electromagnetic radiation. They include radio waves, microwaves, infrared radiation, visible light, ultraviolet radiation, X-rays and gamma rays. In the field of RF and microwave engineering the term RF generally refers to Radio Frequency signals with frequencies in the 3 KHz to 300 MHz range. The term Microwave refers to signals with frequencies from 300 MHz to 300 GHz having wavelengths from 1 meter to 1 millimeter. The RF and microwave frequencies form the spectrum of all radio, television, data, and satellite communications. A typical spectrum chart may show the RF and microwave frequencies up through the extremely high frequency, EHF, range or 300 GHz. This text will focus on the RF and microwave frequencies as the foundation for component design techniques. The application of Linear Technology software will enhance the student's understanding of the underlying principles presented throughout the text. It also helps in the analysis and modeling of many active devices, passive components, and circuits. The practicing engineer will find the text an invaluable reference to the RF and microwave theory and techniques. The numerous examples enable the setup and design of many RF and microwave circuit design problems.

1.2 Straight Wire Inductance

A conducting wire carrying an AC current produces a changing magnetic field around the wire. According to Faraday's law the changing magnetic field induces a voltage in the wire that opposes any change in the current flow. This opposition to change is called self-inductance. At high frequencies even a short piece of straight wire possesses frequency dependent resistance and inductance behaving as a circuit element.

Example 1.1: Calculate the inductance of a three inch length of AWG #28 copper wire in free space.

Solution: The straight wire inductance can be calculated from the empirical Equation (1-2).

$$L = Kl\left(\ln\frac{4l}{D} - 0.75\right) nH \qquad (1\text{-}2)$$

where,

$l =$ Length of the wire

$D =$ Diameter of the wire

$K = 2$ for dimensions in cm and $K=5.08$ for dimensions in inches

The diameter of the AWG#28 wire is 0.0126 inches. Solving Equation (1-2) the inductance is calculated.

$$L = 5.08 \ (3)\left(\ln\frac{4\ (3)}{0.0126} - 0.75\right) = 93.1 \ nH$$

It is interesting to examine the reactance of the wire. We know that the reactance is a function of the frequency and is related to the inductance by the following equation.

$$X_L = 2\pi f L \quad \Omega \qquad (1\text{-}3)$$

where,

f is the frequency in Hz

L is the inductance in Henries

Calculating the reactance at 60 Hz, 1 MHz, and 1 GHz we can see how the reactive component of the wire increases dramatically with frequency. At 60 Hz the reactance is well below 1 Ω while at microwave frequencies the reactance increases to several hundred ohms.

60 Hz:	$X_L = 2\pi\,(60)(93.1\cdot10^{-9}) = 35\ \mu\Omega$
500MHz:	$X_L = 2\pi\,(10^6)(500)(93.1\cdot10^{-9}) = 292\ \Omega$
1 GHz:	$X_L = 2\pi\,(10^9)(93.1\cdot10^{-9}) = 585\ \Omega$

Material	Resistivity Relative to Copper	Actual Resistivity Ω-meters	Actual Resistivity Ω-inches
Copper, annealed	1.00	$1.68\cdot10^{-8}$	$6.61\cdot10^{-7}$
Silver	0.95	$1.59\cdot10^{-8}$	$6.26\cdot10^{-7}$
Gold	1.42	$2.35\cdot10^{-8}$	$9.25\cdot10^{-7}$
Aluminum	1.64	$2.65\cdot10^{-8}$	$1.04\cdot10^{-6}$
Tungsten	3.25	$5.60\cdot10^{-8}$	$2.20\cdot10^{-6}$
Zinc	3.40	$5.90\cdot10^{-8}$	$2.32\cdot10^{-6}$
Nickel	5.05	$6.84\cdot10^{-8}$	$2.69\cdot10^{-6}$
Iron	5.45	$1.00\cdot10^{-7}$	$3.94\cdot10^{-6}$
Platinum	6.16	$1.06\cdot10^{-7}$	$4.17\cdot10^{-6}$
Tin	52.8	$1.09\cdot10^{-7}$	$4.29\cdot10^{-6}$
Nichrome	65.5	$1.10\cdot10^{-6}$	$4.33\cdot10^{-5}$
Carbon	2083.3	$3.50\cdot10^{-5}$	$1.38\cdot10^{-3}$

Table 1-1 Resistivity of common materials relative to copper

It is a common practice in most commercial microwave software programs to specify resistivity in relative terms, compared to copper. Table 1-1 provides a reference of the materials used in microwave engineering.

1.3 Skin Effect in Conductors

At RF and microwave frequencies, due to the larger inductive reactance caused by the increase in flux linkage toward the center of the conductor, the current in the conductor is forced to flow near the conductor surface. As a result the amplitude of the current density decays exponentially with the depth of penetration from the surface. At low frequencies the entire cross sectional area is carrying the current. As the frequency increases to the RF and microwave region, the current flows much closer to the outside of the conductor. At the higher end of microwave frequency range, the current is essentially carried near the surface with almost no current at the central region of the conductor. The skin depth, δ, is the distance from the surface where the charge carrier density falls to 37% of its value at the surface. Therefore 63% of the RF current flows within the skin depth region. The skin depth is a function of the frequency and the properties of the conductor as defined by Equation (1-4). As the cross sectional area of the conductor effectively decreases the resistance of the conductor will increase [7]. Due to skin effect microwave components are often coated with a thin layer of good conductors, such as gold or copper, to improve their electrical performance.

$$\delta = \sqrt{\frac{\rho}{\mu \, \pi \, f}} \qquad (1\text{-}4)$$

where,

δ = skin depth

ρ = resistivity of the conductor

f = frequency

μ = permeability of the conductor

Use caution when solving Equation (1-4) to keep the units of ρ and μ consistent. Table 1-1 contains values of resistivity in units of Ω-meters and Ω-inches. The permeability μ is the permeability of the conductor. It is a property of a material to support a magnetic flux. Some reference tables will show relative permeability. In this case the relative permeability is normalized to the permeability of free space which is: $4\pi 10^{-7}$ Henries per meter. The relationship between relative permeability to the actual permeability is given in Equation (1-5). Most conductors have a relative permeability μ_r very close to one. Therefore conductor permeability μ is often given the same value as μ_o.

$$\mu = \mu_r \ \mu_o \tag{1-5}$$

where,

μ = actual permeability of the material

μ_r = relative permeability of material

μ_o = permeability of free space

Example 1.2: Calculate the skin depth of copper wire at a frequency of 25 MHz.

Solution: From Table 1-1, $\rho = 6.61 \cdot 10^{-7}$ Ω-inches. Using Equation (1-4), and converting the permeability from H/m to H/inch, the skin depth is:

$$\delta = \sqrt{\frac{6.61 \cdot 10^{-7}}{\left(3.19 \cdot 10^{-8}\right) \pi \left(25 \cdot 10^{6}\right)}} = 5.14 \cdot 10^{-4} \ inches$$

As the frequency increases, the current is primarily flowing in the region of the skin depth. It can be visualized that a wire would have greater resistance at higher frequencies due to the skin effect. The resistance of a length of wire is determined by the resistivity and the geometry of the wire as defined by Equation (1-6).

$$R \ = \ \frac{\rho l}{A} \quad \Omega \tag{1-6}$$

where,

ρ = Resistivity of the wire

$l = $ Length of the wire

$A = $ Cross sectional area

Example 1.3: Calculate the resistance of a 12 inch length of AWG #24 copper wire at DC and at 25 MHz.

Solution: Assume the wire diameter is 20.1 mils (0.0201 inches). The DC resistance is then calculated using Equation (1-6).

$$R_{DC} = \frac{(6.61 \cdot 10^{-7}) \cdot 12}{\pi \left(\frac{0.0201}{2}\right)^2} = 0.025 \ \Omega$$

To calculate the resistance at 25 MHz the cross sectional area of the conduction region must be redefined by the skin depth. We can refer to this as the effective area, A_{eff}.

$$A_{eff} = \pi \ (R^2 - r^2) \tag{1-7}$$

where, $r = R - \delta$.

For the AWG#24 wire at 25 MHz the A_{eff} is calculated as:

$$A_{eff} = \pi \left(\frac{0.0201}{2}\right)^2 - \pi \left(\left(\frac{0.0201}{2}\right) - 5.14 \cdot 10^{-4}\right)^2 = 3.14 \cdot 10^{-5} \ in^2$$

Then apply Equation (1-6) to calculate the resistance of the 12 inch wire at 25 MHz.

$$R_{25MHz} = \frac{12 \left(6.61 \cdot 10^{-7}\right)}{3.14 \cdot 10^{-5}} = 0.253 \ \Omega$$

We can see that the resistance at 25 MHz is more than 10 times greater than the resistance at DC.

1.4 Flat Ribbon Inductance

Flat ribbon style conductors are very common in RF and microwave engineering. Flat ribbon conductors are encountered in RF systems in the form of low inductance ground straps. Flat ribbon conductors can also be encountered in microwave integrated circuits (MIC) as gold bonding straps. When a very low inductance is required the flat ribbon or copper strap is a good choice. The flat ribbon inductance can be calculated from the empirical Equation (1-8).

$$L = Kl \left[\ln\left(\frac{2l}{W+T} \right) + 0.223\left(\frac{W+T}{l} \right) + 0.5 \right] nH \qquad (1\text{-}8)$$

where,

l = The length of the wire

K = 2 for dimensions in cm and K = 5.08 for dimensions in inches

W = the width of the conductor

T = the thickness of the conductor

1.5 Ideal and Leaded Resistors

The resistance of a material determines the rate at which electrical energy is converted to heat. In Table 1-1 we have seen that the resistivity of materials is specified in Ω-meters rather than Ω/meter. This facilitates the calculation of resistance using Equation (1-6). When working with low frequency or logic circuits we are used to treating resistors as ideal resistive components.

Example 1.4: Plot the impedance of a 50 Ohm ideal resistor in LTspice over a frequency range of 0 to 2 GHz.

Solution: For the 50 Ohm ideal resistor, the LTspice schematic is shown in Figure 1-1.

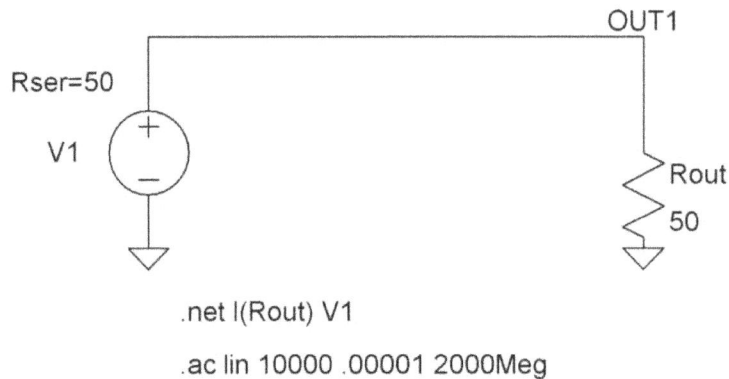

Figure 1-1 Schematic of an Ideal 50 Ω resistor

Simulate the schematic and display the input impedance over a frequency range of 0 to 2 GHz, as shown in Figure 1-2. The plot shows a constant resistance at all frequencies.

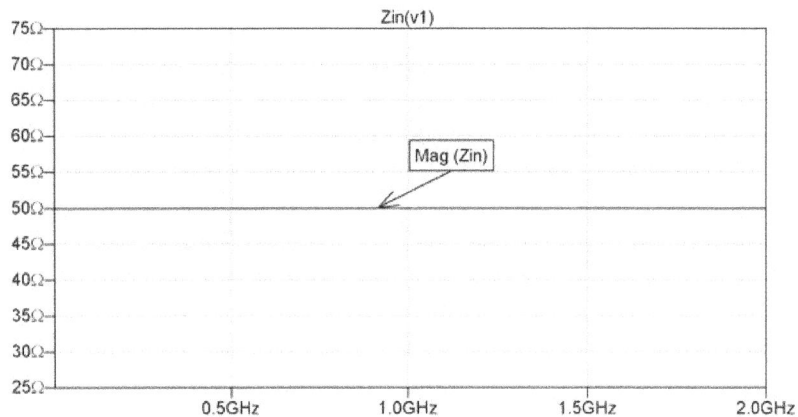

Figure 1-2 Ideal 50Ω resistor impedance versus frequency

Example 1.5: Plot the impedance of a 50 Ohm leaded resistor in LTspice over a frequency range of 0 to 2 GHz.

Solution: At RF and microwave frequencies however, the leaded resistors also possess inductive and capacitive elements. The stray inductance and capacitance associated with a resistor are often called parasitic elements. For a 1/8 watt leaded resistor it is not uncommon for each lead to have about 10 nH of inductance. The body of the resistor may exhibit 0.5 pF capacitance between the leads as shown in Figure 1-3.

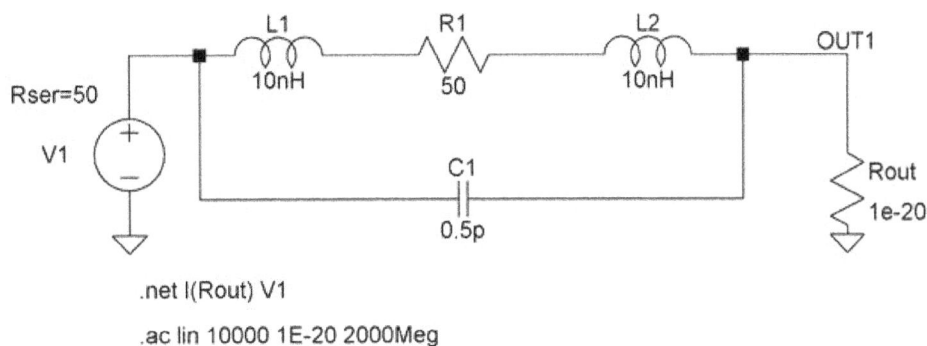

.net I(Rout) V1

.ac lin 10000 1E-20 2000Meg

Figure 1-3 Schematic of the leaded resistor with parasitic elements

Simulate the schematic and plot the input impedance as a function of frequency as shown in Figure 1-4.

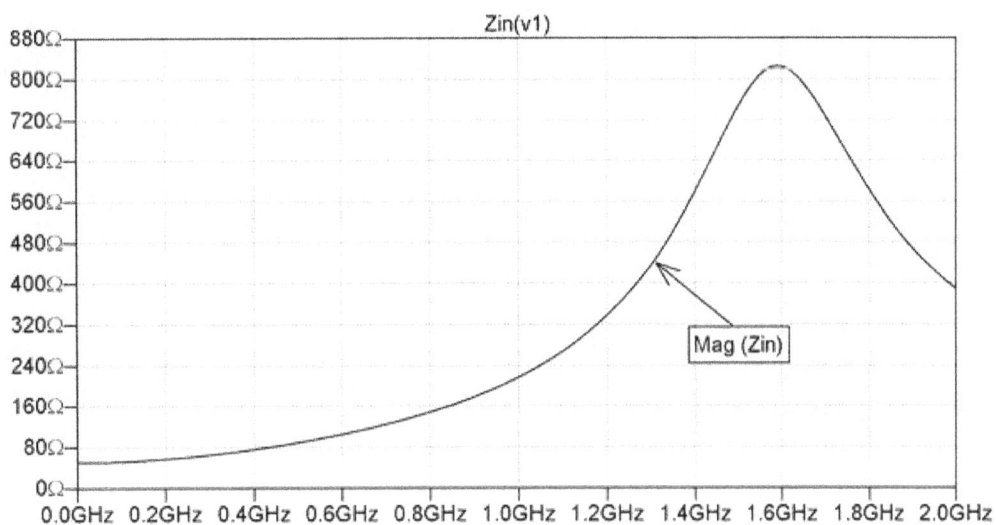

Figure 1-4 Leaded-resistor impedance versus frequency

1.6 Chip Resistors

Thick film resistors are used in most contemporary electronic equipment. The thick film resistor, often called chip resistor, comes close to eliminating much of the inductance that plagues the leaded resistor. The chip resistor works well with popular surface mount assembly techniques preferred in modern electronic manufacturing.

There are many types of chip resistors designed for specific applications. Common sizes and power ratings are shown in Table 1-2.

Size	Length x Width	Power Rating
0201	20mils x 10mils	50mW
0402	40mils x 20mils	62mW
0603	60mils x 30mils	100mW
0805	80mils x 50mils	125mW
1206	120mils x 60mils	250mW
2010	200mils x 100mils	500mW
2512	250mils x 120mils	1W

Table 1-2 Standard thick film resistor size and approximate power rating

The thick film resistor is comprised of a carbon based film that is deposited onto the substrate. Contrasted with a thin film resistor that is typically etched onto a substrate or printed circuit board, the thick film resistor can usually handle higher power dissipation. The ends of the chip have metalized wraps that are used to attach the resistor to a circuit board does not include native models for thick film resistors. Some manufacturers may provide models that can be incorporated into. There are also companies that specialize in developing CAD models of components such as Modelithics, Inc. Modelithics has a wide variety of component model libraries that can be incorporated into.

1.7 Inductor Q Factor

In section 1.2 we introduced the topic of inductance. The inductance of straight cylindrical and flat ribbon conductors was examined. The primary method of increasing inductance is not to simply keep increasing the length of a straight conductor but rather form a coil of wire. Forming a coil of wire increases the magnetic flux linkage and greatly increases the overall inductance. Because of the greater surrounding magnetic flux, inductors store energy in the magnetic field. Lumped element inductors are used in bias circuits, impedance matching networks, filters, and resonators. As we will see throughout this section inductors are realized in many forms including: air-core, toroidal and very small chip inductors. The concept of Q factor is introduced and will come up frequently in RF and microwave circuit design. It is a unit-less figure of merit that is used in circuits in which both reactive and resistive elements coexist.

Basically, the higher the Q factor, the less loss or resistance exists in the energy storage property. The inductor quality factor Q is defined as:

$$Q = \frac{X}{R_s} \qquad\qquad (1\text{-}9)$$

where,

X is the reactance of the inductor

R_s is the resistance in the inductor

At low RF frequencies the resistance comes primarily from the resistivity of the wire and as such is quite low. At higher frequencies the skin effect and inter-winding capacitance begin to influence the resistance and reactance thus causing the Q factor to decrease. In most applications we want as high a component Q factor as possible. We can increase the Q factor of inductors by using larger diameter wire or by silver plating the wire. In a multi-turn coil, the windings can be separated to reduce the inter-winding capacitance which in turn will increase the Q factor. Winding the coil on a magnetic core can increase the Q factor.

1.8 Air Core Inductors

Forming a wire on a removable cylinder is the basic realization of the air core inductor. When designing an air-core inductor, use the largest wire size and close spaced windings to result in the lowest series resistance and high Q. The basic empirical equation is given by Equation (1-10)

$$L = \frac{(17)N^{1.3}(D+D1)^{1.7}}{(D1+S)^{0.7}}$$
(1-10)

where,

N = Number of turns of wire

D = Core form diameter in inches

$D1$ = Wire diameter in inches

L = Coil inductance in nH

S = Spacing between turns in inches

Example 1.6: Calculate the amount of inductance that we can realize in that same three inches of wire if we wind it around a core to form an inductor. Choose a core form of 0.095 inches diameter as a convenient form to wrap the wire around.

Solution: First we need to calculate the approximate number of turns that we can expect to have with the three inch length of wire. We know that the circumference of a circle is related to the diameter by the following equation.

$$Circumference = \pi\,(Diameter) = \pi\,(0.095) = 0.2985 \quad inches$$

With a circumference of 0.2985 inches we can calculate the approximate number of turns that we can wrap around the 0.095 inch core with three inches of wire.

$$N = \frac{3}{0.2985} = approximately\ 10\ turns$$

From Equation (1-10) we can see that the spacing between the turns has a strong effect on the value of the inductance that we can expect from the coil. When hand winding the coil, it may be difficult to maintain an exact spacing of zero inches between the turns. Therefore it is useful to solve Equation (1-10) in terms of a variety of coil spacing so that we can see the effect on the inductance. We can solve Equation (1-10) in MATLAB for a variety of coil spacing, as follows.

1. Enter design parameters

D= 0.095; D1=0.0126; N=10; Spacing = [0;0.002;0.004;0.006;0.008;0.010];

2. Calculate the coil length and inductance in nano Henries

Inductance_nH = (17*(N^1.3)*((D+D1)^1.7))/((D1+Spacing)^.7)

Coil_Length = (D1*N) + (Spacing.*(N-1))

Note that the coil spacing is variable and Spacing has been defined as an array variable. Placing a semicolon between the values makes the array organized in a column format. This is handy for viewing the results in tabular format. Placing a comma between the values would organize the array in a row format. As an aid in forming the coil, the overall coil length is also calculated. The coil length is simply the summation of the overall wire thickness times the number of turns and the spacing between the turns.

Add the calculated inductance, spacing, and coil length to a table as shown in Table1-3.

Index	Inductance_nH	InductanceCalculator.Spacing	InductanceCalculator.Coil_Length
1	163.784	0	0.126
2	147.735	2e-3	0.144
3	135.037	4e-3	0.162
4	124.701	6e-3	0.18
5	116.097	8e-3	0.198
6	108.806	0.01	0.216

Table 1-3 Coil inductance, spacing and coil length

The Table1-3 shows that the coil inductance with no spacing between the turns is 163.78 nH and it is 0.126 inches long. Contrast this to the 93.1 nH inductance with the same three inches of wire in a straight length. We can clearly see the dramatic impact of the magnetic flux linkage in increasing the inductance by forming the wire into a coil. The Figure also shows the strong influence of the inter-winding capacitance in influencing the inductance of the coil. Just 10 mils spacing between the turns reduces the coil's inductance from 163.78 nH to 108.8 nH. In practice this is an effective means to tune the inductor's value in circuit. When designing and building the inductor it is necessary to solve Equation (1-10) for the number of turns given a desired value of inductance. The procedure in MATLAB script follows.

1. Enter design parameters

D = 0.095; D1 = 0.0126; Inductance_nH = 163.784;
Spacing = [0;.002;.004;.006;.008;.010]

2. Calculate the number of turns and coil length

N = ((((D1+Spacing)^.7)*Inductance_nH)/(17*((D+D1)^1.7)))^.7692
Coil_Length = (D1*N) + (Spacing.*(N-1))

Note that there is one subtle difference namely the calculation of the coil length. In this case both variables, Spacing and N, are array variables. When multiplying array variables use a period in front of the multiplication sign to signify that this is an operation on arrays. This makes sure that the correct array index is maintained between the variables. In the previous equations this notation was not necessary because N was a constant. The tabulated results are shown in Table 1-4.

Index	N	Number_of_Turns.Spacing	Number_of_Turns.Coil_Length
1	9.999	0	0.126
2	10.825	2e-3	0.156
3	11.599	4e-3	0.189
4	12.332	6e-3	0.223
5	13.029	8e-3	0.26
6	13.696	0.01	0.3

Table 1-4 Number of turns, coil spacing and length

Example 1.7 Create a simple RLC network in LTspice that gives a resonator response around 1 GHz. Such a circuit is shown in Figure 1-5.

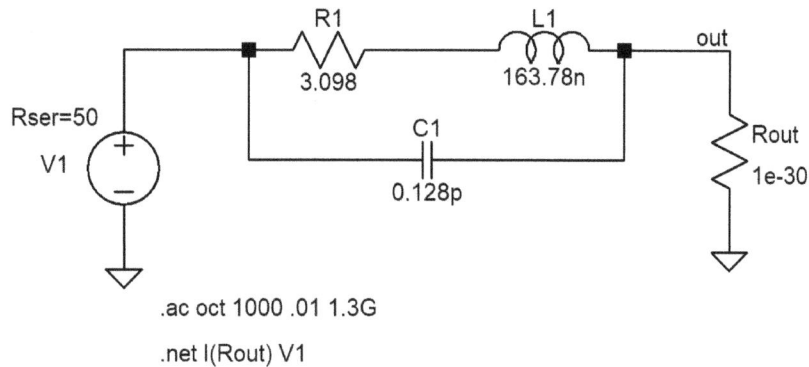

Figure 1-5 Equivalent model of the air core inductor

The simulated response is given in Figure 1-6.

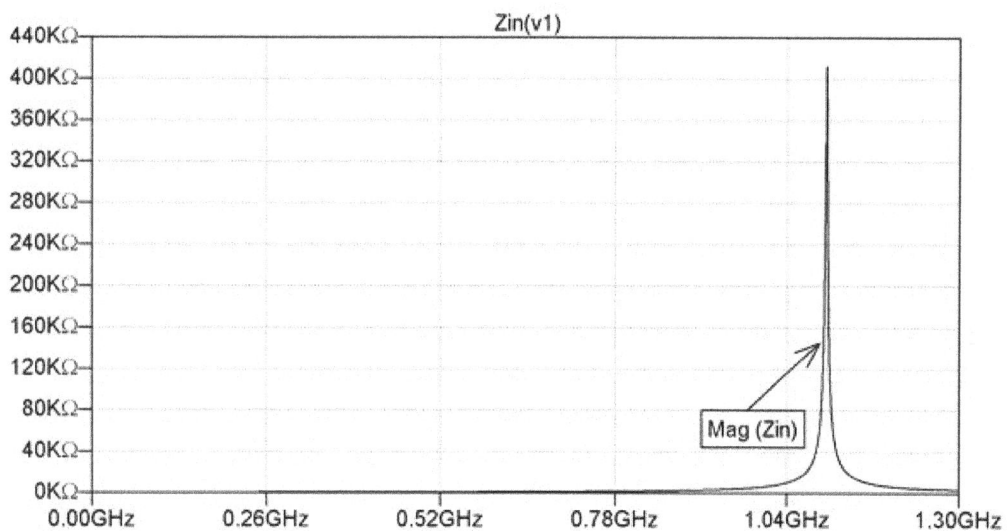

Figure 1-6 Input impedance of the air core inductor

Resonant circuits are covered in detail in chapter 4 but it is important to understand that each individual component such as the air core inductor has its own resonant frequency. The resonant frequency is the frequency at which the inductive reactance and capacitive reactance are equal and cancel one another. When this condition occurs in the inductor it is a parallel resonant circuit which results in a very high real impedance. If we plot the reactance along with the impedance a very interesting response is obtained. This response is shown in Figure 1-7.

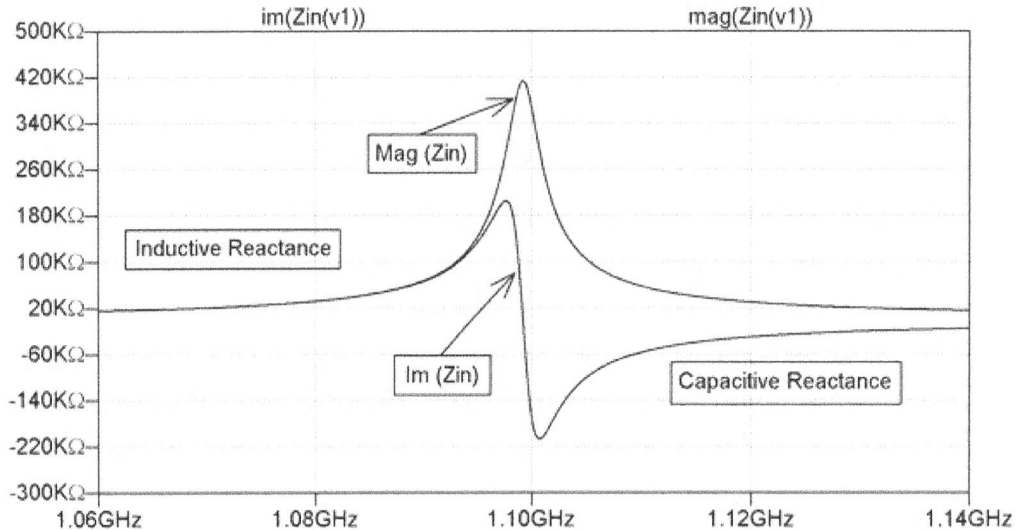

Figure 1-7 Impedance and reactance of the air core inductor

The reactance, up to the resonant frequency, is positive but beyond resonance the reactance becomes negative. From basic circuit theory we know that a negative reactance is associated with a capacitor. Therefore above the Self Resonant Frequency, SRF, the inductor actually becomes a capacitor. In practice we want to make sure that our inductor really behaves like an inductor. A good design practice is to keep this self-resonant frequency about four times higher than the frequency of operation. However using the inductor near its resonant frequency might make a good choke. A choke is a high reactance inductor often used to feed voltage to a circuit in which all RF energy is blocked from the DC side of the circuit. The inductor manufacturer will typically specify SRF of the inductor. It is important to remember that the inductor's SRF is the parallel resonant frequency; not the series resonant frequency.

1.9 Air Core Inductor Q Factor

Example 1.8: Plot the Q factor of the air core inductor in Figure 1-5 versus frequency from 0 to 1.3 GHz.

Solution: The plot of inductor Q factor versus frequency is shown in Figure 1-8.

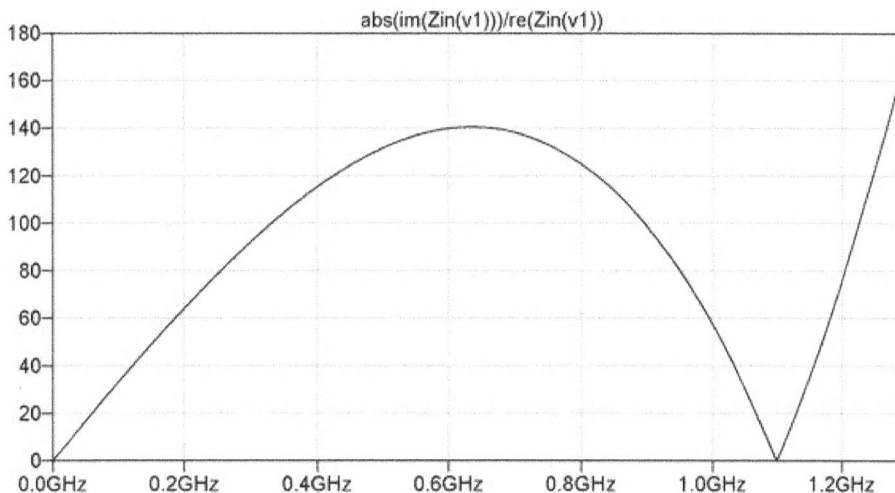

Figure 1-8 Air core inductor Q factor versus frequency

It is interesting to note that the Q factor peaks at a frequency well below the self-resonant frequency of the inductor. The actual frequency at which the Q factor peaks will vary among inductor designs but is usually ranges from 2 to 5 times less than the SRF. Close winding spacing results in inter-winding capacitance, which lowers the self-resonant frequency of the inductor. Thus there is a tradeoff between maximum Q factor and high self resonant frequency. It is also noteworthy that the Q factor goes to zero at the self-resonant frequency. Figure 1-8 shows the Q factor monotonously increasing beyond the self-resonant frequency. This is erroneous and is due to the fact that the model used to simulate the inductor's performance is invalid beyond the self-resonance. The Air Core inductor model uses a simplified network similar to the one shown in Figure 1-5. Beyond self resonance the complexity and number of ideal elements need to increase in order to accurately model the inductor. A resistor needs to be added in series with the capacitor to begin to model the Q factor because the inductor is becoming a capacitor above self resonance. For most practical work however the native model will work fine because we should be using the inductor well below the self-resonant frequency. A technique commonly used by microwave engineers to increase the Q factor of an inductor is to silver plate the wire. This can be modeled by setting Rho = 0.95 in the inductor model.

1.10 Chip Inductors

The inductor core does not have to be air. Other materials may be used as the core of an inductor. Similar in size to the chip resistor there is a large assortment of chip inductors that are popular in surface mount designs. The chip inductor is a form of dielectric core inductor. There are a variety of modeling techniques used for chip inductors. One of the more popular modeling techniques is with the use of S parameter files. The subject of S parameters is covered in chapter 3. At this point consider the S parameter file as an external data file that contains an extremely accurate network model of the component. Most component manufacturers provide S parameter data files for their products. It is a good practice to always check the manufacturer's website for current S parameter data files. CoilCraft, Inc. is one manufacturer of chip inductors.

The characteristic differences among the various sizes are more difficult to quantify than the chip resistors. A careful study of the data sheets is required for the proper selection of a chip inductor. In general the larger chip inductors will have higher inductance values. Often the smaller chip inductors will have higher Q factor. The impedance and self-resonant frequency can vary significantly across the sizes as well as the current handling capability. The plot of the Q factor can be derived from the S parameter file given by manufacturers. Manufacturers will often plot the Q vs frequency on a logarithmic scale. It is very easy to change the x-axis to a logarithmic scale on the rectangular graph properties window. This allows us to have a visual comparison to the manufacturer's catalog plot.

1.11 Magnetic Core Inductors

We have seen that the inductance of a length of wire can be increased by forming the wire into a coil. We can make an even greater increase in the inductance by replacing the air core with a magnetic material such as ferrite or powdered iron. The magnetic field around an inductor is characterized by the magnetic force H, and the magnetic flux B. They are related by the level of the applied signal and the permeability, μ, of the core material. This relationship is given by Equation (1-11).

$$B = \mu \, H \qquad\qquad (1\text{-}11)$$

where,

> B = Flux density in Gauss
> H = Magnetization intensity in Oersteds
> μ = Permeability in Webers/Ampere-turn

This relationship is nonlinear in that as H increases, the amount of flux density will eventually level off or saturate. We will consider the linear region of this relationship throughout the discussion of this text. In an iron core inductor the permeability of the magnetic core is much higher than an air core and produces a high flux density. The rod inductor has magnetic flux outside of the core as well as inside the core. Rod inductors that are used in tuned circuits generally require a metal shield around the inductor to contain this magnetic flux so that it does not interfere or couple to adjacent circuits and other inductors. The toroidal inductor flux remains primarily inside the core material. This suggests that the toroidal inductor experiences less loss and should have higher Q factor. This also gives the toroidal inductor a self-shielding characteristic and does not require a metallic enclosure. Because of its self-shielding properties and high Q factor the toroidal inductor is one of the most popular of all magnetic core inductors. However one advantage of the rod inductor is that it is much easier to tune. The coil can be wound on a hollow plastic cylindrical form in which the magnetic rod can be placed inside. The rod is then free to move longitudinally which can tune the inductance value. Core materials are characterized by their permeability. Core permeability can vary quite a bit with frequency and temperature and can be confusing to specify for a given application. The stability of the permeability can change with the magnetic field due to DC current or RF drive through the inductor. As the frequency increases the permeability eventually reduces to the same value as air. Therefore iron core inductors are used only up to about 200 MHz. In general powdered iron can handle higher RF power without saturation and permanent damage. Ferrite cores have much higher permeability. The higher permeability of ferrite results in higher inductance values but lower Q factors. This characteristic can be advantageous in the design of RF chokes and broad band transformers. For inductors used in tuned circuits and filters, however, the higher Q factor of powdered iron is preferred. The powdered iron cores are manufactured in a variety of mixes to achieve different characteristics. The iron powders are made of hydrogen reduced iron and have greater permeability and lower Q factor. These cores are often used in RF chokes, electromagnetic interference (EMI) filters, and switched mode power supplies. Carbonyl iron tends to have better temperature stability and more constant permeability over a wide range of power. At the same time the Carbonyl iron maintains very good Q factor making them very popular in RF circuits. These characteristics lead to the popularity of toroidal inductors of Carbonyl iron for the manufacture of RF inductors. There is a wide variety of sizes and mixtures of Carbonyl iron that are used in the design of toroidal inductors. A few of the popular sizes that are manufactured by Micrometals Inc. are shown in Table 1-5.

Core Designator	OD, inches	ID, inches	Height, inches
T30	0.307	0.151	0.128
T37	0.375	0.205	0.128
T44	0.440	0.229	0.159
T50	0.500	0.303	0.190
T68	0.690	0.370	0.190
T80	0.795	0.495	0.250
T94	0.942	0.560	0.312
T106	1.060	0.570	0.437
T130	1.300	0.780	0.437
T157	1.570	0.950	0.570
T200	2.000	1.250	0.550
T300	3.040	1.930	0.500
T400	4.000	2.250	0.650

Table 1-5 Partial listing of popular toroidal cores with designators

The inductance per turn of a toroidal inductor is directly related to its permeability and the ratio of its cross section to flux path length as given by Equation (1-12) [4].

$$L = \frac{4 \pi N^2 \mu A}{length} \quad nH \tag{1-12}$$

where,

L_{nH} = inductance

μ = permeability

A = cross sectional area

$length$ = flux path length

N = number of turns

As Equation (1-12) shows the inductance is proportional to the square of the turns. A standard specification used by toroid manufacturers for the calculation of inductance is the inductive index, A_L. The inductive index is typically given in units of nH/turn. The inductance can then be defined by Equation (1-13).

$$L = N^2 A_L \quad nH \tag{1-13}$$

Some manufacturers specify the A_L in terms of uH or mH. To convert among the three quantities use the following guideline.

$$\frac{1 \, nH}{turn} = \frac{10 \, uH}{100 \, turns} = \frac{1 \, mH}{1000 \, turns} \tag{1-14}$$

Example 1.9: Design a 550 nH inductor using the Carbonyl W core of size T30. Determine the number of turns of the inductor.

Solution: From the manufacturer's data sheet the A_L value is 2.5 for a T30-10 toroidal core. Rearranging Equation (1-13) to solve for the number of turns, we find that 14.8 turns are required.

$$N = \sqrt{\frac{L}{A_L}} = \sqrt{\frac{550 \, nH}{2.5}} = 14.8$$

To reduce the winding loss we want to use the largest diameter of wire that will result in a single layer winding around the toroid. Equation (1-15) will give us the wire diameter.

$$d = \frac{\pi \ ID}{N + \pi} \qquad\qquad (1\text{-}15)$$

where,

d = Diameter of the wire in inches
ID = Inner diameter of the core in inches (from Table 1-5)
N = Number of turns

Therefore,

$$d = \frac{\pi \ (0.151)}{14.8 + \pi} \ = \ \frac{0.4744}{19.942} \ = \ 0.0238 \quad inches$$

AWG#23 wire is the largest diameter wire that can be used to wind a single layer around the T30 toroid. Normally AWG#24 is chosen because this is a more readily available standard wire size. The toroidal inductor model requires a few more pieces of information. The model requires that we enter the total winding resistance, core Q factor, and the frequency for the Q factor, F_q. As an approximation, set F_q to about six times the frequency of operation. In this case set F_q to $(6)(25)$ MHz = 150 MHz. Then tune the value of Q to get the best curve fit to the manufacturer's Q curve. We know that we have 14.8 turns on the toroid but we need to calculate the length of wire that these turns represent. The approximate wire length around one turn of the toroid is calculated from the following equation.

$$Length = \left[(2) \, Height + (OD - ID) \right] (\#turns) \qquad\qquad (1\text{-}16)$$

Using the dimensions for the T30 toroid from Table 1-5 we can calculate the total length of the wire as 6.10 inches.

$$\left[(2)(0.128) + (0.307 - 0.151) \right] \cdot 14.8 = 6.10 \quad inches$$

1.12 Single Layer Capacitor

The single layer capacitor is one of the simplest and most versatile of the surface mount capacitors. It is formed with two plates that are separated by a single dielectric layer. Most of the electric field (E) is contained within the dielectric however there is a fraction of the E field that exists outside of the plates. This is known as the fringing field.

The capacitance formed by a dielectric material between two parallel plate conductors is given by Equation 1-17.

$$C = (N-1)\left(\frac{KA\varepsilon_r}{t}\right)(FF) \quad pF \qquad (1\text{-}17)$$

where,

A = plate area

ε_r = relative dielectric constant

t = separation

K = unit conversion factor; 0.885 for cm and 0.225 for inches

FF = fringing factor; 1.2 when mounted on microstrip

N = number of parallel plates.

Example 1.10: Design a single layer capacitor from a dielectric that is 0.010 inches thick and has a dielectric constant of three. Each plate is cut to 0.040 inches square.

Solution: When the capacitor is mounted with at least one plate on a large printed circuit board track, a value of 1.2 is typically used in calculation. The single layer capacitor can be modeled as the Thin Film Capacitor.

$$C = (2-1)\left(\frac{(0.225)(0.04)(0.04)(3)}{0.010}\right)(1.2) = 0.13 \, pF$$

The ceramic dielectrics used in capacitors are divided into two major classifications. Class 1 dielectrics have the most stable characteristics in terms of temperature stability. Class 2 dielectrics use higher dielectric constants which result in higher capacitance values but have greater variation over temperature. The temperature coefficient is specified in either percentage of nominal value or parts per million per degree Celsius (ppm/°C). Ceramic materials with a high dielectric constant tend to dominate RF applications with a few exceptions. NPO (negative-positive-zero) is a popular ceramic that has extremely good stability of the nominal capacitance versus temperature.

Dielectric Material	Dielectric Constant
Vacuum	1.0
Air	1.004
Mylar	3
Paper	4 - 6
Mica	4 - 8
Glass	3.7 - 19
Alumina	9.9
Ceramic (low ε_r)	10
Ceramic (high ε_r)	100 – 10,000

Table 1-6 Dielectric constants of materials

1.13 Capacitor Q Factor

RF losses in the dielectric material of a capacitor are characterized by the dissipation factor. The dissipation factor is also referred to as the loss tangent and is the ratio of energy dissipated to the energy stored over a period of time. It is essentially the capacitor's efficiency rating. The dissipation factor and other ohmic losses lead to a parameter known as the Equivalent Series Resistance, ESR. The dissipation factor is the reciprocal of the Q factor. Just as we have seen with resistors and inductors, the physical model of a capacitor is a network of R, L, and C components.

Example 1.11: Calculate the Q factor versus frequency for the physical model of an 8.2 pF chip capacitor shown in Figure 1-9.

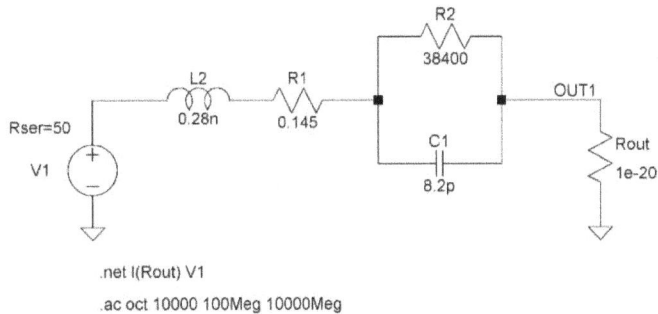

Figure 1-9 Physical model of the 8.2 pF chip capacitor

The input impedance of the physical model is shown in Figure 1-10.

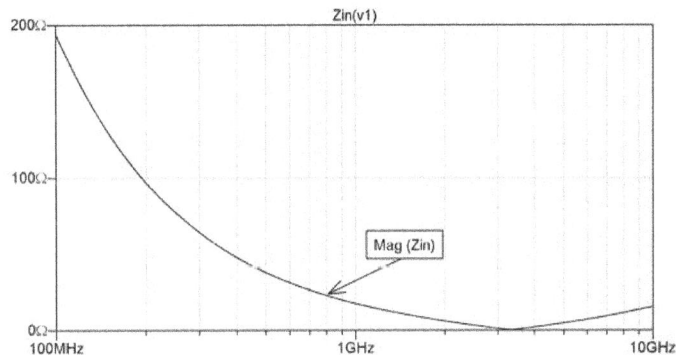

Figure 1-10 Physical 8.2 pF chip capacitor impedance

The capacitor has a series inductance and resistance component along with a resistance in parallel with the capacitance. The parallel resistor sets the losses in the dielectric material. The series resistance and inductance represent any residual lead inductance and ohmic resistances.

Solution: The values entered for the physical model and the Q factor can be obtained from the capacitor manufacturer. The plot of Figure 1-10 shows the impedance of the capacitor versus frequency. Note that the impedance decreases as would be expected until the self-resonant frequency is reached. Above the self-resonant frequency the impedance begins to increase suggesting that the capacitor is behaving as an inductor. The self-resonant frequency is due to the series inductance resonating with the capacitance. At resonance the reactance cancels leaving only the resistances *R1* and *R2*. The parallel resistance, *R2*, can be converted to an equivalent series resistance by Equation (1-18).

$$R2' = \frac{1}{1+Q^2} \ R2 \qquad (1\text{-}18)$$

These two series resistances can then be added to find the equivalent series resistance, ESR, as defined by Equation (1-19).

$$ESR = R1 + R\acute{2} \qquad (1\text{-}19)$$

The capacitor Q factor is then calculated by Equation (1-20). X_T is the total series reactance of the inductive and capacitive reactance.

$$Q = \frac{X_T}{ESR} \qquad (1\text{-}20)$$

As Figure 1-10 shows, the 8.2 pF chip capacitor has a series resonant frequency of over 3 GHz. An improved model for analyzing the capacitor's characteristics over a wide frequency range is to use the model for capacitor with Q. Using this model the Q factor of the capacitor can be made proportional to the square root of the applied frequency. The inductance, capacitance, and Q factor can be calculated from the impedance using the following equations.

X_T = Reactance = im(Zin1)

Resistance = re(Zin1)

Inductance = abs((Reactance)/(freq*2*pi))

Qfactor = abs(Inductance/Resistance)

Capacitance = 1/(abs(Reactance)*freq*2*pi)

Example 1.12 Calculate the Q factor versus frequency for the modified model of an 8.2 pF chip capacitor shown in Figure 1-11.

Solution: The modified model is shown in Figure 1-11.

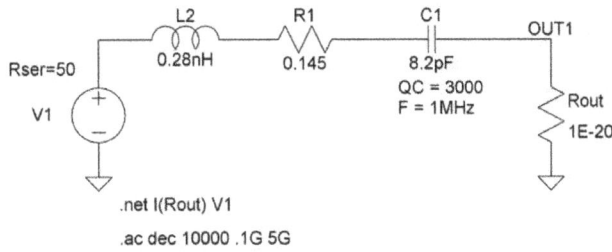

Figure 1-11 Modified physical model of 8.2 pF chip capacitor

The capacitor Q factor of the modified model versus frequency is plotted in Figure 1-12.

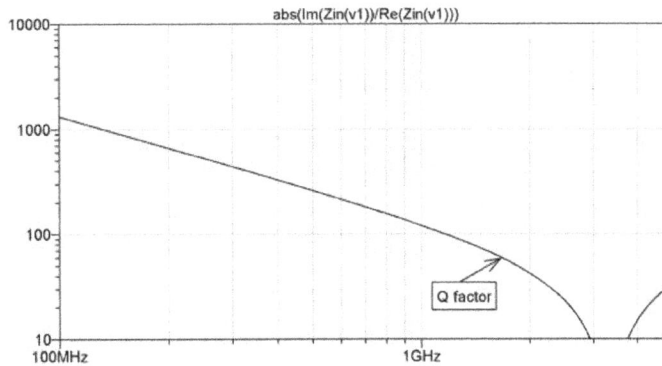

Figure 1-12 Calculated Q factor of 8.2 pF chip capacitor

Figure 1-12 shows the large dependence of the capacitor Q on frequency. The Q factor goes to zero at the self-resonant frequency. Above the self-resonant frequency the Q factor is undefined.

Calculate the effective capacitance from the total reactance of the model using Equation (1-21).

$$C = \frac{1}{2\pi F X_T} \qquad (1\text{-}21)$$

The plot of Figure 1-13 reveals some interesting characteristics about the chip capacitor. From 100 MHz to 300 MHz the capacitance value is fairly constant. As the frequency increases we see that the capacitance actually increases. The parasitic inductive reactance of the capacitor package actually makes the effective capacitance greater than its nominal value.

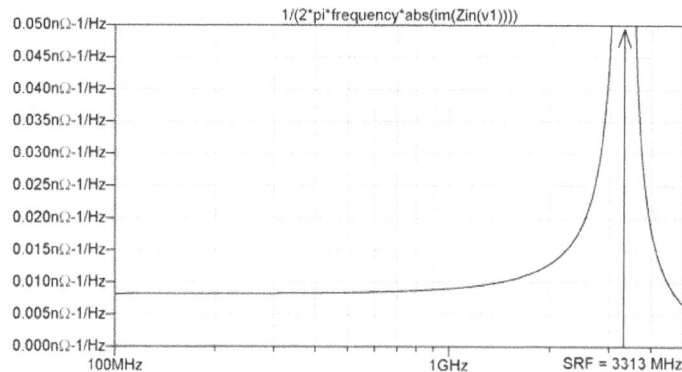

Figure 1-13 Effective capacitance of the 8.2 pF chip capacitor

This is a property of the capacitor that is not always intuitive. As the frequency approaches the self-resonant frequency (SRF) the capacitance rapidly approaches infinity. The capacitor actually becomes nearly a short circuit to RF at the self-resonant frequency. This is an important property of the capacitor that is used frequently in RF and microwave design. In RF coupling or bypass capacitor applications, capacitors are very often used at or near their self resonant frequency. The capacitor's self resonant frequency is due to the series resonant circuit. In applications requiring bypassing over a wide range of frequencies it is often necessary to use several capacitors each selected to have a uniquely spaced self resonant frequency. In filter and other tuned circuit applications where we want the chip capacitor to appear as an 8.2 pF capacitor, we clearly must stay well below the self-resonant frequency of the capacitor. A typical rule-of-thumb is to use the capacitor over a frequency range up to 35% of the self-resonant frequency. Therefore the 8.2 pF chip capacitor with a self-resonant frequency of 3321 MHz would be used as a capacitor in tuned circuits up to a frequency of about 1006 MHz.

1.14 Wideband Transformers

Linear power amplifiers (PAs) and push-pull amplifiers are often required to operate over large range of frequencies [10-12]. For certain applications such as for

Class B and D PAs, transformers are required to have good frequency response over three times the operating output frequency. For above applications as well as in push-pull amplifiers, conventional transformers cannot be used due to their inherent bandwidth limitations. Instead, wideband transformers and *Baluns* made of transmission lines are used for proper impedance transformation between the device output and the load. Figure 1-14 shows configurations for two transmission line transformers [10]. Figure 1-14(a) depicts a 1:1 balun, in which a balanced differential voltage is transformed to an unbalanced single-ended output. Equal impedance R looking into the differential port is being transformed to the load R connected to ground. In Figure 1-14(b) a unbalanced impedance equal to 4R is being transformed to a load impedance equal to R. In both figures, currents and voltages at various nodes are shown.

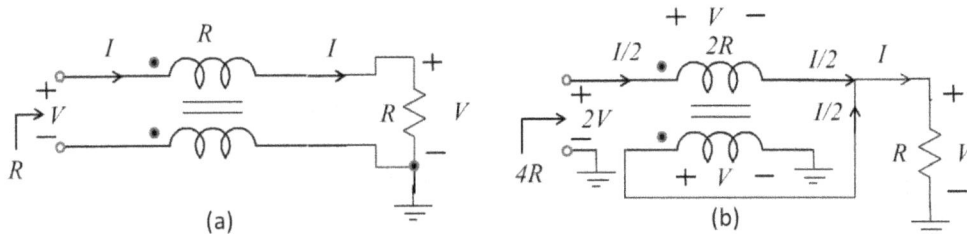

Figure 1-14 Transmission-line transformer configurations. (a) Balun (1:1): (b) 4:1

For impedance transformation between differential outputs and the single-ended load, baluns of various ratios are used. An example for a 4:1 Guanella balun is shown in Figure 1-15. One simplified construction method for such a wideband transformer is depicted in Figure 1-16. In practice, a third wire (not connected to either input or output) is used between the other two wires to achieve desired line impedance. The core material of high permeability material provides isolation for the transformer and immunity to radiation effects.

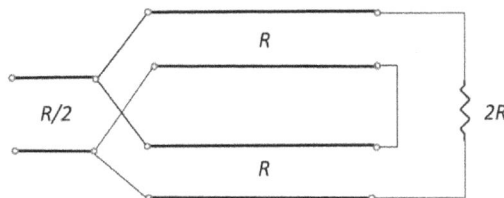

Figure 1-15 A simplified schematic diagram of a 4:1 Guanella balun [12]

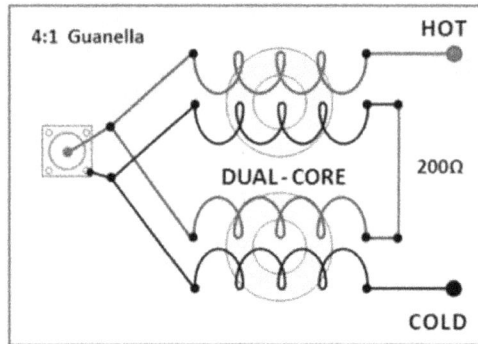

Figure 1-16 A more detailed schematic diagram of a Dual-Core 4:1 Guanella balun [Courtesy of Amateur Radio Practical Solutions]

Finally, high frequency integrated circuit technologies allow for design and fabrication of these transformers on chip using their corresponding process.

Problems

1-1.   Calculate the wavelength of an electromagnetic wave operating at a frequency of 1000MHz.

1-2.   Calculate the inductance of a 1 inch length of AWG #30 straight copper wire.

1-3.   Calculate the resistance of a 12 inch length of AWG #24 copper wire at DC and at 50MHz

1-4.   Find the skin depth and the resistance of a 1 meter length of copper coaxial line at 1 GHz. The inner conductor radius is 1 mm and the outer conductor is 4 mm.

1-5.   Calculate the inductance of a 1 inch length of copper flat ribbon conductor. The dimensions of the ribbon are 0.100 inches in width and 0.002 inches thick.

1-6.   Design an air core inductor with an inductance value of 56 nH. Use a copper wire of 0.050 inch diameter wound on a core diameter or 0.100

inch. Determine the number of turns required assuming a tight spaced winding.

1-7. Using the inductor from Problem 1-6, determine the self-resonant frequency of the inductor and comment on the maximum frequency in which the inductor may be used in a tuned circuit application.

1-8. Using the inductor from Problem 1-6, determine the maximum Q factor of the inductor and the frequency at which the maximum Q factor is obtained.

1-9. 1-13. Design a 1mH toroidal inductor on a Carbonyl W core size T30. Determine the maximum wire size that could be used to realize a single layer winding.

1-10. Using the inductor from Problem 1-9, model the inductor and determine the approximate self-resonant frequency. Comment on the maximum usable frequency of the inductor in the front end of a radio receiver.

1-11. A 0.05pF capacitor is required to couple a transistor to the resonator of a microwave oscillator. Design a single layer capacitor using a 0.020 inch thick dielectric with ε_r =2.2. Determine the dimensions of the capacitor assuming square footprint is desired.

1-12. For the single layer capacitor of Problem 11, determine the dimensions of the capacitor with a dielectric constant ε_r =10.2.

References and Further Readings

[1]   Ali A. Behagi, *RF and Microwave Circuit Design*, A Design Approach Using (**ADS**) Software, Techno Search, Ladera Ranch, CA 2015

[2]   Ali Behagi and Manou Ghanevati, *Fundamentals of RF and Microwave Circuit Design,* Ladera Ranch, CA 2017

[3]   Paul Lorrain, Dale P. Corson, and Francois Lorrain, *Electromagnetic Fields and Waves*, W.H. Freeman and Company, New York, 1988

[4]   *Design Guide, Microwave Components* Inc., P.O. Box 4132, South Chelmsford, MA 01824

[5]   *Iron Powder Cores for High Q Inductors*, Micrometals, Inc.

[6]   *Capacitors for RF Applications,* Dielectric Laboratories, Inc., 2777 Rt.20 East, Cazenovia, NY. 13035

[7]   R. Ludwig and P. Bretchko, *RF Circuit Design -Theory and Applications*, Prentice Hall, New Jersey, 2000

[8]   M.F. "Doug" DeMaw, *Ferromagnetic Core Design & Application Handbook*, MFJ Publishing Co., Inc. Starkville, MS. 39759, 1996

[9]   *The RF Capacitor Handbook,* American Technical Ceramics, One Norden Lane, Huntington, New York 11746

[10]   Herbert L. Kraus, Charles W. Bostian, and Fredrick H. Raab, *Solid State Radio Engineering*, John Wiley & Sons, Inc., New York, 1980.

[11]   Jerry Sevick, *Transmission Line Transformers*, The American Radio Relay League, Newington, CT., 1990

[12]   M. Ghanevati, A. V. Thangavelu, J. H. Lee, R. Gutierrez and A. S. Daryoush, "An efficient SOM for front-end UHF electronics," *1999 IEEE MTT-S International Microwave Symposium Digest (Cat. No.99CH36282)*, Anaheim, CA, USA, 1999, pp. 1769-1772 vol.4. doi: 10.1109/MWSYM.1999.780314

Chapter 2

Transmission Lines

2.1 Introduction

In this chapter we present the lumped-element equivalent model of a transmission line and define several important transmission line parameters such as transmission and reflection coefficients, characteristic impedance, propagation constant, attenuation constant, phase constant, voltage standing wave ratio, return loss, velocity factor and group delay [1-11]. It is demonstrated how to simulate and measure these parameters using the software. Popular types of transmission lines, such as: coaxial lines, microstrip lines, strip lines, and waveguides are briefly discussed. Several methods of characterizing reflection coefficients and the characteristic impedance of these transmission lines are examined. Field coupling between adjacent (coupled) transmission lines is introduced. The chapter concludes with the design of a microstrip directional coupler.

2.2 Plane Waves in a Lossless Medium

Plane waves are among the simplest form of electromagnetic waves in which the electric field intensity, E, is perpendicular to magnetic field intensity, H, and both are perpendicular to the direction of propagation (E and H are represented by vectors). Such a wave is also called transverse electromagnetic or TEM wave.

In rectangular coordinates, assuming that electric field E is a vector in the x direction and varies as it moves along the z direction, the Wave Equation for the electric field in a lossless medium can be written as:

$$\frac{\partial^2 E_x}{\partial z^2} + k^2 E_x = 0 \tag{2-1}$$

where,

$k = \omega\sqrt{\mu\varepsilon}$ is the wave number

ω is the angular frequency in radians per second

μ is permeability of the medium in Henries per meter

ε is permittivity of the medium in Far per meter

In a lossless medium μ and ε are positive real numbers, therefore k is also positive and real. The solution to Equation (2-1) is of the following form.

$$E_x(z) = E^+ e^{-jkz} + E^- e^{+jkz} \qquad (2\text{-}2)$$

Where E^+ and E^- are arbitrary constants determined by the boundary conditions. For the sinusoidal waveforms at frequency ω, Equation (2-2) can be written as:

$$\mathcal{E}_x(z,t) = E^+ \cos(\omega t - kz) + E^- \cos(\omega t + kz) \qquad (2\text{-}3)$$

Where,

$$E^+ \cos(\omega t - k z) \quad \text{is the wave traveling in the forward direction}$$
$$E^- \cos(\omega t + k z) \quad \text{is the wave traveling in the reverse direction}$$

Some of the wave characteristics are as follows:

1. Phase velocity is the velocity of a fixed point on the wave that is obtained by setting the derivative of the phase, with respect to t, equal to zero:

$$\omega - k \frac{\partial z}{\partial t} = 0$$

$$v_p = \frac{\partial z}{\partial t} = \frac{\omega}{k} = \frac{1}{\sqrt{\mu \varepsilon}} \qquad (2\text{-}4)$$

In free space: $\mu = \mu_0 = 4\pi\ (10^{-7})$ H/m and $\varepsilon = \varepsilon_0 = 8.854(10^{-12})$ F/m, therefore,

$$v_p = \frac{1}{\sqrt{\mu_0 \epsilon_0}} = 2.998(10^8) = c \quad meters\,/\,second$$

where c is the velocity of light in free space.

2. The wavelength, λ, is defined as the distance between two successive maximum or minimum points on the wave at a fixed instant of time.

Therefore $k\,\lambda = 2\pi$ which leads to Equation (2-5).

$$\lambda = \frac{2\,\pi}{k} = \frac{v_p}{f} \tag{2-5}$$

3. The plane wave impedance, η, is defined as the ratio between electric and magnetic field components travelling in the same direction. Therefore, if

$$E_x^+ = E^+ \cos\left(\omega\,t - k\,z\right)$$
$$H_y^+ = H^+ \cos\left(\omega\,t - k\,z\right)$$

then,

$$\eta = \frac{E_x^+}{H_y^+} = \frac{E^+}{H^+} \tag{2-6}$$

Here E^+ and H^+ represent the electric and magnetic field amplitude travelling in the positive z direction in units of Volt/meter and Ampere/meter. Based on Maxwell's curl equations the plane wave impedance is given by:

$$\eta = \sqrt{\frac{\mu}{\varepsilon}} \tag{2-7}$$

In free space the plane wave impedance is,

$$\eta_0 = \sqrt{\frac{\mu_0}{\varepsilon_0}} = 377 \ \Omega$$

2.3 Plane Waves in a Good Conductor

Metallic conductors used in microwave networks are not perfect but they are considered to be very good conductors. In a material with conductivity σ the current density due to conduction is given by:

$$J = \sigma E \qquad (2\text{-}8)$$

Where σ is the conductivity of the conductor in S/m.

In a good conductor the conductive current is much greater than the displacement current, therefore, by ignoring the displacement current, the propagation constant can be written as [2]:

$$\gamma = \alpha + j\beta = \sqrt{\frac{\omega\mu\sigma}{2}} + j\sqrt{\frac{\omega\mu\sigma}{2}} \qquad (2\text{-}9)$$

The skin depth for a good conductor is defined as the inverse of the attenuation constant:

$$\delta = \frac{1}{\alpha} = \frac{1}{\sqrt{\pi f \mu \sigma}} \qquad (2\text{-}10)$$

Where,

   δ is the skin depth in meters

   f is the frequency in Hertz

   μ is the permeability in H/m

   σ is the conductivity in S/m

In a good conductor the positive traveling wave is of the form:

$$e^{-\alpha z}\cos(\omega t - \beta z) \qquad (2\text{-}11)$$

Note that when the wave travels a distance equal to, $Z = \delta = 1/\alpha$, the amplitude drops to $e^{-1}\cos(\omega t - \beta z) = 0.368\cos(\omega t - \beta z)$ which is 36.8% of the original signal's amplitude at $z = 0$. Also from Equation (2-9) it can be seen that the phase of the propagation constant in a good conductor is 45 degrees.

2.4 Representation of Transmission Lines

A lumped model of a small section of parallel wire transmission line, of physical length Δz, is shown in Figure 2-1. Any transmission line can be considered a cascade of many sections of the length Δz [1-4].

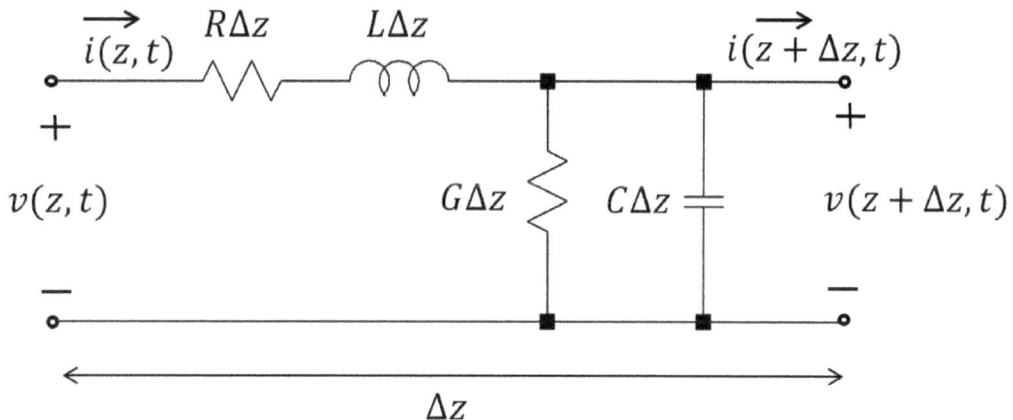

Figure 2-1 Lumped element model of the transmission line section

In Figure 2-1, R is the per unit length series resistance of both lines in Ω/m, L is the per unit length series inductance of both lines in Henries/meter, G is the per unit length shunt conductance of the line in Siemens/meter, and C is the per unit length capacitance of the line in Farads/meter. At higher frequencies, where the wavelength of the signal is smaller than the physical dimension of the network, the voltages and currents along a uniform transmission line are functions of position and time. For a TEM wave traveling in the z direction, the voltage and current are given by:

$$V(z,t)=re\left[V(z)e^{j\omega t}\right] \qquad (2\text{-}12)$$

$$I(z,t)=re\left[I(z)e^{j\omega t}\right] \qquad (2\text{-}13)$$

The quantities *V(z)* and *I(z)* are complex functions of *z* along the transmission line and ω is the frequency of the source in radians per second.

2.5 Transmission Line Equations and Parameters

Please refer to Appendix A for detailed derivation of transmission line equations and parameters. For sinusoidal steady-state excitations, Kirchhoff's voltage and current laws along the line yield the following equations:

$$\frac{dV(z)}{dz}+(R+j\omega L)I(z)=0 \qquad (2\text{-}14)$$

$$\frac{dI(z)}{dz}+(G+j\omega C)V(z)=0 \qquad (2\text{-}15)$$

By taking the derivative of both sides of Equation (2-14) with respect to z, and substituting for $\frac{dI(z)}{dz}$ from Equation (2-15), we have:

$$\frac{d^2V(z)}{dz^2}-(R+j\omega L)(G+j\omega C)\,V(z)=0 \qquad (2\text{-}16)$$

Similarly, by taking the derivative of both sides of Equation (2-15) with respect to z, and substituting for $\frac{dV(z)}{dz}$ from Equation (2-14), we have:

$$\frac{d^2I(z)}{dz^2}-(R+j\omega L)(G+j\omega C)\,I(z)=0 \qquad (2\text{-}17)$$

By defining the complex propagation constant:

$$\gamma = \sqrt{(R+j\omega L)(G+j\omega C)} \qquad (2\text{-}18)$$

Equations (2-16) and (2-17) can be redefined as:

$$\frac{d^2V(z)}{dz^2} - \gamma^2 V(z) = 0 \tag{2-19}$$

$$\frac{d^2I(z)}{dz^2} - \gamma^2 I(z) = 0 \tag{2-20}$$

To satisfy equations (2-19) and (2-20), the voltage and current wave equations can be given as

$$V(z) = A\,e^{-\gamma z} + B\,e^{\gamma z} \tag{2-21a}$$

$$I(z) = \bar{A}\,e^{-\gamma z} + \bar{B}\,e^{\gamma z} \tag{2-21b}$$

where,

A, B, $\bar{A}$, and $\bar{B}$ are constants

$Ae^{-\gamma z}$ and $\bar{A}e^{-\gamma z}$ are the incident voltage traveling in the +z direction

$B\,e^{\gamma z}$ and $\bar{B}\,e^{\gamma z}$ are the reflected voltage traveling in the -z direction

The relation between A and $\bar{A}$ can be found using the steps below:

$$\frac{dI(z)}{dz} = -\gamma\bar{A}\,e^{-\gamma z} + \gamma\,e^{\gamma z}$$

Also,

$$\frac{dI(z)}{dz} + (G + j\omega C)(R + j\omega L)V(z) = 0$$

Therefore,

$$-\gamma\bar{A}\,e^{-\gamma z} + \gamma\bar{B}e^{\gamma z} + (G + j\omega C)(Ae^{-\gamma z} + Be^{\gamma z}) = 0$$

From which we can obtain the relationship between A and $\bar{A}$:

$$\bar{A} = \frac{A\,(G + j\omega C)}{\gamma} = A\frac{\gamma}{(R + j\omega L)}$$

Following the same procedure as above, the relationship between B and $\bar{B}$ can be determined resulting in

$$I(z) = A\frac{\gamma}{(R + j\omega L)} e^{-\gamma z} - B\frac{\gamma}{(R + j\omega L)} e^{\gamma z} \qquad (2\text{-}22)$$

where,

$\frac{\gamma}{R + j\omega L} A e^{-\gamma z}$ is the incident current traveling in the $+z$ direction

$\frac{\gamma}{R + j\omega L} B e^{\gamma z}$ is the reflected current traveling in the $-z$ direction

The quantity $\frac{R + j\omega L}{\gamma}$ has a significant importance for the medium the wave is traveling in, and as will be shown later, it is related to the ratio of traveling voltage and current in an unbounded medium. That is the wave is only moving in one direction. The quantities A and B can be determined using voltage and current boundary conditions (see Appendix A, Fig. A-2). Assume that the above incident and reflected waves are generated as a result of connecting a finite transmission line of length d to a voltage signal source having internal source impedance Z_s on one end and a load impedance Z_L on the other end. Assume further that the source is situated at $Z = 0$ and that the load is located at $Z = d$. The values of voltage and current at $Z = d$ are simply voltage and current at the load, namely V_L and I_L, and they are related to load impedance as $Z_L = V_L / I_L$. Evaluation of load voltage V_L and load current I_L result in two equations from which A and B can be determined (See Appendix A for detailed characteristics of voltage and current along a finite transmission line that is terminated in a load impedance Z_L).

$$I(z = d) = \frac{A}{Z_0} e^{-\gamma d} - \frac{B}{Z_0} e^{\gamma d} = I_L$$

and

$$V(z = d) = A e^{-\gamma d} + B e^{\gamma d} = V_L$$

Solving the above equations simultaneously for A and B and noting that voltage at the load $V_L = I_L Z_L$, we obtain

$$A = \frac{e^{\gamma d}}{2} I_L (Z_L + Z_0)$$

$$B = \frac{e^{-\gamma d}}{2} I_L (Z_L - Z_0)$$

2.6 Definition of Attenuation and Phase Constant

The complex propagation constant is written as:

$$\gamma = \alpha + j\beta \tag{2-23}$$

Where the real part α is defined as the attenuation constant in Nepers per meter (1 Neper = abs($20\log(e^{-1})$) = 8.686 dB) and the imaginary part β is defined as the phase constant in radians per meter.

2.7 Definition of Characteristic Impedance

The characteristic impedance of a transmission line, Z_0, is defined as the ratio of the incident voltage to incident current. Therefore by dividing the incident voltage by the incident current, in Equations (2-21) and (2-22), and replacing γ from Equation (2-18), the transmission line characteristic impedance is written as:

$$Z_o = \frac{(R + j\omega L)}{\gamma} = \sqrt{\frac{R + j\omega L}{G + j\omega C}} \tag{2-24}$$

2.8 Definition of Reflection Coefficient

It is important to understand the concepts of incident (forward) and reflected (backward) wave propagation in transmission lines. In high power RF systems there can be potentially dangerous high voltage peaks that can occur at points along the transmission line when the incident and reflected waves are in phase and add together. The voltage reflection coefficient, $\Gamma(z)$, of a transmission line along the z axis is defined as the ratio of reflected to incident voltage as shown in Equation (2-25).

$$\Gamma_{in}(z) = \frac{B\, e^{\gamma z}}{A\, e^{-\gamma z}} = \frac{\frac{e^{-\gamma d}}{2}\, I_L\,(Z_L - Z_0) e^{\gamma z}}{\frac{e^{\gamma d}}{2}\, I_L\,(Z_L + Z_0) e^{-\gamma z}}$$

$$= \frac{Z_L - Z_0}{Z_L + Z_0}\, e^{-2\gamma d}\, e^{2\gamma z} \qquad (2\text{-}25a)$$

At the load end z = d, and

$$\Gamma_{in}(z = d) = \frac{Z_L - Z_0}{Z_L + Z_0} = \Gamma_L = |\Gamma_L| e^{j\theta_L} \qquad (2\text{-}25b)$$

where, Γ_L is the load reflection coefficient, and in general it is a complex quantity. Note that $\Gamma_L = 0$ when $Z_L = Z_0$.

At the source end z = 0, and

$$\Gamma_{in}(z = 0) = \frac{Z_L - Z_0}{Z_L + Z_0}\, e^{-2\gamma d} = |\Gamma_L| e^{j\theta_L}\, e^{-2\gamma d} \qquad (2\text{-}25c)$$

For a lossless line

$$\Gamma_{in}(z = 0) = \frac{Z_L - Z_0}{Z_L + Z_0}\, e^{-2j\beta d} = |\Gamma_L| e^{j\theta_L}\, e^{-2j\beta d} \qquad (2\text{-}26)$$

2.9 Definition of Voltage Standing Wave Ratio

VSWR is the ratio of the maximum to minimum value of the standing wave. The VSWR is a very common quantity for describing the percentage of power reflected by a given load impedance. The forward and reflected waves travel in opposite directions to form a standing wave pattern. The maximum value of the standing wave (for a lossless line, see Appendix A) is given by Equation (2-27) while the minimum value is given by Equation (2-28).

$$|V(z)|_{max} = |A|\,(1 + |\Gamma_L|) \qquad (2\text{-}27)$$

$$|V(z)|_{min} = |A| (1 - |\Gamma_L|) \qquad (2\text{-}28)$$

The ratio of the maximum to minimum voltage is related to the reflection coefficient as shown in Equation (2-29).

$$VSWR = \frac{|V(\acute{z})_{max}|}{|V(\acute{z})_{min}|} = \frac{1 + |\Gamma_L|}{1 - |\Gamma_L|} \qquad (2\text{-}29)$$

Solving the equation for the magnitude of the reflection coefficient, Γ_L results in Equation (2-30).

$$|\Gamma_L| = \frac{VSWR - 1}{VSWR + 1} \qquad (2\text{-}30)$$

Notice that when the reflection coefficient is zero, VSWR = 1. This VSWR is commonly presented as 1:1 ratio.

2.10 Definition of Return Loss

When the transmission line is mismatched to the load, a portion of the incident power is reflected back to the source. This can be considered a loss of power absorbed by the load. Therefore the return loss, RL, in dB, is defined as:

$$RL\,(dB) = -20\log\left|\Gamma\right| \qquad (2\text{-}31)$$

For a matched line:

$$|\Gamma| = 0 \text{ and RL} = \infty \text{ dB whereas for } |\Gamma| = 1, \text{ RL} = 0 \text{ dB}$$

2.11 Lossless Transmission Line Parameters

From Figure 2-1 we can see that a transmission line is considered lossless when R = G = 0. For a lossless transmission line the propagation constant reduces to:

$$\gamma = j\beta = j\omega \sqrt{L\,C} \qquad (2\text{-}32)$$

Therefore,

$$\beta = \omega \sqrt{L\,C} \tag{2-33}$$

and

$$\alpha = 0 \tag{2-34}$$

The characteristic impedance of a lossless transmission line i.e. R = G = 0 is obtained from Equation (2-24) as:

$$Z_0 = \sqrt{\frac{L}{C}} \tag{2-35}$$

Similarly the voltage reflection coefficient in Equation (2-25) reduces to:

$$\Gamma_{in}(z) = \frac{Z_L - Z_0}{Z_L + Z_0}\, e^{-2j\beta d}\, e^{2j\beta z} = |\Gamma_L| e^{j\theta_L}\, e^{2j\beta(z-d)} \tag{2-36}$$

where, Γ_L was defined as the load reflection coefficient.

Notice from Equations (2-32) through Equation (2-34) for lossless transmission lines the attenuation constant is zero and the phase constant is linearly proportional to frequency. The propagation constant is purely imaginary, and the characteristic impedance is a positive real number.

2.12 Lossless Transmission Line Terminations

For detailed discussion of the following, please see Appendix A. The input impedance of a transmission line having length d and characteristics Z_0 and γ, that is terminated in a given load impedance Z_L, can be determined by dividing the voltage equation of (2-21) to current Equation of (2-22) and by evaluating the equations at a distance d away from the load. The input impedance of the transmission is then given by Equation (2-37a) and for a lossless line ($\gamma = j\beta$) by Equation (2-37b)

$$Z_{in} = z_0 \frac{Z_L + Z_0 \tanh \gamma d}{Z_0 + Z_L \tanh \gamma d}, \text{for a lossy line} \tag{2-37a}$$

$$Z_{in} = z_0 \frac{Z_L + jZ_0 \tan \beta d}{Z_0 + jZ_L \tan \beta d}, \text{for a lossless line} \qquad (2\text{-}37\text{b})$$

where γ and β $(= 2\pi/\lambda)$ are the propagation and phase constant of the line.

In the following sections we discuss the input impedance of terminated transmission lines.

2.12-1 Transmission Line Terminated in Z_0

When a lossless transmission line of characteristic impedance Z_0 is terminated in a load equal to Z_0, Equation (2-37) shows that the input impedance becomes equal to Z_0. Such a line behaves like an infinitely long transmission line with no reflection. In this case all of the incident power is absorbed by the load.

2.12-2 Transmission Line Terminated in a Short Circuit

When a lossless transmission line of characteristic impedance Z_0 is terminated in a short circuit, $Z_L = 0$, Equation (2-38) shows that the input impedance becomes a purely imaginary number equal to:

$$Z_{in} = jZ_0 tan \beta d \qquad (2\text{-}38)$$

Depending on the length of the line the input impedance takes any possible reactive value from minus to plus infinity.

2.12-3 Transmission Line Terminated in an Open Circuit

When a lossless transmission line of characteristic impedance Z_0 is terminated in an open circuit, $Z_L = \infty$, Equation (2-37) shows that the input impedance becomes equal to:

$$Z_{in} = -jZ_o cot \beta d \qquad (2\text{-}39)$$

In this case Z_{in} is also purely imaginary. Depending on the length of the line the input impedance takes any possible reactive value from minus to plus infinity.

2.12-4 Half Wavelength Transmission Lines

For a lossless transmission line of characteristic impedance Z_0 with the length $d = \lambda/2$, the input impedance from Equation (2-37) is equal to:

$$Z_{in} = Z_L \qquad (2\text{-}40)$$

This means that the input impedance of a transmission line of one half-wavelength is equal to the load impedance regardless of the line characteristic impedance.

2.12-5 Quarter Wavelength Transmission Lines

For a lossless transmission line with the length $d = \lambda/4$, the input impedance from Equation (2-37) becomes equal to:

$$Z_{in} = \frac{Z_0^2}{Z_L} \qquad (2\text{-}41)$$

In this case the transmission line transforms the load impedance to a different impedance as defined by Equation (2-41). Such a line is called a quarter-wave transformer. Equation (2-41) can be rewritten to solve for the characteristic impedance of a quarter-wave matching section as:

$$Z_0 = \sqrt{Z_{in} Z_L} \qquad (2\text{-}42)$$

2.13 Simulation of Reflection Coefficient and VSWR

The software has built-in functions to directly display the common measurements of VSWR, return loss, and reflection coefficient. The following example explores the simulation of input reflection coefficient, S_{11} and VSWR.

Example 2.1: For the series RLC elements in Figure 2-2 measure the reflection coefficients and VSWR from 100 to 1000 MHz in 100 MHz steps.

Solution: To generate a table of tabulated numbers, follow the procedure outlined below:

1. Construct the schematic diagram as shown as shown in Fig. 2-2.

2. From the main window select: Simulate > Edit Simulation Cmd to open the Edit Simulation Command window.

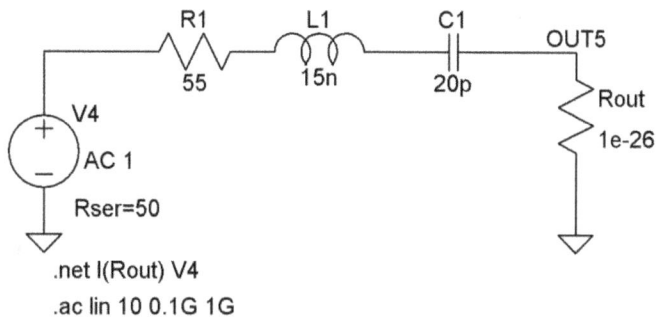

Figure 2-2 Schematic for simulating of S parameters.

3. Select AC Analysis and complete the simulation box as shown in Figure 2-3. Then click OK and run simulation.

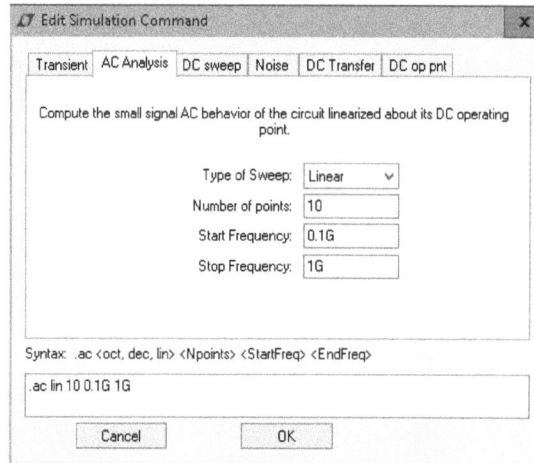

Figure 2-3 AC analysis setup

4. Go to the waveform and Add Trace by typing the required formula to compute VSWR under "Expressions to add," as shown in Figure 2-4.

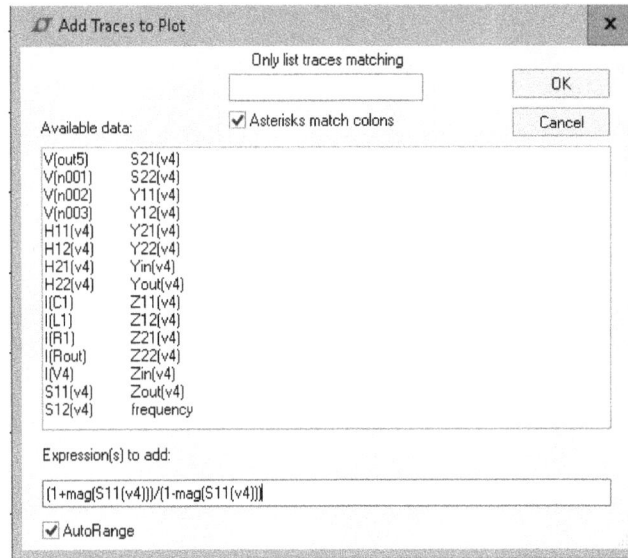

Figure 2-4 Window Add Traces to Plot

a. Press Ok to generate Figure 2-5.

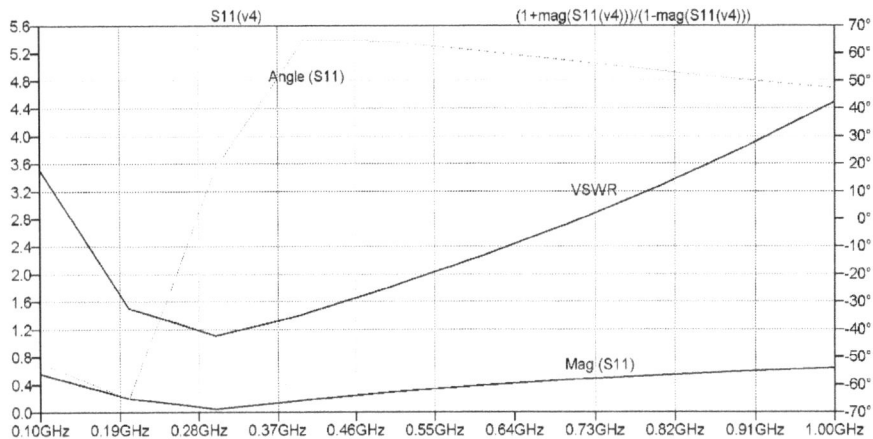

Figure 2-5 S_{11} Magnitude and Angle plus VSWR

5. Select the desired traces and format [Real, Imaginary or Polar: (dB,deg), as shown in Figure 2-6.

6. Press Ok to save the text file. Use the Browse tab to save the text file in the desired directory.

7. Go to the above directory and open the file to access data.

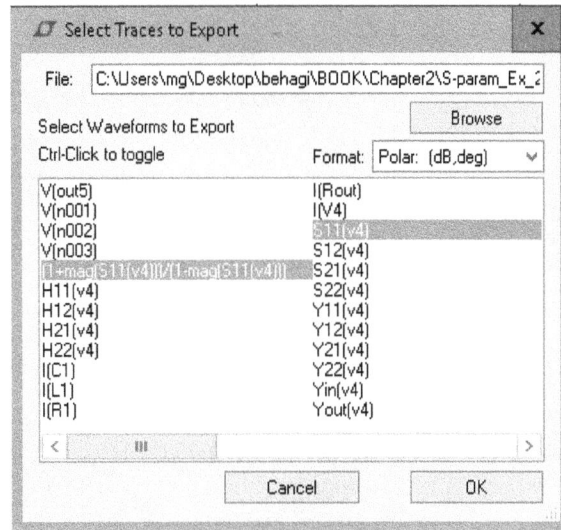

Figure 2-6 Select traces to add in polar format

8. Organize the tabulated data in a proper Table.

Freq. (Hz)	VSWR(dB)		Mag,S_{11}(dB)		Angle,S_{11}(Deg)
1.00E+08	10.92		-5.0837		-52.175
2.00E+08	3.541		-13.933		-65.29
3.00E+08	0.877		-25.943		18.317
4.00E+08	3.048		-15.206		64.687
5.00E+08	5.159		-10.797		64.343
6.00E+08	7.024		-8.321		61.006
7.00E+08	8.712		-6.682		57.291
8.00E+08	10.26		-5.506		53.694
9.00E+08	11.71		-4.62		50.344
1.00E+09	13.05		-3.931		47.27

Table 2-1(a): Tabulated Magnitude,S_{11} and VSWR in dB

Freq.(Hz)	VSWR		Real,S_{11}		Imag.,S_{11}
1.00E+08	3.51401		0.34154659		-0.43992229
2.00E+08	1.50334		0.084053375		-0.18265703
3.00E+08	1.10626		0.047892137		0.015854803
4.00E+08	1.42028		0.074246399		0.156977608
5.00E+08	1.81111		0.124931429		0.260087866
6.00E+08	2.24497		0.185967242		0.335577977
7.00E+08	2.72658		0.250364556		0.389844097
8.00E+08	3.26009		0.314124589		0.427531726
9.00E+08	3.84853		0.37493197		0.452313137
1.00E+09	4.49418		0.431541452		0.467160375

Table 2-1 (b): Tabulated VSWR and S_{11} in real and imaginary

2.14 Return Loss, VSWR, and Gamma

Return Loss, VSWR, and Reflection Coefficient are all different ways of characterizing the wave reflection. These definitions are often used interchangeably in practice. Therefore it is important to be able to convert from one form of reflection to another. Return Loss is often used to characterize components such as filters, amplifiers, and networks. VSWR is normally used in systems such as radio and TV transmitters. Reflection coefficient is normally used in device characterization such as transistors, capacitors, inductors, etc. Equations (2-29), (2-30), and (2-31) give the conversions among these three parameters Note, however, that the reflection coefficient is a vector quantity whose magnitude is all that is required to determine the VSWR or return loss. Therefore when calculating Γ from the VSWR or return loss we can only find the magnitude and not the angle of the reflection coefficient. Another useful parameter is the mismatch loss. The mismatch loss is a measure of power that is reflected by the load.

$$Mismatch\ Loss\ (dB) = -10\log\left(1 - \left|\Gamma\right|^2\right) \qquad (2\text{-}43)$$

Example 2.2: Generate a table showing the return loss, the reflection coefficient, and the percentage of reflected power as a function of VSWR.

Solution: Create a schematic with as shown in Figure 2-7. Make the resistance value a tunable variable. Set the Linear AC Analysis at any fixed frequency for one point only. Add a Parameter Sweep to sweep the value of the resistor.

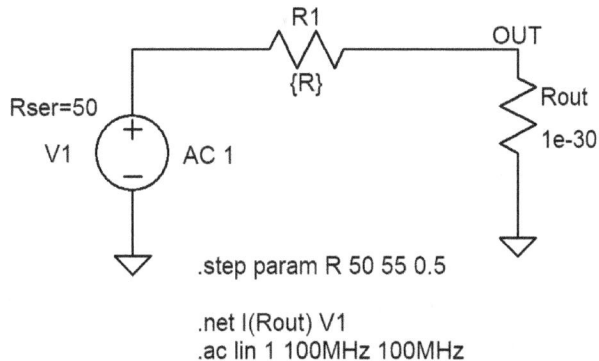

Figure 2-7 Schematic to generate VSWR table

Simulate (Run) the schematic and follow below instructions:

1. Right click on the waveform window > Add trace > Expression Editor opens up. Type below expressions in the Expression Editor to plot reflection coefficient, VSWR, and % reflected power, respectively:

mag(S11(v1))

(1+mag(S11(v1)))/(1-mag(S11(v1)))

(mag(S11(v1))^2)*100

2. Click File > Exports to open the "Select Traces to Export" window

3. Hold the Ctrl key on your keyboard

4. In the Select Traces to Export window, select above expressions.

5. In the above window, select the proper format for the tabulated data

6. Use the top-down arrow to select Real and Imaginary > OK.

7. Go to the directory where the generated .txt file is stored. Open the file and organize the data as shown in Table 2-2. Simulate Figure 2-7 for other range of resistor values R to complete Table 2-2 for all desired VSWR values. Following are the equations for the calculation of mismatch loss.

ReflCoef = (VSWR - 1)/(VSWR + 1)

Mismatch = 1 − ((ReflCoef)^2)

Powerloss = (1 − Mismatch)*100

We will present the mismatch loss as a percentage of the available power that is reflected by the load. As we can see from the table if we can keep the VSWR less than 1.25:1 we will have less than 1% power loss due to reflective impedance mismatch.

2.15 Characteristic Impedance and Velocity of Propagation

There are many physical transmission lines that are encountered in RF and microwave circuits and systems. In addition it should not be overlooked that free space is also a transmission medium. Therefore Equation (2-35) could be used to describe the characteristic impedance, Z_o, of free space. In order to define L and C we need to consider the inductive and capacitive properties of free space. Permeability is the ability of a transmission media to support a magnetic field. Considered as a density in free space the permeability is related to inductance and defined by Equation (2-44).

$$\mu_0 = 4\pi.\,10^{-7} \quad Henries/meter \qquad (2\text{-}44)$$

VSWR	$\lvert\Gamma\rvert$	Return Loss (dB)	% Reflected Power	VSWR	$\lvert\Gamma\rvert$	Return Loss (dB)	% Reflected Power
1.01	0.005	46.06	0.002	2.40	0.412	7.71	16.955
1.02	0.010	40.09	0.010	2.50	0.429	7.36	18.367
1.10	0.048	26.44	0.227	3.00	0.500	6.02	25.000
1.20	0.091	20.83	0.826	3.50	0.556	5.11	30.864
1.30	0.130	17.69	1.701	4.00	0.600	4.44	36.000
1.40	0.167	15.56	2.778	4.50	0.636	3.93	40.496
1.50	0.200	13.98	4.000	5.00	0.667	3.52	44.444
1.60	0.231	12.74	5.325	6.00	0.714	2.92	51.020
1.70	0.259	11.73	6.722	7.00	0.750	2.50	56.250
1.80	0.286	10.88	8.163	8.00	0.778	2.18	60.494
1.90	0.310	10.16	9.631	9.00	0.800	1.94	64.000
2.00	0.333	9.54	11.111	10.00	0.818	1.74	66.942
2.10	0.355	9.00	12.591	20.00	0.905	0.87	81.859
2.20	0.375	8.52	14.063	200.00	0.990	0.09	98.020
2.30	0.394	8.09	15.519	2000.00	0.999	0.01	99.800

Table 2-2 Relationship among VSWR, Γ, Return Loss, and Refl. Power

Permittivity is the ability of a transmission media to support an electric field. Considered as a density in free space the permittivity is closely related to capacitance in equation (2-45).

$$\varepsilon_o = \frac{1}{36\pi} \cdot 10^{-9} \quad Farads/meter \qquad (2\text{-}45)$$

Therefore we can rewrite equation (2-35) as:

$$Z_o = \sqrt{\frac{\mu_o}{\varepsilon_o}} = \sqrt{\frac{4\pi \cdot 10^{-7}}{\frac{1}{36\pi} \cdot 10^{-9}}} = 377\ \Omega \qquad (2\text{-}46)$$

Equation (2-46) shows that the characteristic impedance is related to the inductance and capacitance of the transmission media. The velocity of propagation is also related to inductance and capacitance. It can be shown that the time required for a sine wave to propagate through a unit length of lossless transmission line is related by Equation (2-47) [8].

$$t = \sqrt{LC} \qquad (2\text{-}47)$$

The velocity of propagation is related to the wavelength in one period T,

$$v = \frac{\lambda}{T} = \frac{1}{t} = \frac{1}{\sqrt{LC}} \qquad (2\text{-}48)$$

where, T is the time period of the sine wave. Using the free space permeability and permittivity the velocity of propagation through free space is:

$$v_0 = \frac{1}{\sqrt{\mu_0 \varepsilon_0}} = 2.998 \cdot 10^8 \ meter/second \qquad (2\text{-}49)$$

From Equations (2-46) and (2-49) we can see that the characteristic impedance and velocity of propagation can be defined for any transmission media based on the inductive and capacitive properties of that media.

2.16 Physical Transmission Lines

Free space can be used for the propagation of radio signals across long distances but may not be as useful for the point to point connection and isolation of specific RF signals. For this purpose engineers use a variety of different transmission line media. This text is focused on a few of the most commonly used transmission media in RF and microwave circuit and system design including, microstrip, stripline, coaxial, and waveguide transmission lines. As a general differentiator, transmission media may be divided among pure TEM, quasi TEM, and non TEM propagation modes. TEM refers to the transverse electromagnetic mode of wave propagation. Signals traveling through space propagate in the TEM mode. This simply means that the magnetic field, electric field, and the direction of propagation are all orthogonal to one another. The velocity with which a wave travels will almost always be slower in a physical transmission line than it is in free space. Engineers frequently need to trim a transmission line for a specific wavelength so it is essential that they know the velocity of propagation in the transmission line. Knowing the distributed inductive and capacitive properties of the media, Equation (2-48) could be used to determine the propagation velocity. However it is common for many transmission line manufacturers to specify a velocity factor. The velocity factor is a ratio of the actual transmission line velocity v, to the velocity of free space.

$$v_f = \frac{v}{v_0} \tag{2-50}$$

Similar to the velocity factor, manufacturers of transmission line media typically express the permittivity of the dielectric material as a relative dielectric constant. The relative dielectric constant, ε_r, is the ratio of the actual material dielectric constant to the dielectric constant of free space.

$$\varepsilon_r = \frac{\varepsilon}{\varepsilon_0} \tag{2-51}$$

Depending on the type of transmission media the manufacturer may choose to specify either the velocity factor, v_f, or relative dielectric constant ε_r. The two quantities are related by Equation (2-52).

$$v_f = \frac{1}{\sqrt{\varepsilon_r}}$$ (2-52)

There are many types of physical transmission lines ranging from twisted wire pairs to fiber optic cables. There are many good reference sources to examine the characteristics and application of transmission lines [4]. In this chapter we will focus on just a few of these transmission lines that the RF and microwave engineer will frequently encounter. They include:

Coaxial Transmission Lines
Microstrip Transmission Lines
Stripline Transmission Lines
Waveguides

Note that coaxial, microstrip, and stripline transmission lines require two conductors to transfer power from a given source to its load while waveguides require only a single hollow conductor. A brief introduction to each of the above transmission media is given here.

2.17 Coaxial Transmission Lines

Coaxial transmission line, also known as coaxial cable, is one of the most common types of transmission line used in electrical equipment interconnection. A typical coaxial line consists of an inner conductor inside another cylindrical conductor and a dielectric material in between. The outer conductor, or the shield, is usually grounded to minimize RF interference and radiation loss. The coaxial transmission line supports pure TEM propagation. The dielectric material is typically a form of Teflon. Teflon has very good physical strength as well as low loss and high temperature operation. For very high power and low loss applications the dielectric material may be air. In this case there must be some dielectric spacer installed at certain intervals to support the center conductor and maintain concentricity. It is also common to introduce nitrogen into the air dielectric to help prevent condensation from forming inside the cable. Condensation would greatly increase the loss and lower the voltage breakdown. Coaxial cables that are used in very high power applications generally require a larger diameter to increase the voltage breakdown.

Table 2-3 gives a brief listing of some commonly used flexible coaxial cables.

Loss in dB/100ft. @ frequency in MHz										
Coaxial Cable	100 MHz	200 MHz	400 MHz	700 MHz	900 MHz	Z_o	V_f	Capac. per ft.	O.D. inch	Shield
Belden 9913	1.30	1.80	2.70	3.60	4.20	50	.84	24.6 pF	0.405	100%
Belden 9914	1.60	2.40	3.50	5.00	5.70	50	.82	24.8 pF	0.403	100%
Belden 8214 RG-8/U	1.80	2.70	4.20	5.80	6.70	50	.78	26.0 pF	0.405	97%
Belden 8238 RG-11/U	2.00	2.90	4.20	5.80	6.70	75	.66	20.5 pF	0.405	97%
Belden 8267 RG-213/U	1.90	2.70	4.10	6.50	7.60	50	.66	30.8 pF	0.405	97%
Belden 8242 RG-9/U	2.10	3.00	4.80	6.50	7.60	51	.66	30.0 pF	0.420	98%
Belden 9258 RG-8/X	3.70	5.40	8.00	11.10	12.80	50	.80	25.3 pF	0.242	95%
Belden 84142 RG-142	3.90	5.60	8.20	11.00	12.50	50	.695	29.2 pF	0.195	98%
Belden 9273 RG-223/U	4.10	6.00	8.80	12.00	13.80	50	.66	30.8 pF	0.212	95%
Belden 8240 RG-58/U	4.50	6.80	10.00	14.00	16.00	51.5	.66	29.9 pF	0.195	95%
Belden 8259 RG-58A/U	4.90	7.30	11.50	17.00	20.00	50	.66	30.8 pF	0.193	95%
Belden 9259 RG-59/U	3.00	4.50	6.60	8.90	10.1	75	.78	17.3 pF	0.242	95%
Belden 8241 RG-59/U	3.40	4.90	7.00	9.70	11.1	75	.66	20.5 pF	0.242	95%
Belden 8216 RG-174/U	8.40	12.50	19.00	27.00	31.00	50	.66	30.8 pF	0.101	90%
Belden 9228 RG-62A/U	2.70	3.80	5.30	7.30	8.20	93	.84	13.5 pF	0.242	95%

Table 2-3 Sample table of coaxial cable specifications

Manufacturers typically specify the cable loss in dB per 100 foot lengths at various frequencies. This table gives the manufacturer's (Belden) part number along with an RG designation. The RG designation is an attempt by the U.S. government under MIL-C-17 specification to have a standard designation for the manufacture of coaxial cables. The R means that the cable is intended for RF frequency usage. The G means that the cable is manufactured to general specifications. The number then identifies the unique specifications that the cable is designed to meet. Minor differences in the specification have a letter appended to the numerical designator. The characteristic impedance and velocity factor are also specified. It is not uncommon to have the cable's capacitance, per unit length, specified as well. Two physical characteristics included in Table 2-3 are the outer dimension (O.D.) and the shield percentage. The shield percentage is only applicable in flexible cables in which the outer conductor is actually a braid of several discrete wires that form a tight mesh. A 100% rating means that there are no air gaps in the wires that comprise the braid and therefore a perfect outer conductor is formed. The term O.D. is referred to as either the outer dimension or overall diameter of the cable including the outer jacket; not the diameter of the dielectric itself. The characteristic impedance of a coaxial line is determined by the diameter of the inner conductor (d) and the outer conductor (b) along with the relative dielectric constant, as given by Equation (2-53).

$$Z_o = \frac{60}{\sqrt{\varepsilon_r}} \cdot \ln\left(\frac{b}{d}\right) \qquad (2\text{-}53)$$

2.18 Microstrip Transmission Lines

Microstrip [9] is a planar transmission line media in which the transmission line is etched onto the top side of a printed circuit board. In a single layer PCB the bottom side of the printed circuit board is completely metalized and grounded. The printed circuit board has a low loss dielectric material that is suitable for microwave transmission. Unlike the coaxial cable the center conductor is not shielded. This means that a portion of the electric and magnetic field is in the air space above the microstrip line. Thus the propagation in microstrip is not purely TEM but rather quasi TEM. This also leads to the fact that the dielectric constant beneath the microstrip line is slightly less than the relative dielectric constant of the material. This is known as the effective dielectric constant, ε_{eff}, and is a function of the width of the microstrip line, W and the height of the substrate, h. Figure 2-8 shows the geometry of a microstrip line as well as the electric and magnetic field lines in the substrate and in the air. Due to quasi TEM mode of propagation, the phase velocity can be expressed as

$$v_p = \frac{c}{\sqrt{\varepsilon_{eff}}} \tag{2-54}$$

where $c = 3 \times 10^8$ m/s is speed of light in vacuum. Therefore, the wavelength inside the microstrip can be written as

$$\lambda = \frac{v_p}{f} = \frac{\lambda_0}{\sqrt{\varepsilon_{eff}}} \tag{2-55}$$

The characteristic impedance of the microstrip line Z_0 is related to capacitance per unit length and the phase velocity, and it is given by

$$Z_0 = \sqrt{\frac{L}{C}} = \sqrt{\frac{LC}{CC}} = \frac{1}{v_p C} \tag{2-56}$$

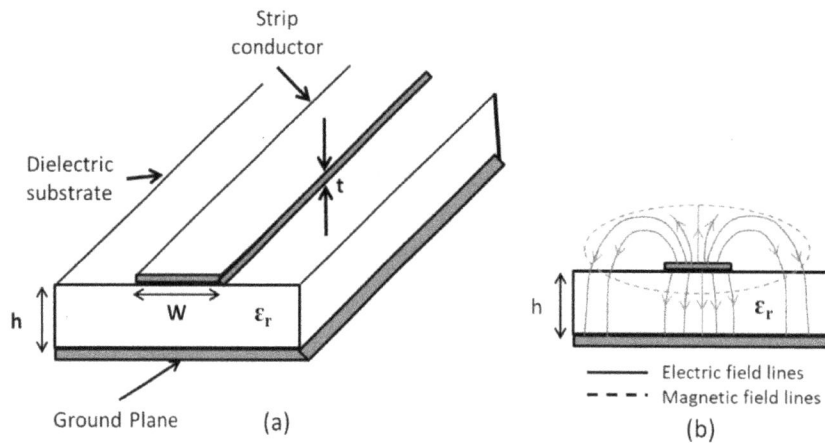

Figure 2-8 The geometry of a microstrip line, (a) and (b) the electric and magnetic field lines

When the thickness of the strip conductor is negligible compare to the substrate height (i.e., $t < 0.005h$), some useful expressions for characteristic impedance and wavelength can be given as follows:

For W/h ≤ 1:

$$Z_0 = \frac{60}{\sqrt{\varepsilon_{ff}}} \ln\left(8\frac{h}{W} + 0.25\frac{W}{h}\right) \tag{2-57}$$

where

$$\varepsilon_{ff} = \frac{\varepsilon_r + 1}{2} + \frac{\varepsilon_r - 1}{2}\left[\left(1 + 12\frac{h}{W}\right)^{-\frac{1}{2}} + 0.04\left(1 - \frac{W}{h}\right)^2\right] \tag{2-58}$$

For W/h ≥ 1:

$$Z_0 = \frac{120\pi/\sqrt{\varepsilon_{ff}}}{\frac{W}{h} + 1.393 + 0.667\ln\left(\frac{W}{h} + 1.444\right)} \tag{2-59}$$

where

$$\varepsilon_{ff} = \frac{\varepsilon_r + 1}{2} + \frac{\varepsilon_r - 1}{2}\left(1 + 12\frac{h}{W}\right)^{-\frac{1}{2}} \tag{2-60}$$

For W/h ≥ 0.6:

$$\lambda = \frac{\lambda_0}{\sqrt{\varepsilon_r}}\left[\frac{\varepsilon_r}{1 + 0.63\,(\varepsilon_r - 1)\,(W/h)^{0.1255}}\right]^{1/2} \tag{2-61}$$

For W/h ≤ 0.6:

$$\lambda = \frac{\lambda_0}{\sqrt{\varepsilon_r}}\left[\frac{\varepsilon_r}{1 + 0.6\,(\varepsilon_r - 1)\,(W/h)^{0.0297}}\right]^{1/2} \tag{2-62}$$

When the strip thickness t cannot be neglected (i.e., t < h and t < W/2), the conductor width W should be replaced with an effective width W_{eff} given by

For W/h ≥ 1/2 π :

$$\frac{W_{eff}}{h} = \frac{W}{h} + \frac{t}{\pi h}\left(1 + \ln\frac{2h}{t}\right) \tag{2-63}$$

For W/h ≤ 1/2 π :

$$\frac{W_{eff}}{h} = \frac{W}{h} + \frac{t}{\pi h}\left(1 + \ln\frac{4\pi W}{t}\right) \tag{2-64}$$

As frequency increases, the microstrip line becomes dispersive and the quasi TEM mode of operation does not hold. In that case, phase velocity decreases and line impedance increases with increase in frequency. Microstrip parameters become function of frequency. In general, dispersion may be neglected below frequency given by

$$f_0(GHz) = 0.3\sqrt{\frac{Z_0}{h\sqrt{\varepsilon_r - 1}}} \tag{2-65}$$

where h is in centimeter. As an example, for a FR4 material having $\varepsilon_r = 4.6$ and h = 30 mils, the above frequency is 5.7 GHz.

The quality factor Q of the microstrip line for a fractional bandwidth around a given center frequency can be determined similar to the method discussed in Appendix A, and it is written by

$$Q = \frac{\beta}{2\alpha} = \frac{2\pi/\lambda}{2\alpha} = \frac{\pi}{\alpha\lambda} \tag{2-66}$$

The quality factor per wavelength can be calculated as

$$\frac{Q}{\lambda} = \frac{(\pi)(8.686)}{\alpha} = \frac{27.3}{\alpha} \quad dB/\lambda \tag{2-67}$$

2.19 Stripline Transmission Lines

Stripline is a transmission line using planar dielectric material similar to microstrip. The major difference is that the top half of the line also consists of a dielectric of the same material as the bottom conductor. Therefore the transmission line is shielded in much the same way as the coaxial transmission line. As such the stripline transmission line supports a pure TEM propagation mode. Because of the presence of the top dielectric it is much more difficult to integrate discrete components such as transistors, and chip inductors and capacitors. As a pure TEM transmission line, stripline does offer superior performance in distributed filters, directional couplers, and power combiners.

Because the propagation mode is pure TEM, there is no effective dielectric constant, just the relative dielectric constant of the substrate material. Soft substrates are almost entirely used for stripline transmission circuits. One popular variation is when an air dielectric is used. This is referred to as suspended substrate stripline (SSS) and is characterized by very low loss. Suspended substrate stripline is often used in filter and multiplexer circuit designs. The characteristic impedance of a stripline transmission line is given by the empirical Equation (2-68) [10]. Here, b is the separation between upper and lower ground planes. W and t are the width and thickness of the metal conductor, respectively.

$$Z_0 = \frac{94.2}{\sqrt{\varepsilon_r}} \ln \left[\frac{1 + \frac{W}{b}}{\frac{W}{b} + \frac{t}{b}} \right] \qquad (2\text{-}68)$$

In practice a 60 mil stripline media is realized by clamping or fusing two 30 mil dielectric halves together. The circuit pattern is etched on one of the 30 mil dielectrics similar to a microstrip circuit pattern. The top half dielectric would then have all metal removed on the side which meets the bottom half circuit patterns. A special bonding film can be used to effectively glue the halves together or the dielectrics could be clamped together using sufficient pressure.

2.20 Waveguide Transmission Lines

A waveguide is a special form of transmission line that is used primarily at microwave frequencies above 2 GHz. The most important difference between waveguide and other forms of transmission line is the low loss and capability to transmit very high power. Waveguides do not have a separate conductor, ground plane, or shield as the previous transmission line structures. Rather waveguide is a hollow tube in which the wave propagates through. Even though the propagation is in the air space, enclosing an RF wave in a metallic boundary causes the wave to propagate quite differently than it would in free space. Waveguide tubes can be circular or rectangular but the rectangular tube is much more popular due to its ease of manufacture. The flanges allow a connection to be made to a mating piece of waveguide or waveguide component. The broad dimension of the waveguide is labeled as the 'a' dimension while the narrow side is labeled the 'b' dimension. The TEM propagation mode does not exist in waveguide. The propagation must be either transverse electric (TE) or transverse magnetic (TM). The propagation modes are quite complex with many subcategories of TE and TM propagation in existence. In TE mode the electric field is transverse to the direction of propagation. This means that there is no component of the electric field in the direction of propagation. In the TM mode there is no magnetic field component in the direction of propagation. Depending on how the RF energy is launched into the waveguide there are many variations of both TE and TM propagation modes. These sub-modes are designated with the subscripts m and n such as $TE_{m,n}$. The sub-mode notation describes the field patterns that exist in the waveguide. The dominant sub-mode is the $TE_{1,0}$ mode. This is the lowest frequency that the waveguide will support. The remaining discussion of rectangular waveguide properties is based on the $TE_{1,0}$ mode. The wave pattern in a waveguide leads to an interesting condition where at or below a certain frequency the wave bounces from side-to-side or top-to-bottom in the waveguide tube and no longer travels in the direction of propagation. This frequency is known as the cutoff frequency. Below the cutoff frequency the waveguide transmits very little energy. As such the waveguide has a natural high pass filter characteristic. The cutoff frequency of the $TE_{1,0}$ in rectangular waveguide is given by Equation (2-69),

$$f_{c_{1,0}} = \frac{c}{2a} = \frac{3 \cdot 10^8}{2a}$$

(2-69)

Where, c is the speed of light and a is the wider dimension of the waveguide.

The wavelength in a rectangular waveguide is then defined as.

$$\lambda_g = \frac{\lambda}{\sqrt{1 - \left[\frac{\lambda}{(2a)^2}\right]}}$$

(2-70)

where λ is the wavelength of the signal in free space.

The characteristic impedance of waveguide is based on the propagation mode. The characteristic impedance must be defined separately for the transverse electric and magnetic fields as given by equations (2-71) and (2-72).

$$Z_{o\,(TE_{m,n})} = \frac{\eta}{\sqrt{1-\left(\dfrac{f_c}{f}\right)^2}} \qquad \text{m, n=1, 0} \qquad (2\text{-}71)$$

$$Z_{o\,(TM_{m,n})} = \eta\sqrt{1-\left(\dfrac{f_c}{f}\right)^2} \qquad \text{m, n=1, 0} \qquad (2\text{-}72)$$

The term, η is the intrinsic impedance of the medium. In air $\eta = 377\ \Omega$. Z_0 is rarely used in practice. Waveguide is typically characterized by its cross section dimensions as shown in Table 2-4. Table 2-4 shows a listing of some of the more popular waveguides by WR designator.

Frequency Band, GHz	U.S. (EIA) Designator	British WG Designator	Cut Off Freq. in GHz $TE_{1,0}$	a dimension inches	b dimension inches
1.12 - 1.70	WR 650	WG 6	0.908	6.500	3.250
1.45 - 2.20	WR 510	WG 7	1.158	5.100	2.550
1.70 - 2.60	WR 430	WG 8	1.375	4.300	2.150
2.20 - 3.30	WR 340	WG 9A	1.737	3.400	1.700
2.60 - 3.95	WR 284	WG 10	2.080	2.840	1.340
3.30 - 4.90	WR 229	WG 11A	2.579	2.290	1.145
3.95 - 5.85	WR 187	WG 12	3.155	1.872	0.872
4.90 - 7.05	WR 159	WG 13	3.714	1.590	0.795
5.85 - 8.20	WR 137	WG 14	4.285	1.372	0.622
7.05 - 10.0	WR 112	WG 15	5.260	1.122	0.497
8.2 - 12.4	WR 90	WG 16	6.560	0.900	0.400
9.84 - 15.0	WR 75	WG 17	7.873	0.750	0.375
11.9 - 18.0	WR 62	WG 18	9.490	0.622	0.311
14.5 - 22.0	WR 51	WG 19	11.578	0.510	0.255
17.6 - 26.7	WR 42	WG 20	14.080	0.420	0.170
21.7 - 33.0	WR 34	WG 21	17.368	0.340	0.170
26.4 - 40.0	WR 28	WG 22	21.100	0.280	0.140
32.9 - 50.1	WR 22	WG 23	26.350	0.224	0.112
39.2 - 59.6	WR 19	WG 24	31.410	0.188	0.094
49.8 - 75.8	WR 15	WG 25	39.900	0.148	0.074
60.5 - 91.9	WR 12	WG 26	48.400	0.122	0.061
73.8 - 112	WR 10	WG 27	59.050	0.100	0.050

Table 2-4 Standard rectangular waveguide characteristics

2.21 Group Delay in Transmission Lines

A frequently encountered concept related to the transmission line velocity factor is group delay. Group delay is a measure of the time that it takes a signal to traverse a transmission line, or its transit time. It is a strong function of the length of the line, and usually a weak function of frequency. It is expressed in units of time, picoseconds for short distances or nanoseconds for longer distances. Remember that in free space all electromagnetic signals travel at the speed of light, c, which is approximately 300,000 kilometers per second. Therefore, in free space, electromagnetic radiation travels one foot in one nanosecond, unless there is something to slow it down such as a dielectric. Mathematically the group delay is the derivative of phase versus frequency. In communication systems, the ripple in the group delay creates a form of distortion.

2.22 Transmission Line Components

There are many useful components that can be realized using transmission lines. These include power splitters, directional couplers, voltage and current insertion networks, as well as various filter networks. These networks can be realized with any of the physical transmission line structures. However, because of its popularity, we will explore a few of these components in microstrip transmission line. These components are referred to as distributed components. It can be shown that the series inductance and shunt capacitance can be realized with distributed microstrip transmission lines. We will begin this section with an examination of the short and open-circuited microstrip transmission lines.

2.23 Short-Circuited Transmission Line

Equation (2-38) shows that the input impedance of a lossless short-circuited transmission line is a pure imaginary function; therefore, the input reactance is given by the following equation.

$$X_{in} = Z_O \tan \theta \qquad (2\text{-}73)$$

where $\theta = \beta d$ is the electrical length of the transmission line in degrees.

From Equation (2-86) we can see that this reactance can change from inductive to capacitive depending on the length of the transmission line.

Example 2.3: Plot the input impedance (or reactance) of a lossless short-circuited transmission line as a function of the electrical length of the line.

Solution: To plot the reactance of the short-circuited transmission line create a schematic with a grounded transmission line. Make the length of the transmission line a variable with any starting value in degrees. Setup a new S parameter simulation with a single frequency at 1500 MHz.

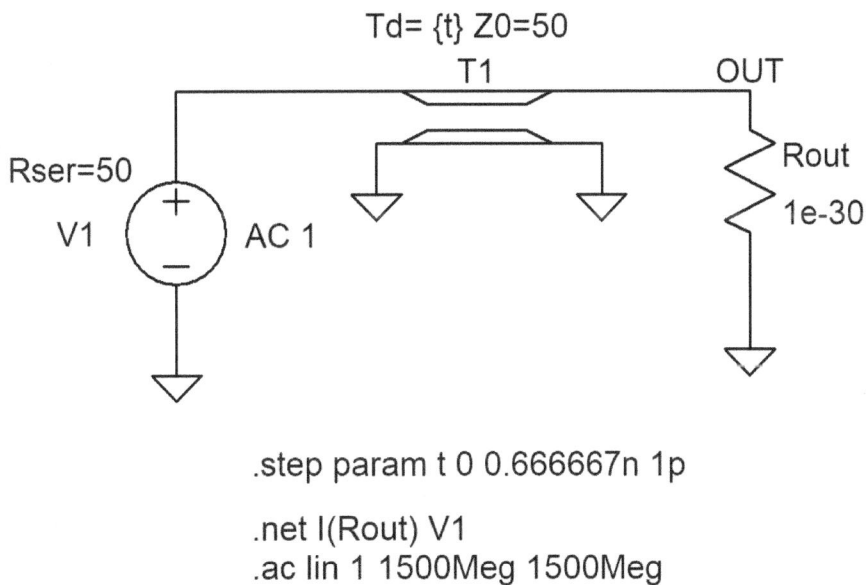

Td= {t} Z0=50

T1   OUT

Rser=50

V1   AC 1   Rout
                     1e-30

.step param t 0 0.666667n 1p

.net I(Rout) V1
.ac lin 1 1500Meg 1500Meg

Figure 2-9 Schematic to simulate reactance of short-circuited transmission line

Then use a Parameter Sweep to vary the electrical length of the transmission line from 0 to 360 degrees. Setup a graph to plot the reactance of the shorted transmission line vs. the electrical length (Group Delay) from the Parameter sweep data set as shown in Fig. 2-9.

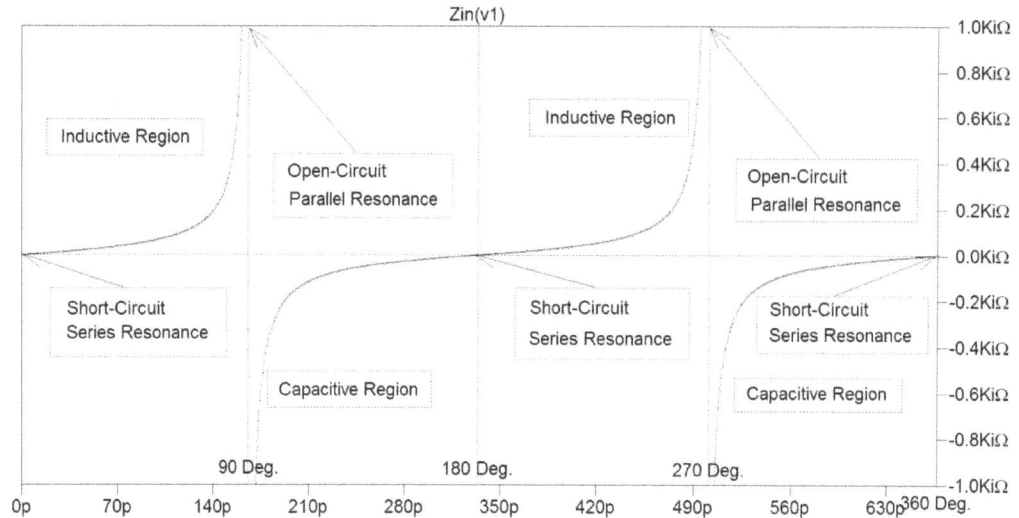

Figure 2-10 Short-circuited line reactance versus electrical length

Normally the independent variable in most linear simulations is frequency. From the reactance plot of Figure 2-10 we can see that at one quarter wavelength, $\lambda_g/4$, the short-circuited line looks like an open circuited line. From 0 to 90 degrees the line looks like an inductor. From $90°$ to $180°$ the circuit looks like a capacitor. The pattern then repeats from $180°$ to $360°$.

2.24 Modeling Short-Circuited Microstrip Lines

The short-circuited microstrip line can be modeled as a microstrip transmission line connected to a grounded via hole. A via hole is made by drilling a hole in the dielectric and metalizing the inside of the hole to form a conductive path to the ground side of the dielectric. To create a 90 degree line in the microstrip substrate we must know the effective dielectric constant so that the wavelength in the dielectric, λ_g, can be calculated. The relationships between line length and electrical degrees are as follows.

$$\theta = \frac{2\pi l}{\lambda_g} \tag{2-74}$$

$$l = \frac{\theta \, \lambda_0}{360 \sqrt{\varepsilon_{eff}}} \qquad (2\text{-}75)$$

Example 2.4: Calculate the time delay required to realize an ideal 50-ohm quarter-wavelength lossless transmission line at 1.5 GHz. Compute the input impedance of this quarter-wave when short–circuited at 1.5 GHz.

Solution: The schematic diagram of the short-circuited transmission line is shown in Figure 2-11. The required time delay is $\tau = \frac{l}{c} = \frac{\lambda_0/4}{c} = \frac{c/f}{4c} = \frac{1}{4f} = \frac{1}{4(1.5x10^9)} \cong$ $0.16666ns$. Here, c is the speed of light in vacuum and inside ideal transmission line.

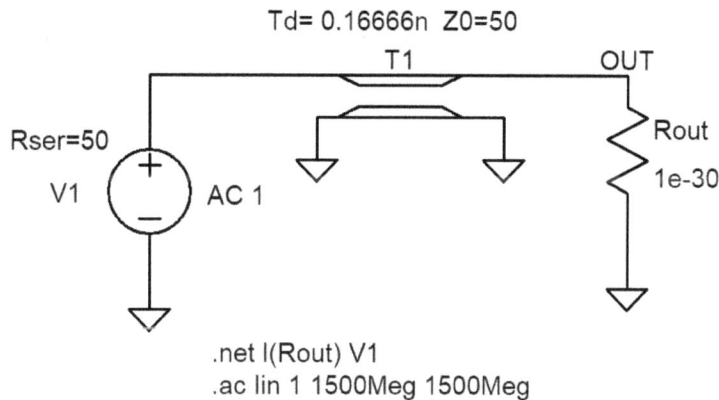

Figure 2-11 Quarter wave short circuited line

Frequency	Zin	Zin
1.5 GHz	0 + j 796 kΩ	796 KΩ / 90°

Figure 2-12 Quarter wave short-circuited line impedance in two formats

As the Figure 2-12 shows, the impedance of a quarter-wave section of short circuit line is quite high, close to an open circuit. Note that depending on the time delay approximation, Zin value can change significantly at 1.5 GHz. This type of line section could be used as a parallel resonant circuit.

2.25 Open-Circuited Transmission Line

Equation (2-20) showed that the input impedance of a lossless open circuited transmission line is a pure imaginary function; therefore, the input reactance is given by the following equation.

$$X_{in} = Z_O \cot\theta \qquad (2\text{-}76)$$

where θ is the electrical length of the transmission line in degrees.

Use a parameter sweep to observe the behavior of this reactance as the length of the open circuit transmission line is varied from 0 to 360 degrees (Figure 2-13). Note again that at 1.5 GHz the length of a quarter wavelength translates into about 167 ps of delay. This delay is equivalent to 90 degrees of electrical length. As can be seen the transmission line is terminated in a 10^6 Ω load to emulate an open circuit transmission line. However, for increased accuracy, a higher value for Rout may be used.

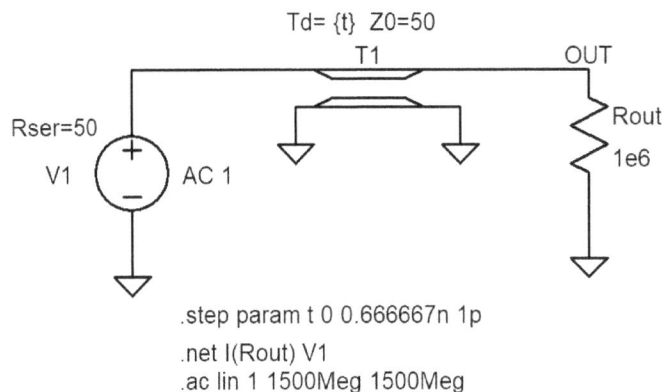

Figure 2-13 Schematic to simulate reactance of open-circuited transmission line

Comparing the open circuit reactance to the short-circuited line reactance we can see that a 90°, $\lambda_g/4$, offset is present. Here, the input impedance is 0 Ω when transmission line delay is about 167 ns or 90 degrees.

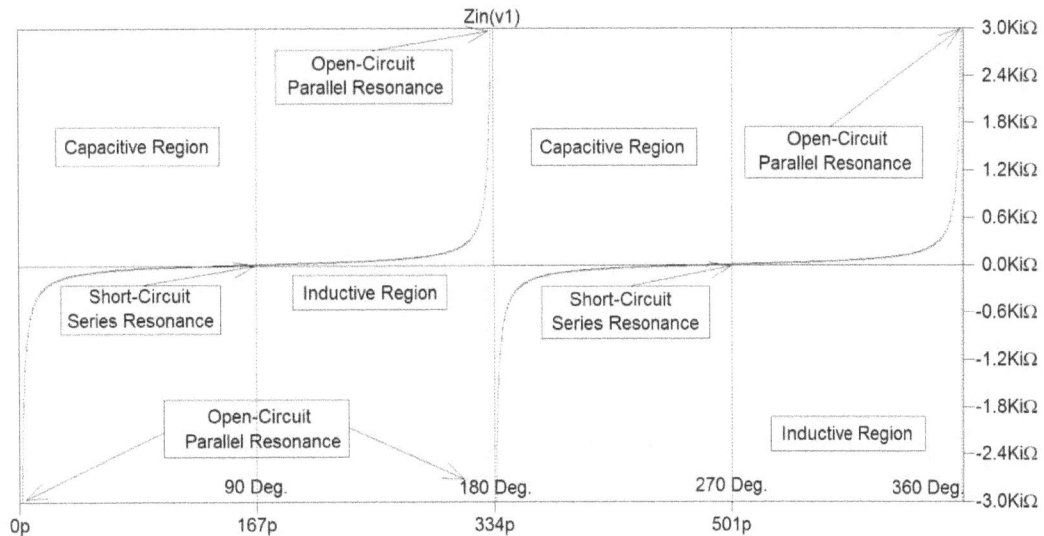

Figure 2-14 Open circuit transmission line reactance

2.26 Modeling Open-Circuited Microstrip Lines

Care must be used when modeling the open circuit microstrip line due to the radiation effects from the end of the transmission line. The E fields that exist in the air space of the microstrip line add capacitance to the microstrip transmission line. On an open circuit microstrip line this fringing capacitance is referred to as an end effect.

Example 2.5: Calculate the input impedance of a quarter-wave open–circuited transmission line at 1.5 GHz. Assume an ideal lossless transmission line as in Example 2.4.

Solution: Place an open-circuited transmission line on the schematic. Figure 2-15 shows that we have used a 1 Mega Ohm resistor to act as an open circuit for the transmission line.

Simulate the schematic and display the input impedance as shown in Figure 2-16.

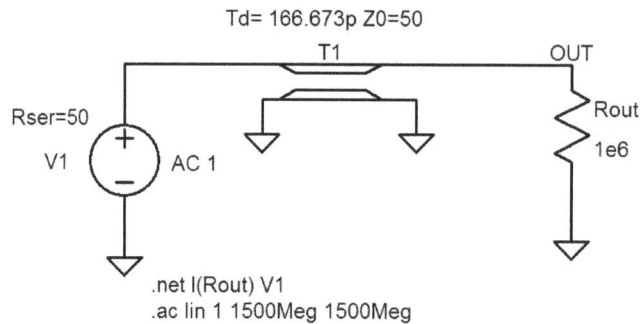

Figure 2-15 Quarter wave open circuited line schematic

freq	Zin1	Zin1
1.500 GHz	0.045 + j0.835	0.837 / 86.935

Figure 2-16 Quarter wave open circuited line impedance

2.27 Distributed Inductive and Capacitive Elements

Thus far we have dealt with only 50 Ω transmission lines. It is possible to synthesize series inductance by using short lengths of transmission lines that have considerably higher impedance than 50 Ω. It is possible to synthesize shunt capacitors by using short lengths of transmission lines that have considerably lower impedance than 50 Ω. Typical impedances would range from approximately 20 Ω for capacitive elements and 80 Ω for inductive elements. The actual impedance used is a compromise between the substrate height and dielectric constant and the ability to physically realize the distributed element. For example an 80 Ω line on a thin substrate or a high dielectric constant substrate may be too narrow to etch on a printed circuit board. In such a case it may be necessary to use a 70 Ω impedance to realize the inductive element. These elements can be used successfully in narrow bandwidth applications. These distributed elements are very useful in the design of filters, bias feed networks, and impedance matching networks.

2.28 Microstrip Inductance and Capacitance

For short lengths of high impedance transmission line use the following equation to calculate the length of microstrip line to synthesize a specific value of inductance.

$$Inductive\ Line\ Length = \frac{f\,\lambda_g\,L}{Z_L} \qquad (2\text{-}77)$$

$$Capacitive\ Line\ Length = f\,\lambda_g\,Z_C\,C \qquad (2\text{-}78)$$

where,

f = frequency at which inductance is calculated

L = nominal inductance value

C = nominal capacitance value

Z_L = impedance of the inductive transmission line

λ_g = wavelength using the effective dielectric constant, ε_{eff}

Z_C = impedance of the capacitive transmission line

Example 2.6: Convert the lumped elements capacitors and inductors (Figure 2-17) to distributed elements. Assume frequency of operation is 100 MHz.

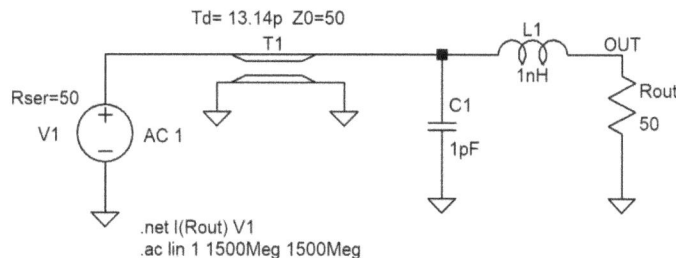

Figure 2-17 Lumped capacitive and inductive lines with PCB layout

Solution: Assume using microstrip on Roger's RO3003 substrate having ε_r=3, t = 1.4 mils, H = 30 mils. We use W = 250/20 mils (i.e, 20.5/96.7 Ω) to realize 1 pF capacitor and 1 nH inductor elements, respectively. Use equations 2-69 and 2-70 to compute the capacitive and inductive line lengths. Use AppCAD (see Figure 2-18(c)) for line impedance / electrical length calculations. Convert the electrical length to delay in seconds (= length of line/speed of light) and enter in LTspice schematic (Figure 2-18(a)). The printed circuit layout (Figure 2-19) shows the line width relationship among the 50 Ω, 20.5 Ω, and 96.7 Ω microstrip lines.

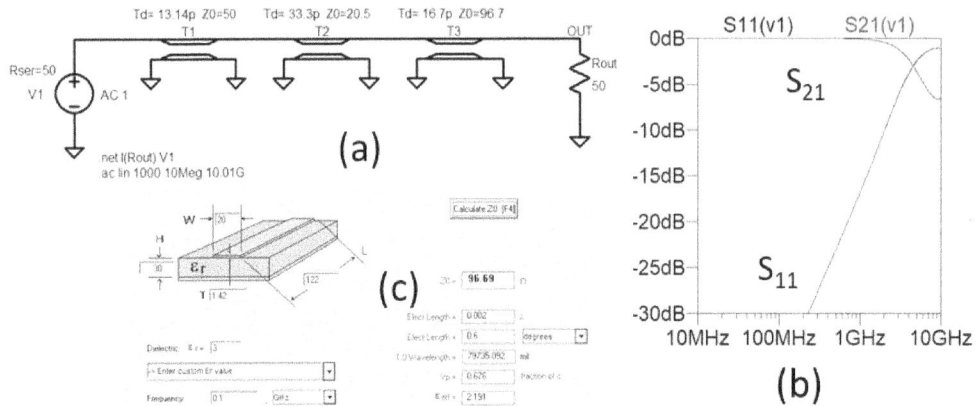

Figure 2-18 Schematic for distributed capacitive and inductive lines in LTs (a) Simulated response (b), Impedance / electrical length calculations in AppCAD (c)

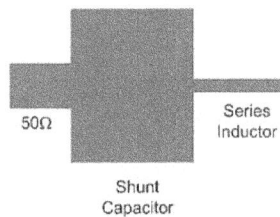

Figure 2-19 Capacitive and inductive lines in PCB layout

2.29 Microstrip Bias Feed Networks

Another useful purpose for high impedance and low impedance microstrip transmission lines is the design of bias feed networks. Often it is necessary to insert voltage and current to a device that is attached to a microstrip line. Such a device could be a transistor, MMIC amplifier, or diode. The basic bias feed or "bias decoupling network" consists of an inductor (used as an "RF Choke") and shunt capacitor (used as a bypass capacitor). At lower RF frequencies (less than 200 MHz) these networks are almost entirely realized with lumped element components. Even at these low frequencies it is very important to account for the parasitic in the components.

Fig. 2-19 shows a typical series inductor, shunt capacitor, lumped element bias feed and its effect on a 50 Ω transmission line.

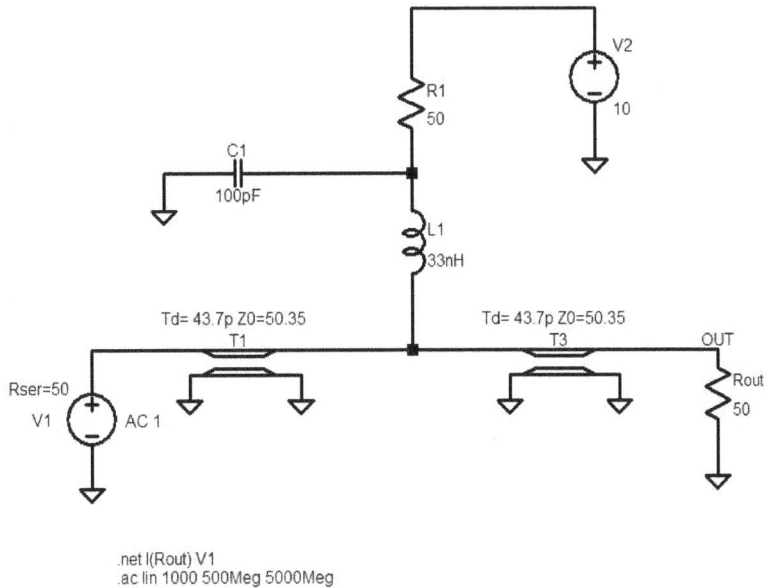

Figure 2-20 Inductor and bypass capacitor bias feed network

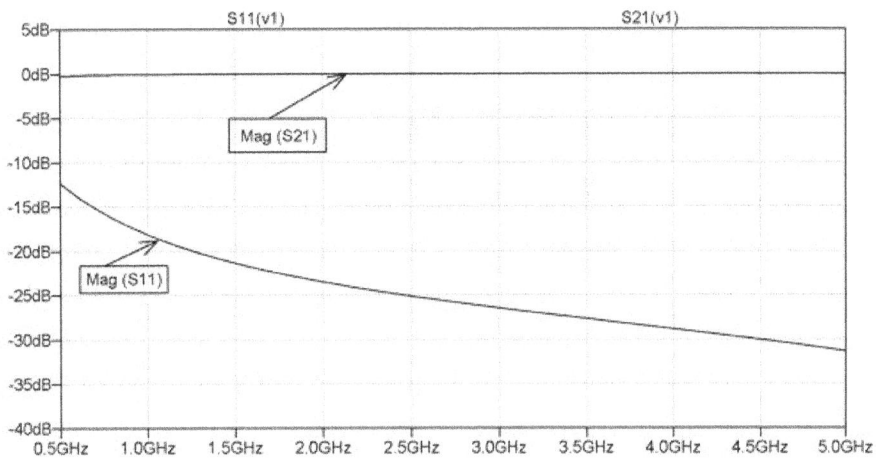

Figure 2-21 Response of inductor and capacitor bias feed network

2.30 Distributed Bias Feed Design

A high impedance microstrip line of $\lambda_g/4$ can be used to replace the lumped element inductor. Similarly a $\lambda_g/4$ of low impedance line can be used to model the shunt capacitor.

Example 2.7: Use the Advanced TLine utility to calculate the physical line length of the $\lambda_g/4$ sections of 80 Ω and 20 Ω microstrip lines at a frequency of 2 GHz. Create a schematic of a distributed bias feed network.

Solution: Use the 80 Ω high impedance quarter wave section and a shunt capacitance as shown in Figure 2-22. A tee junction, if available, can be used to accurately model the electrical length of each junction.

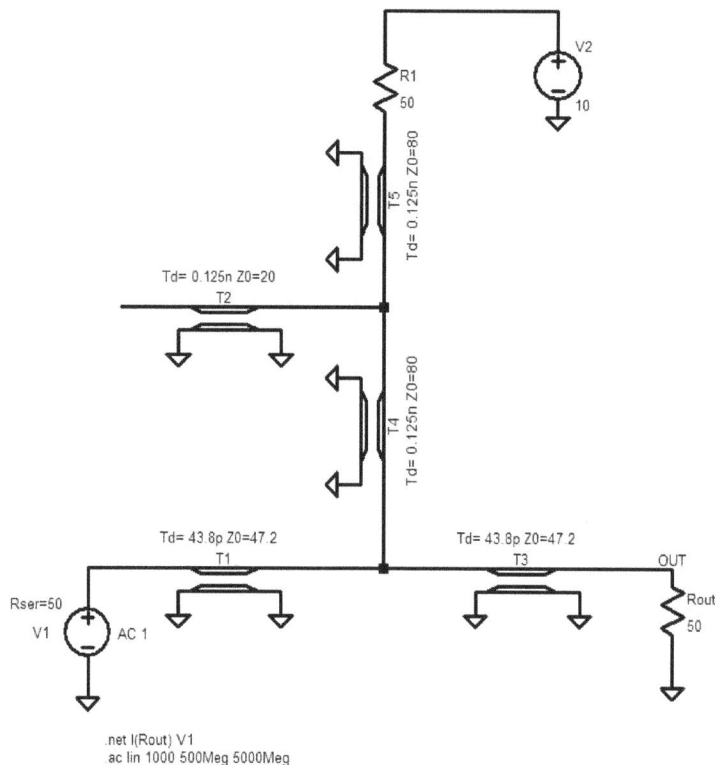

Figure 2-22 Distributed Bias feed network

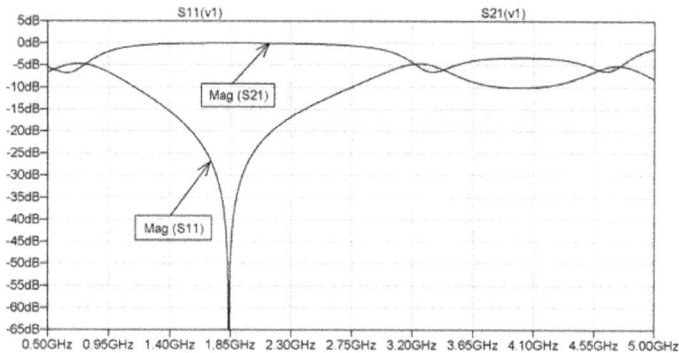

Figure 2-23 Bias feed response

2.31 Coupled Transmission Lines

There are three primary methods in which coupled lines are used in microwave circuit design. These are end-coupled, edge-coupled, and broadside coupled line structures. End coupled lines are often used to realize microstrip resonators and filters. Edge coupled lines are used in both coupler and filter designs. Broadside coupled lines are popular with various coupler designs. Because of the coupling between the lines there exist two modes of impedance required to characterize the circuit. These are known as the even mode, Z_{oe}, and odd mode, Z_{oo}, impedance. Figure 2-24 shows the field distribution on the edge coupled lines for both conduction modes. The magnitude of the even and odd mode impedance is strongly dependent on the separation between the lines which also determine the electrical coupling between the lines. The coupling between the lines, in dB, is defined by Equation (2-79).

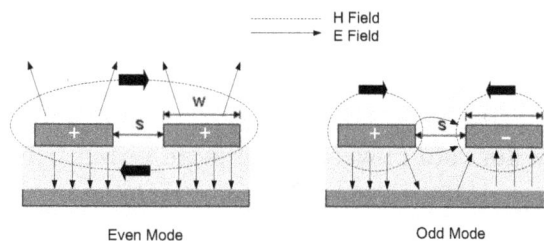

Figure 2-24 Edge coupled microstrip line field distribution

$$C = 20 \ \log \left| \frac{Z_{oe} - Z_{oo}}{Z_{oe} + Z_{oo}} \right|$$

(2-79)

The even and odd mode impedances are then defined by the following equations [6].

$$Z_{oo} = Z_o \sqrt{\frac{1 - 10^{\left(\frac{-C}{20}\right)}}{1 + 10^{\left(\frac{-C}{20}\right)}}} \tag{2-80}$$

$$Z_{oe} = Z_o \sqrt{\frac{1 + 10^{\left(\frac{-C}{20}\right)}}{1 - 10^{\left(\frac{-C}{20}\right)}}} \tag{2-81}$$

$$Z = \sqrt{Z_{oo} \, Z_{oe}} \tag{2-82}$$

The even and odd mode impedance can be calculated for a given characteristic impedance, Z_o, and coupling ratio in dB.

2.32 Directional Coupler

One important use of coupled lines is the design of directional couplers. Directional couplers are useful components for sampling an RF signal without significantly loading or perturbing the input signal. Simple directional couplers are often used to provide a sample of an RF signal for measurement. High quality, precision, directional couplers can separate incident and reflected signals and are the fundamental component used for the measurement of VSWR and return loss. Two directional couplers can be placed back-to-back to form a dual directional coupler (Figure 2-24). This type of coupler forms a four port network that can provide a sample of the forward power and reflected power between a source and load. Some basic properties of directional couplers include:

Insertion Loss:
Insertion Loss is simply the ratio of the output power at P2 to the input power at P1. Expressed in dB:

$$Insertion\ Loss\ (dB) = P2_{dBm} - P1_{dBm} \tag{2-83}$$

Coupling:

The coupling factor is the ratio of the output power at P3 to the input power at P1. In microstrip and stripline circuits the coupled port is adjacent to the input line port. In waveguide couplers the coupled port is furthest from the input port.

$$Input\ Port\ Coupling\ (dB) = P1_{dBm} - P3_{dBm} \qquad (2\text{-}84)$$

Isolation:

Isolation is the ratio of the output power at P4 to the input power at P1.

$$Isolation\ (dB) = P1_{dBm} - P4_{dBm} \qquad (2\text{-}85)$$

Directivity:

Directivity is the difference between the isolation and the coupling when P2 is perfectly terminated in 50 Ω. Another way to think of a coupler's directivity is its ability to properly separate the forward and reflected waves.

$$Directivity\ (dB) = P3_{dBm} - P4_{dBm} \qquad (2\text{-}86)$$

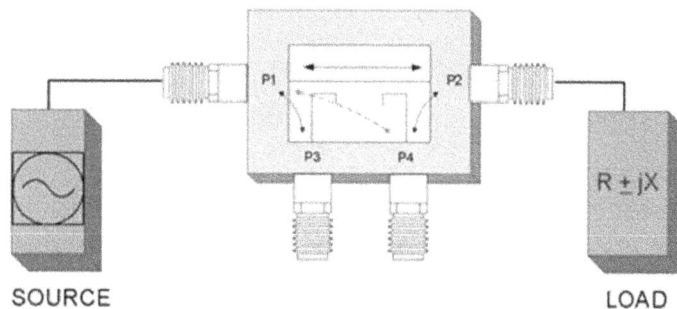

Figure 2-25 Dual directional coupler

A coupler will always have a finite amount of isolation. Ideally, all of the power input to P1 should be directed to P2 and P3. However some finite amount of power will show up at P4. This power will then add with any reflected power coming from P2 and being directed to P4. It is this finite isolation that limits the directivity of the directional coupler. Thus the power at P3 is the (Incident power at P1 – coupling factor) + the (Reflected power from P2– isolation). For simple RF power sampling applications, the directivity is not that critical. But if we are using the directional coupler to measure VSWR, the directivity is very important.

In VSWR measurement applications it is important to know the directivity of the directional coupler that is used to perform the measurement. A significant measurement error can exist when the coupler directivity is less than 40 dB.

Problems

2-1.   The input reflection coefficient of a transistor is measured to be 0.1 at an angle of $30°$. Determine the input VSWR of the device.

2-2.   A *Low-Loss Transmission Line* is defined as one having R $\ll$ ωL and G $\ll$ ωC. Show that below conditions hold:

$$\alpha = \frac{1}{2}\left(R\sqrt{\frac{C}{L}} + G\sqrt{\frac{L}{C}}\right), \quad \beta \cong \omega\sqrt{LC}, \quad v_p = \frac{\omega}{\beta} \cong \frac{1}{\sqrt{LC}}, \quad Z_0 \cong \sqrt{\frac{L}{C}}$$

2-3.   A line is *Distortionless* if R/L = G/C. Show that under this condition, below relations hold:

$$\gamma = \alpha + j\beta = R\sqrt{\frac{L}{C}} + j\omega\sqrt{LC} \ and$$

$$v_p = \frac{\omega}{\beta} = \frac{1}{\sqrt{LC}} \ and \ Z_0 = \sqrt{\frac{L}{C}}$$

2-4.   A voltage source $v_s = 0.75\cos(2\pi 10^9 t)$ (V) having internal source resistance of 5 ohms is connected to a 20-meter long lossless transmission line. The characteristic impedance of the line is 75 ohms. The line is terminated in a matched load. Assume that the velocity of wave propagation on the line $v_p = 85\%$ of speed of light in vacuum. Determine

(a) The input impedance to the line as seen by the voltage source, Z_{in}.
(b) The peak amplitude of the phasor voltage and phasor current at the input of the transmission line.
(c) The propagation constant γ.
(d) V(z) at any point on the transmission line.
(e) I(z) at any point on the transmission line.
(f) The instantaneous voltage $v(z, t)$ at any point on the transmission line.

(g) The instantaneous current $i(z, t)$ at any point on the transmission line.

(h) The instantaneous voltage and current at the load.

(i) The average power at the input of the line and the average power delivered to the load. What can be said about the two powers?

2-5.   Determine the impedance of a quarter-wave transformer to match a 30 Ω load to a 50 Ω source.

2-6.   Design the quarter-wave transformer using a microstrip transmission line given the conditions in Problem 5. The frequency of operation is 3 GHz. The dielectric constant is 3.4 with a thickness of 0.030 in. Determine the length and the width of the microstrip line.

2-7.   A radio transmitter is operating into a transmission line that measures a 2:1 VSWR. Determine the percentage of power that would be expected to reflect back into the transmitter.

2-8.   A series RLC load, R = 70 Ω, L = 10 nH, C = 25 pF is connected to a 50 Ω transmission line. Setup a Linear Analysis to sweep the frequency from 200 MHz to 2000 MHz in 200 MHz steps. Display the input reflection Coefficient, S11, and VSWR in a Table.

2-9.   Calculate the cutoff frequency of the TE1,0 mode in a rectangular waveguide with a height of 0.200 inches and a width of 0.45 inches. Also calculate the waveguide wavelength, λ_g.

2-10.   Design a distributed bias feed network for a C Band amplifier operating at 6.0 GHz. Use a microstrip substrate with a dielectric constant of 10.2 and a thickness of 0.025 inches. Plot the insertion loss and return loss from 2 GHz to 10 GHz.

2-11.   Determine the physical length of a $\lambda_g/4$ open circuit microstrip line with an impedance of 25 Ω. The frequency of operation is 10 GHz. Use a microstrip dielectric constant of 2.2 and a thickness of 0.010 inches. Determine whether an end-effect model element should be used.

References and Further Readings

[1]   Robert E. Collin, *Foundations for Microwave*, Second Edition, McGraw-Hill, Inc., New York, 1996.

[2]   David M. Pozar, *Microwave Engineering*, Third Edition, John Wiley and Sons, Inc. 2005

[3]   Guillermo Gonzales, *Microwave Transistor Amplifiers – Analysis and Design*, Second Edition, Prentice Hall Inc., Upper Saddle River, NJ.

[4]   David K. Cheng, *Field and Wave Electromagnetics*, Second Edition, Addison-Wesley, Inc. 1989

[5]   Herbert L. Kraus, Charles W. Bostian, and Fredrick H. Raab, *Solid State Radio Engineering*, John Wiley & Sons, Inc., New York, 1980.

[6]   *Foundations for Microstrip Circuit Design*, T.C. Edwards, John Wiley & Sons, New York, 1981

[7]   Ali A. Behagi, *100 ADS Design Examples*, Based on the Textbook: *RF and Microwave Circuit Design,* Techno Search, Ladera Ranch, CA 2016

[8]   *UHF/Microwave Experimenters Manual*, American Radio Relay League, Newington, CT.1990

[9]   Reference: I. J. Bahl and D. K. Trivedi, "A Designer's Guide to Microstrip Line", Microwaves, May 1977, pp. 174-182.

[10]   *Microwave Handbook Volume 1*, Radio Society of Great Britain, The Bath Press, Bath, U.K., 1989.

[11]   *Tatsuo Itoh, Planar Transmission Line Structures*, IEEE Press, New York, NY, 1987.

Chapter 3

Network Parameters and the Smith Chart

3.1 Development of Network S Parameters

For RF and microwave networks, a form of the transmission matrix has been defined based on power measurements into the system's characteristic impedance. These parameters are known as S parameters, named after their scattering matrix form. Consider a two-port network, shown in Figure 3-1, where Z_{01} is the real characteristic impedance and V_1^+ and V_1^-, respectively, are the incident and reflected voltage waveforms at the input port. Similarly, Z_{02} is the real characteristic impedance and V_2^+ and V_2^-, respectively, are the incident and reflected voltage waveforms at the output port.

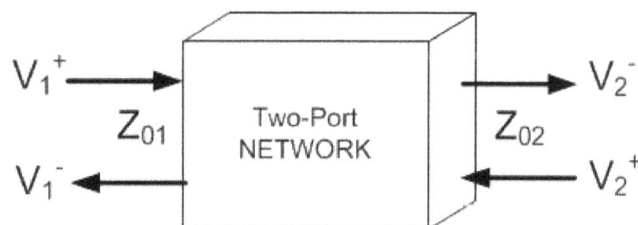

Figure 3-1 Two-port network with incident and reflected voltage waveforms

In order to obtain measurable power relations in terms of wave amplitudes, we need to define a new set of waveforms by normalizing the voltage amplitudes with respect to the square root of the respective characteristic impedances, namely:

$$a_1 = \frac{V_1^+}{\sqrt{Z_{01}}} \tag{3-1}$$

$$b_1 = \frac{V_1^-}{\sqrt{Z_{01}}} \tag{3-2}$$

$$a_2 = \frac{V_2^+}{\sqrt{Z_{02}}}$$ (3-3)

$$b_2 = \frac{V_2^-}{\sqrt{Z_{02}}}$$ (3-4)

The two-port network with normalized waveforms is shown in Figure 3-2.

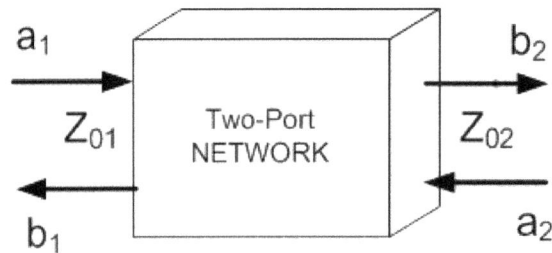

Figure 3-2 Two-port network with incident and reflected waveforms

Notice that:

$$|a_1|^2 = \frac{|V_1^+|^2}{Z_{01}} = \text{Incident power at the network input}$$

$$|b_1|^2 = \frac{|V_1^-|^2}{Z_{01}} = \text{Reflected power at the network input}$$

$$|a_2|^2 = \frac{|V_2^+|^2}{Z_{02}} = \text{Incident power at the network output}$$

$$|b_2|^2 = \frac{|V_2^-|^2}{Z_{02}} = \text{Reflected power at the network output}$$

The S parameters relate b_1 and b_2 to a_1 and a_2 by the following Equations.

$$b_1 = S_{11}a_1 + S_{12}a_2 \tag{3-5}$$

$$b_2 = S_{21}a_1 + S_{22}a_2 \tag{3-6}$$

In matrix form the scattering matrix is written as:

$$\begin{bmatrix} b_1 \\ b_2 \end{bmatrix} = \begin{bmatrix} S_{11} & S_{12} \\ S_{21} & S_{22} \end{bmatrix} \cdot \begin{bmatrix} a_1 \\ a_2 \end{bmatrix}$$

At RF and microwave frequencies the normalized voltage waveforms a_1, a_2, b_1, and b_2 represent vectors having both magnitude and phase. Terminating the output of the two-port network with a real load impedance that is equal to the real system characteristic impedance, $Z_{01} = Z_{02}$, forces $a_2=0$. Solving for the individual S parameters then gives the following relationships.

$$S_{11} = \frac{V_1^-}{V_1^+} = \left. \frac{b_1}{a_1} \right|_{a_2=0} \tag{3-7}$$

$$S_{21} = \frac{V_2^-}{V_1^+} = \left. \frac{b_2}{a_1} \right|_{a_2=0} \tag{3-8}$$

Terminating the input of the two-port network with a real load impedance that is equal to the real system characteristic impedance, $Z_{01} = Z_{02}$, forces $a_1 = 0$. The S parameters S_{22} and S_{12} can be solved by Equations (3-9) and (3-10).

$$S_{22} = \frac{V_2^-}{V_2^+} = \left. \frac{b_2}{a_2} \right|_{a_1=0} \tag{3-9}$$

$$S_{12} = \frac{V_1^-}{V_2^+} = \frac{b_1}{a_2}\bigg|_{a_1=0} \tag{3-10}$$

The relationship between Scattering parameters and incident and reflected voltages V^+ and V^- can be summarize as

$$V_1^- = S_{11}V_1^+ + S_{12} V_2^+ \tag{3-11}$$
$$V_2^- = S_{21} V_1^+ + S_{22} V_2^+ \tag{3-12}$$

In matrix form, for any number of ports, we can write

$$[V^-] = [S][V^+]. \tag{3-13}$$

For a two-port network, S-parameters matrix is given by

$$\begin{bmatrix} V_1^- \\ V_2^- \end{bmatrix} = \begin{bmatrix} S_{11} & S_{12} \\ S_{21} & S_{22} \end{bmatrix}\begin{bmatrix} V_1^+ \\ V_2^+ \end{bmatrix}. \tag{3-14}$$

The input and output reflection coefficients are given by

$$\Gamma_{IN} = \frac{V_1^-}{V_1^+} \tag{3-15}$$

$$\Gamma_{OUT} = \frac{V_2^-}{V_2^+}. \tag{3-16}$$

Given that port two is terminated in a load impedance Z_L, we can assume that

$$\Gamma_L = \frac{V_2^+}{V_2^-} \tag{3-17}$$

Using equations (3-11), (3-12), and (3-17), we can write [1-2]

$$V_2^- = S_{21} V_1^+ + S_{22} \Gamma_L V_2^- \tag{3-18a}$$

$$V_2^-(1 - \Gamma_L S_{22}) = S_{21}V_1^+ \tag{3-18b}$$

$$V_2^- = \frac{S_{21}V_1^+}{(1 - \Gamma_L S_{22})} \tag{3-19}$$

$$V_1^- = S_{11} V_1^+ + S_{12} V_2^+ = S_{11}V_1^+ + S_{12} \Gamma_L V_2^- \tag{3-20}$$

$$V_1^- = S_{11}V_1^+ + S_{12} \Gamma_L \frac{S_{21}V_1^+}{(1 - \Gamma_L S_{22})} \tag{3-21}$$

$$\Gamma_{IN} = \frac{V_1^-}{V_1^+} = S_{11} + \frac{S_{12}S_{21}\Gamma_L}{(1 - \Gamma_L S_{22})} \tag{3-22}$$

Similarly, when port one is terminated in a source impedance Z_s, it can be shown (Problem 3-1) that

$$\Gamma_{OUT} = \frac{V_2^-}{V_2^+} = S_{22} + \frac{S_{12}S_{21}\Gamma_s}{(1 - \Gamma_s S_{11})}. \tag{3-23}$$

When port 2 is terminated in a load resistance equal to line characteristic impedance, $\Gamma_L = 0$, and therefore $\Gamma_{IN} = S_{11}$. Similarly, when port 1 is terminated in a source resistance equal to line characteristic impedance, $\Gamma_S = 0$, and hence $\Gamma_{OUT} = S_{22}$. Note that $\Gamma_S = \frac{Z_S - Z_{01}}{Z_S + Z_{01}} = \frac{V_1^+}{V_1^-}$.

S_{11} is often referred to as the **input reflection coefficient** and S_{22} as the **output reflection coefficient** of the network. S_{21} is the **forward transmission** and is often expressed as a gain or loss depending on whether S_{21}, in dB, is positive or negative. S_{12} is the **reverse transmission** or isolation of the network. Because the S parameters are complex entities they must be measured with a Vector Network Analyzer capable of measuring both amplitude and phase. A Scalar Network Analyzer is used to measure the magnitude of two-port networks. When cascading multiple S parameters, S parameter of each stage needs to be converted to ABCD or T parameters. The ABCD or T matrices (discussed in Appendix B along with other two-port parameters) can then be multiplied together and the resultant matrix can be converted back to S matrix to obtain the overall S parameters.

S parameters of a two port network can also be calculated by constructing the Thevenin equivalent circuits [3] at port 1 and 2.

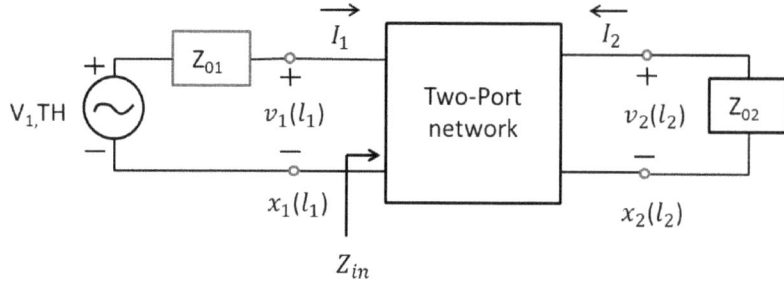

Figure 3-3. Two-port network with Thevenin equivalent at ports 1 and 2

$$S_{11} = \frac{Z_{in} - Z_{01}}{Z_{in} + Z_{01}}$$

(3-24)

$$S_{21} = \frac{2 v_2(l_2) . \sqrt{Z_{01}}}{V_{1,}TH . \sqrt{Z_{02}}}$$

(3-25)

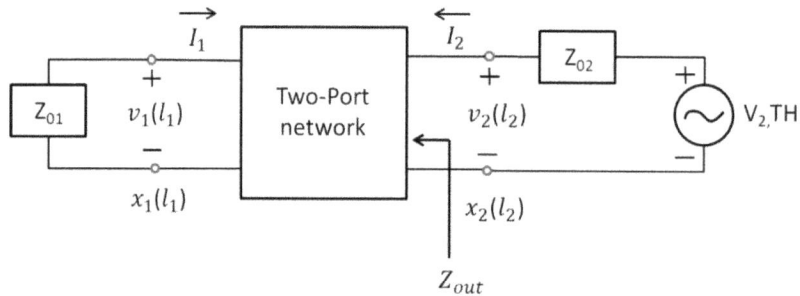

Figure 3-4. Two-port network with Thevenin equivalent at ports 1 and 2

$$S_{22} = \frac{Z_{out} - Z_{02}}{Z_{out} + Z_{02}}$$

(3-26)

$$S_{12} = \frac{2 v_1(l_1) . \sqrt{Z_{02}}}{V_{2,}TH . \sqrt{Z_{01}}}$$

(3-27)

Example 3-1: Find the S parameters of a series impedance Z with line impedances Z_{01} and Z_{02}, as shown below.

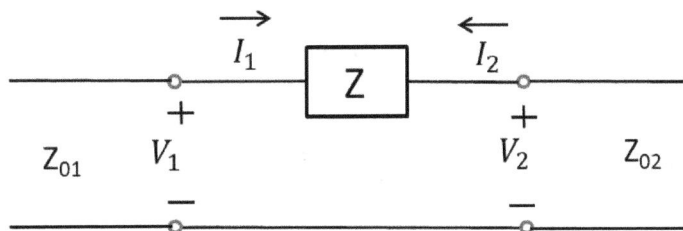

Figure 3-5 A series impedance Z in Z_{01} and Z_{02} system

Solution: Following the analysis of Figures 3-3 and 3-4, we can write

$$S_{11} = \frac{Z_{in} - Z_{01}}{Z_{in} + Z_{01}} = \frac{Z + Z_{02} - Z_{01}}{Z + Z_{02} + Z_{01}}$$

$$S_{22} = \frac{Z_{out} - Z_{02}}{Z_{out} + Z_{02}} = \frac{Z + Z_{01} - Z_{02}}{Z + Z_{01} + Z_{02}}$$

$$S_{21} = \frac{2 V_2 \sqrt{Z_{01}}}{V_{1,TH.} \sqrt{Z_{02}}} = \frac{2\sqrt{Z_{01}}}{\sqrt{Z_{02}}} \frac{Z_{02}}{Z + Z_{01} + Z_{02}} = \frac{2\sqrt{Z_{01} Z_{02}}}{Z + Z_{01} + Z_{02}}$$

It can be shown that $S_{12} = S_{21}$ (Problem 3-2).

When $Z_{01} = Z_{02} = Z_0$, the S parameters reduce to

$$S_{11} = S_{22} = \frac{Z}{Z + 2Z_0}$$

$$S_{21} = S_{12} = \frac{2Z_0}{Z + 2Z_0}$$

Alternately, S_{21} can be determined using continuity of current at port 1 and port 2. At port 1, I_1 is given as

$$I_1 = \frac{V_1^+}{Z_{01}} - \frac{V_1^-}{Z_{01}} = \frac{V_1^+}{Z_{01}}\left(1 - \frac{V_1^-}{V_1^+}\right) = \frac{V_1^+}{Z_{01}}(1 - S_{11})$$

The current at port 2 can be written as: $I_2 = I_2^+ + I_2^- = I_2^- = \frac{V_2^-}{Z_{02}}$.

Above is true since port two is properly matched. Therefore, $I_1 = I_2$, resulting in

$$S_{21} = \frac{V_2^-}{V_1^+} = (1 - S_{11})\frac{Z_{02}}{Z_{01}}$$

If $Z_{01} = Z_{02} = Z_0$, S_{21} then simplifies as

$$S_{21} = \frac{V_2^-}{V_1^+} = (1 - S_{11}) = 1 - \frac{Z}{Z + 2Z_0} = \frac{2Z_0}{Z + 2Z_0}$$

Finally, for simplification, we can apply a source with peak phasor amplitude of 1 V in determining V_2^- and V_1^+. Note that V_1^+ can be determined under the assumption that the source has not "sensed" a mismatch at the input of the two-port terminal (i.e., $V_1^+ = \frac{(1V)Z_{01}}{Z_{01} + Z_{01}} = \left(\frac{1}{2}\right)V$).

Example 3-2: Find the S parameters of admittance Y connected across the transmission line with characteristic impedance and admittance Z_0 and Y_0.

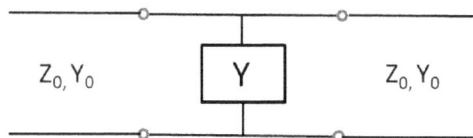

Figure 3-6 A shunt admittance Y in Z_0 / Y_0 system

Solution: It is more convenient to use characteristic admittance of the transmission line in determining S_{11}.

$$S_{11} = \frac{Z_{in} - Z_0}{Z_{in} + Z_0} = \frac{Y_0 - Y_{in}}{Y_0 + Y_{in}} = \frac{Y_0 - (Y + Y_0)}{Y_0 + (Y + Y_0)} = \frac{-Y}{Y + 2Y_0}$$

Since $Y_0 = 1/Z_0$, S_{11} becomes

$$S_{11} = \frac{-Z_0 Y}{2 + Z_0 Y}$$

S_{21} can be easily determined using voltage relations at ports 1 and 2 since $V_1 = V_2$. With port two properly matched, V_2 is simplified as

$$V_2 = V_2^-$$

$$V_1 = V_1^+ + V_1^- = V_1^+ (1 + S_{11})$$

Since $V_1 = V_2$, we have

$$V_1 = V_1^+ + V_1^- = V_1^+ (1 + S_{11}) = V_2^-$$

$$S_{21} = \frac{V_2^-}{V_1^+} = 1 + S_{11} = 1 + \frac{-Z_0 Y}{2 + Z_0 Y} = \frac{2}{2 + Z_0 Y}$$

By symmetry, we can conclude that $S_{22} = S_{11}$ and $S_{12} = S_{21}$.

Example 3-3: The symmetrical attenuator network T consisting of resistors R_1 and R_2 provide voltage reduction factor K, when the attenuator is terminated in a source or load resistance equal to characteristic impedance of the transmission line. For the T attenuator network shown below, R_1 and R_2 are given by

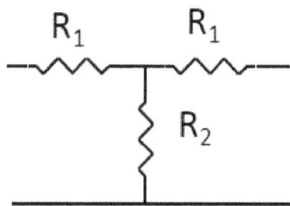

Figure 3-7 The T attenuator network

$$R_1 = \frac{1-K}{1+K} Z_0$$

$$R_2 = \frac{2K}{1-K^2} Z_0$$

where,

Z_0 = characteristic line impedance of a given system

K = voltage attenuation factor.

In a 50-Ω system, for example, Z_0 = 50 Ω, and for a 3-dB attenuator, K can be determined from 10 log_{10} (K^2) = -3. Therefore K = $\sqrt{0.5}$, resulting in R_1 = 8.58 Ω and R_2 = 141.4 Ω. **(a)** Find the S parameters of the T network in a Z_0 system in terms of R_1 and R_2, and Z_0. **(b)** Compute R_1, R_2, and the S parameters for a 10-dB attenuator in a 50-ohm system. Design and simulation of this 10-dB attenuator circuit in LTspice is discussed in Example 3-4.

Solution (a): S_{11} and S_{21} can be determined from the figure below.

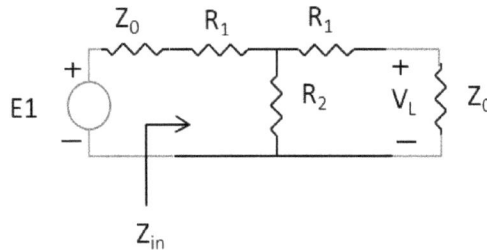

Figure 3-8 Determination of scattering parameters for a T attenuator

$$S_{11} = \frac{Z_{in}-Z_0}{Z_{in}+Z_0}, \text{ where}$$

$$Z_{in} = R_1 + \frac{R_2(R_1+Z_0)}{R_1+R_2+Z_0}$$

$$S_{21} = \frac{2V_2\sqrt{Z_{01}}}{V_{1,TH.}\sqrt{Z_{02}}} = \frac{2V_L}{E1} \text{ since } Z_{01} = Z_{02} = Z_0$$

V_L = (I_L) (Z_0) and I_L can be obtained from

$$I_L = \frac{E1}{Z_{in} + Z_0} \frac{R_2}{R_1 + R_2 + Z_0} = \frac{E1}{2Z_0} \frac{R_2}{R_1 + R_2 + Z_0}.$$

Therefore,

$$S_{21} = \frac{R_2}{R_1 + R_2 + Z_0}.$$

This expression for S_{21} is equal to K for a T attenuator, as expected since both S_{21} and K are voltage reduction factors for the attenuator.

S_{22} and S_{12} can be determined by symmetry, namely, $S_{22} = S_{11}$ and $S_{12} = S_{21}$.

Solution (b): Since for a 10-dB attenuator $10\ log_{10}\ (K^2) = -10$, we obtain K$=\sqrt{0.1}$.

R1 and R2 can be computed using K and Z_0. Thus, $R_1 = 25.97\Omega$ and $R_2 = 35.14\Omega$. Using the expressions for S_{11} and S_{21}, it can be shown that $Z_{in} = 50\Omega$ and therefore $S_{11} = 0$. $S_{21} = 0.316 = \sqrt{0.1}$.

The S matrix for the 10-dB attenuator is given by $S = \begin{bmatrix} 0 & \sqrt{0.1} \\ \sqrt{0.1} & 0 \end{bmatrix}$.

In general, the S parameters matrix of an attenuator is given by

$$S = \begin{bmatrix} S_{11} & S_{12} \\ S_{21} & S_{22} \end{bmatrix} = \begin{bmatrix} 0 & K \\ K & 0 \end{bmatrix}.$$

Example 3-4: Draw a schematic diagram for the 10-dB attenuator in example above. Simulate S_{11} and S_{21} from low frequency to 1 GHz. Comment on S_{22} and S_{12} values.

Solution: The schematic diagram of the attenuator is shown in Fig. 3-9. The simulated values for S_{11} and S_{21} both in linear and decibel are shown in figures 3-9b and 3-9c, respectively. As expected, S_{11} is equal to 0, and S_{21} is equal to $\sqrt{0.1}$ = 0.316. It can be shown that $S_{22} = S_{11}$ and $S_{12} = S_{21}$. This is expected due the symmetry of the attenuator circuit.

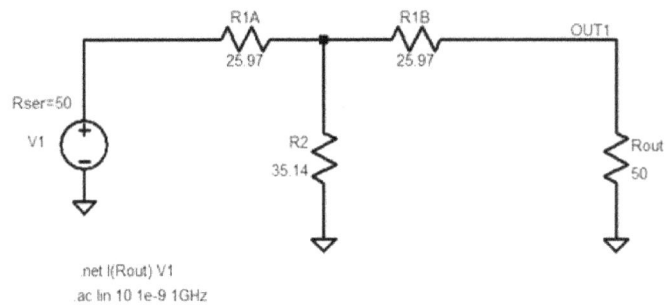

(a)

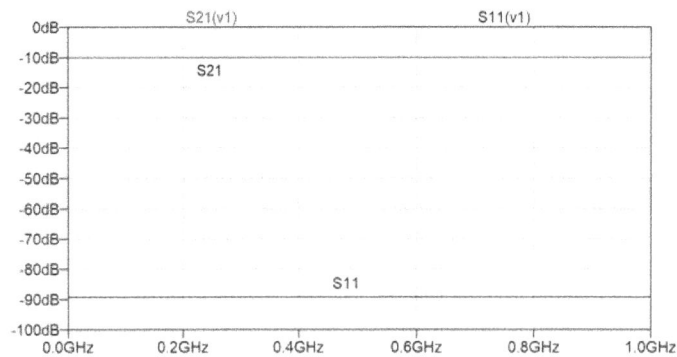

(b)

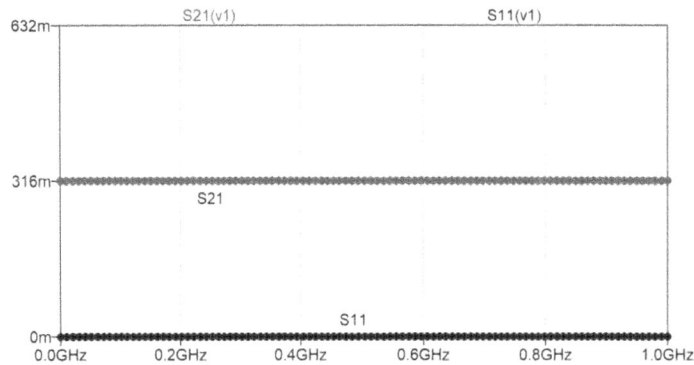

(c)

Figure 3-9 (a) Schematic diagram of the 10-dB attenuator and S-parameter simulation setup. (b) Simulated S11 and S21 in dB. (c) Simulated S11 and S21 showing the linear unit less values.

3.2 Using S Parameter Files

As Table 3-1 shows the definition line contains four descriptive parameters for the data file. The available options are summarized below. Only a single space is required between each entry on a line of the S parameter file. For better visual presentation a tab space can be used between entries.

GHz: Units for the swept frequency data column. Frequency can be GHz, MHz, kHz, or Hz.

S: Defines network parameter type. Parameters can be S, Y, Z, or h parameters format.

MA: Magnitude-Angle format for the parameter data. Available formats are DB for dB-angle, MA for magnitude-angle, or RI for real-imaginary format.

R 50: Reference resistance. This is the characteristic impedance in which the parameters have been measured.

A linear analysis can be setup to analyze the amplifier's S parameter data file over a frequency range of 5800 MHz to 5820 MHz in 1 MHz steps. Setup a tabular output and display each of the four S parameters. On the Table Properties window, set the units to absolute (Abs) and magnitude-angle format. The resulting table is shown in Table 3-1[4]. Note that even though the S Parameters in the data file are recorded in increments greater than 1 MHz, the software has interpolated the values between each data point and can output the S parameters in 1 MHz increments.

Most if not all microwave software are capable of importing files containing S parameter data. The file containing S parameter data is called a **.S2P** file. Similarly, most microwave software are capable of exporting simulated S parameter data in the form of a .S2P file.

LTspice can output S parameter data into a text file, with .txt extension. This file can be easily modified and turned into a .S2P file.

	F (MHz)	mag(S[1,1]) (dB)	ang(S[1,1])	mag(S[2,1]) (dB)	ang(S[2,1])	mag(S[1,2]) (dB)	ang(S[1,2])	mag(S[2,2]) (dB)	ang(S[2,2])
1	5800	0.07	-52.367	65.436	59.242	1e-3	49.912	0.151	-69.52
2	5800.2	0.07	-52.676	65.414	58.75	1.05e-3	42.841	0.151	-69.53
3	5800.4	0.07	-52.985	65.392	58.257	1.1e-3	35.77	0.152	-69.54
4	5800.6	0.07	-53.294	65.37	57.765	1.15e-3	28.699	0.152	-69.55
5	5800.8	0.069	-53.603	65.348	57.272	1.2e-3	21.628	0.153	-69.561
6	5801	0.069	-53.913	65.326	56.78	1.25e-3	14.557	0.153	-69.571
7	5801.2	0.069	-54.222	65.304	56.287	1.3e-3	7.486	0.153	-69.581
8	5801.4	0.069	-54.531	65.282	55.795	1.35e-3	0.415	0.154	-69.591
9	5801.6	0.069	-54.84	65.26	55.302	1.4e-3	-6.656	0.154	-69.601
10	5801.8	0.069	-55.149	65.238	54.81	1.45e-3	-13.727	0.155	-69.611
11	5802	0.068	-55.458	65.217	54.318	1.5e-3	-20.798	0.155	-69.621
12	5802.2	0.068	-55.767	65.195	53.825	1.55e-3	-27.869	0.155	-69.632
13	5802.4	0.068	-56.076	65.173	53.333	1.6e-3	-34.94	0.156	-69.642
14	5802.6	0.068	-56.385	65.151	52.84	1.65e-3	-42.011	0.156	-69.652
15	5802.8	0.068	-56.694	65.129	52.348	1.7e-3	-49.082	0.157	-69.662
16	5803	0.068	-57.004	65.107	51.855	1.75e-3	-56.153	0.157	-69.672
17	5803.2	0.068	-57.313	65.085	51.363	1.8e-3	-63.224	0.157	-69.682
18	5803.4	0.067	-57.622	65.063	50.87	1.85e-3	-70.295	0.158	-69.693
19	5803.6	0.067	-57.931	65.041	50.378	1.9e-3	-77.366	0.158	-69.703
20	5803.8	0.067	-58.24	65.019	49.885	1.95e-3	-84.437	0.159	-69.713
21	5804	0.067	-58.549	64.997	49.393	2e-3	-91.508	0.159	-69.723
22	5804.2	0.068	-58.46	64.969	48.727	2.067e-3	-84.362	0.158	-69.803
23	5804.4	0.068	-58.371	64.941	48.061	2.133e-3	-77.215	0.158	-69.883
24	5804.6	0.069	-58.282	64.912	47.395	2.2e-3	-70.069	0.157	-69.963
25	5804.8	0.07	-58.193	64.884	46.729	2.267e-3	-62.922	0.157	-70.043

Table 3-1 Tabular display of the S Parameter data file

3.3 Scalar Representation of the S Parameters

When working with a two-port network, such as an amplifier, the engineer is usually more interested in the magnitude in (dB's) of the S parameters rather than the absolute units. This is referred to as the Scalar representation of the S parameters and is readily measured on scalar network analyzers. Because the S parameters are based on voltage waveforms they must be multiplied by 20log to convert to dB format.

$$S_{11} \text{ (dB)} = 20log\ |S_{11}|$$ Input return loss [= -S_{11} (dB)]
$$S_{21} \text{ (dB)} = 20log\ |S_{21}|$$ Insertion gain (+) or insertion loss (-)
$$S_{12} \text{ (dB)} = 20log\ |S_{12}|$$ Reverse isolation [= -S_{12} (dB)]
$$S_{22} \text{ (dB)} = 20log\ |S_{22}|$$ Output return loss [= -S_{22} (dB)]

With the C Band amplifier, setup a rectangular graph and display the S Parameters in Scalar format. Assign each S parameter to the graph using the dB-magnitude format. The display is shown in Figure 3-10.

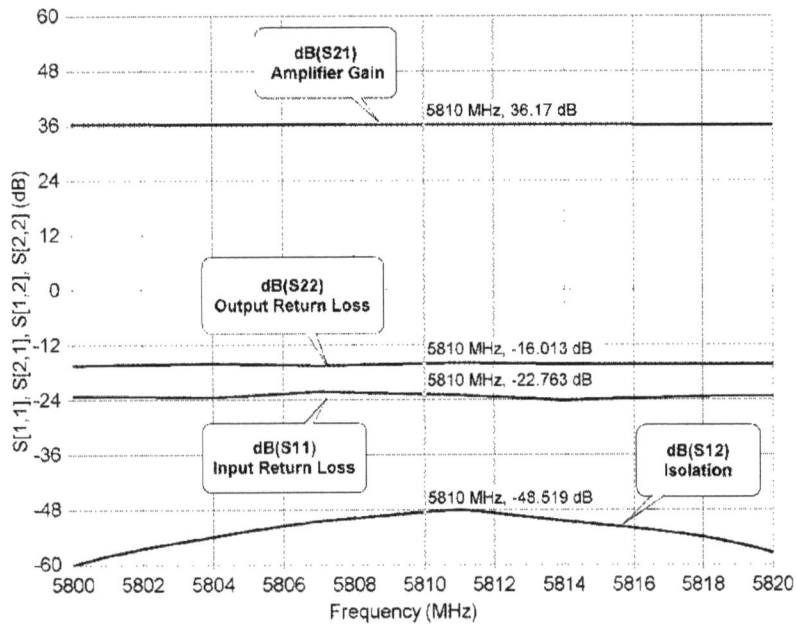

Figure 3-10 Scalar display of frequency swept amplifier S Parameters

3.4 Development of the Smith Chart

The most frequently used graphical tool used to visualize the vector properties of S parameters is the Smith Chart. The Smith Chart conformally maps the familiar rectangular impedance coordinates onto a polar plane. Essentially the reactive, normally the vertical axis, has been bent around in such a way that $\pm$ infinity is included within the boundary of the graph. Therefore any positive complex impedance can be plotted on the standard Smith Chart shown in Figure 3-13. Negative impedances or gain is outside of the standard Smith Chart. A compressed Smith Chart must be used to plot negative impedances. Figure 3-13 shows the standard Smith Chart graph with impedance coordinates. A rectangular axis has been overlaid to show the relationship to the rectilinear grid system. The Smith Chart can be normalized to any characteristic impedance. The normalized impedance is a pure resistance that is a single point at the center of the chart. The purely real impedances exist along the horizontal axis from 0 Ω to infinity. Note the locations of the short circuit (0 Ω) and open circuit (infinity) on the real axis. The family of circles that intersect the real axis are known as the constant resistance circles, and they are centered at $U = r/(r+1)$ and $V=0$. Here U and V are the real/imaginary axis of the reflection coefficient plane Γ, and r is the real part of the normalized impedance $(\bar{Z} = \frac{Z}{Z_0} = r + jx)$. The constant reactance circles appear as arcs on the standard Smith Chart. As shown in Figure 3-13 the reactance circles on the top half of the chart represent inductive reactance while the circles on the bottom half of the chart represent capacitive reactance. Any impedance defined by rectangular coordinates (R + jX) can be plotted as a point where the R value on the constant resistance circle intersects the X value on the reactance circle.

Example 3-5: Consider the load $Z_L = R_L + jX_L$ in Figure 3-11 that results in load impedance equal 25 + j25 Ω at 1 GHz. Use Smith V3.10 to show the impedance at 1 GHz on Smith Chart.

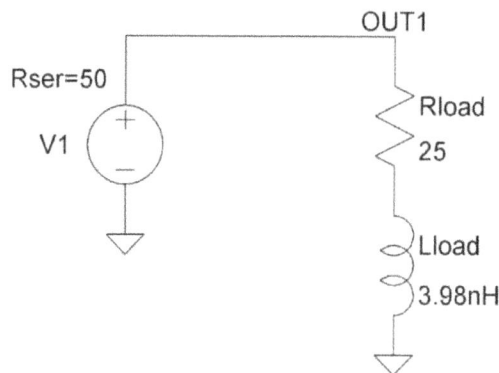

Figure 3-11 Plotting impedance on the Smith chart

Solution: Open Smith V3.10 and click on "Keyboard" Tab. Data Point window opens up > Select impedance (Ω) > Insert 25 for "re" and 25 for "im". Insert 1 under "frequency" and select "GHz" for proper frequency unit (Figure 3-12a). Press OK. Impedance 25 + j25 Ω will show up on Figure 3-12b. Note that actual impedance is shown on Smith Chart utilizing Smith V3.10.

The input impedance in Figure 3-11 can be simulated to verify that impedance of 25 + j25 Ω is achieved at 1 GHz.

The quality factor (Q) is the ratio X_L/R_L on Smith Chart equal to 1 (i.e., 25/25=1).

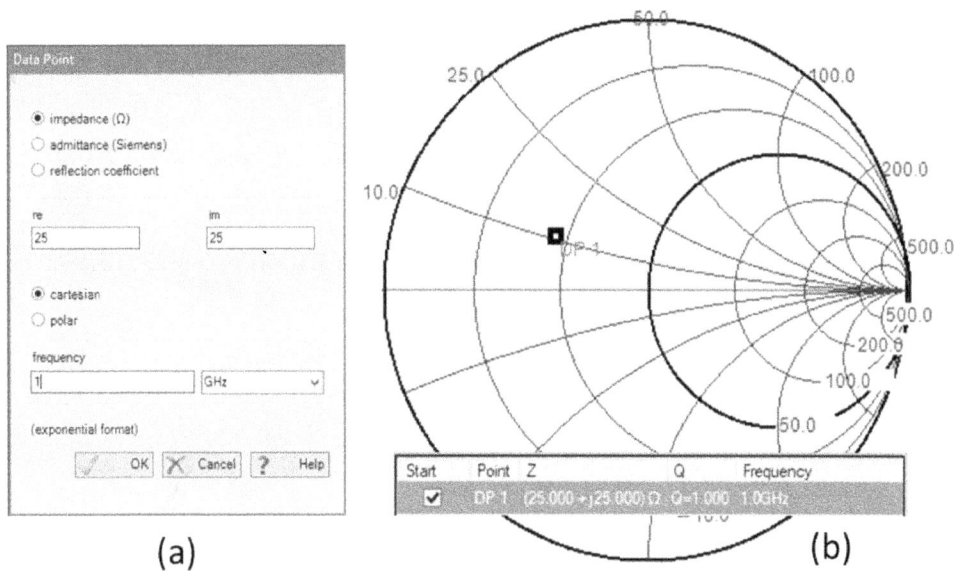

Figure 3-12 Plotting Z=25+j25 on the Smith Chart (utilizing Smith V3.10)

3.5 Normalized Impedance on the Smith Chart

The impedance on a Smith Chart is often presented in its normalized form. This means that the actual impedance is divided by the value of the characteristic impedance. The Smith Chart allows the selection of either normalized or actual impedance via the Properties window of the Smith Chart. Equation (3-11) shows how the reflection coefficient, Γ, is related to the normalized impedance on the Smith Chart.

$$\Gamma = \frac{\left(Z_{actual} - Z_o\right)}{\left(Z_{actual} + Z_o\right)} \tag{3-28}$$

Redefining Equation (3-11) in terms of normalized impedance where; $z = \dfrac{\left(Z_{actual}\right)}{\left(Z_o\right)}$, results in Equation (3-12).

$$\Gamma = \frac{\left(z - 1\right)}{\left(z + 1\right)} \tag{3-29}$$

The normalized impedance of the Smith Chart is the vector from the center of the chart to the normalized impedance. The transmission coefficient is the vector from the origin ($Z = 0$) to the normalized impedance. The reflection S parameters, S_{11} and S_{22}, are measured as a reflection coefficient on the Smith Chart. The transmission S parameters, S_{21} and S_{12}, are measured as transmission parameters on the Smith Chart.

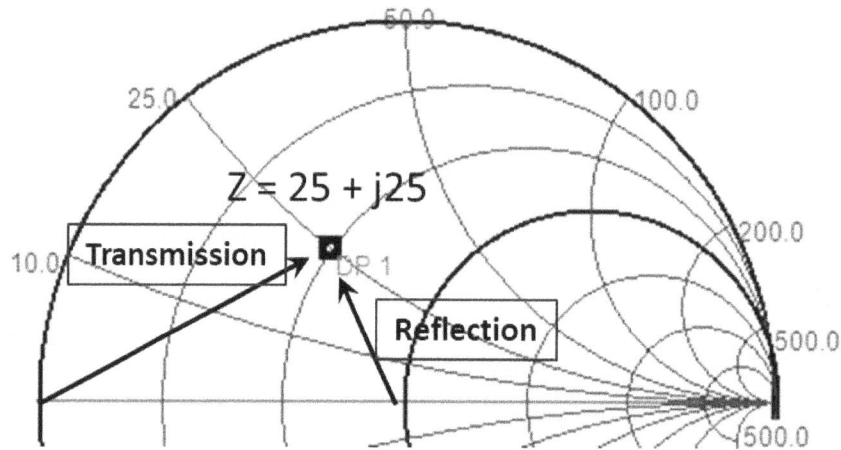

Figure 3-13 Impedance transmission and reflection on the Smith Chart

Knowing that the reflection coefficients and S parameters are vector quantities, there must be a method to measure the angular portion of the vector. Figure 3-14 shows the angular measurement convention on the Smith Chart. Note that the reflection coefficient of a 50 Ω resistance (center of the chart) is equal to zero. A total reflection, like that due to a perfect short or open circuit, has a reflection coefficient equal to one. Therefore all positive impedances result in a reflection coefficient between 0 and 1.

The reflection coefficient of the 25 + j25 Ω impedance can be determined by displaying the magnitude and angle of the S parameter, S11. As Figure 3-14 shows the reflection coefficient of this impedance is 0.447 at an angle of +116.565 degrees.

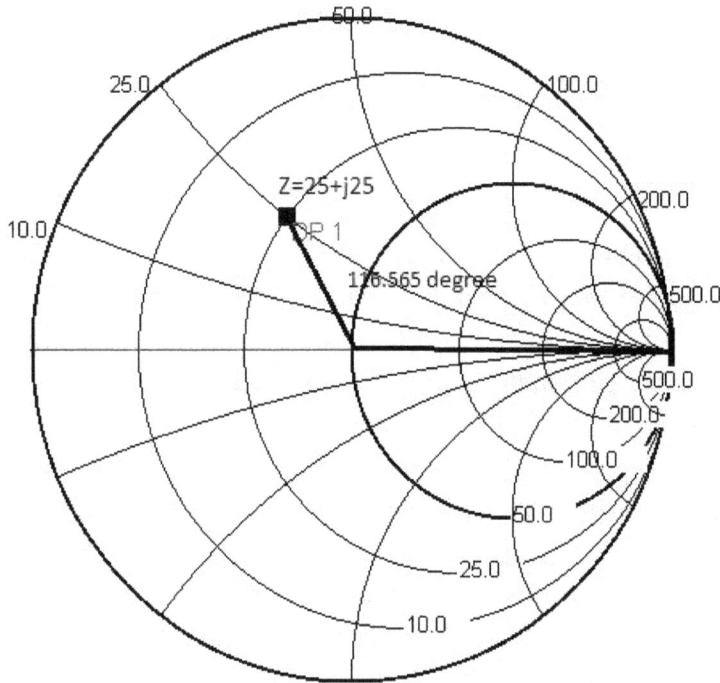

Figure 3-14 Angular measurement of reflection coefficients

3.6 Admittance on the Smith Chart

Admittance circles can also be displayed on the Smith Chart. The admittance circles can be enabled on the Smith Chart by their selection on the graph's Properties page. The admittance circles consist of constant conductance and susceptance circles which are inverted from the impedance circles. Subsequent chapters dealing with the subject of impedance matching will make frequent use of the admittance parameters. Having both impedance and admittance parameters displayed on the Smith Chart makes it very easy to design impedance matching networks that include both series and parallel (shunt) elements. It also becomes very easy to convert series impedance to its parallel admittance equivalent. The admittance of a network is the inverse of the impedance as given by Equation (3-30).

$$Y = \frac{1}{Z} = G \pm jB \qquad (3\text{-}30)$$

where,

G = conductance in mhos

B = susceptance in mhos

The equivalent admittance of a network is read directly from the admittance circles on the Smith Chart. For the normalized $0.5 + j0.5\ \Omega$ series impedance the admittance is read directly from the chart as $1.0 - j1.0$ mho. It is important to note that there is an inversion in the sign of the imaginary component when converting from impedance to admittance or vice versa.

3.7 Lumped Element Movements on the Smith Chart

Lumped element movements on the Smith Chart form the basis for impedance matching. The Smith Chart is a wonderfully intuitive tool, for visualization of moving from one impedance to another, without involving circuit synthesis mathematics. Understanding the basic movements around the Smith Chart will build a foundation for the circuit designs covered in this text. It is helpful to display both the impedance and admittance coordinates simultaneously on the Smith Chart.

3.8 Adding a Series Reactance to an Impedance

Adding a series reactance to an impedance point on the Smith Chart causes the resulting impedance to move along the constant resistance circle in which the impedance intersects. A series inductance will move the impedance in a clockwise direction while a series capacitance will move the resulting impedance in a counter-clockwise direction on the constant resistance circle. The reactance that is added to the impedance by the series element can be read from the Smith Chart by finding the difference between the lines of reactance that intersect the start point and end point on the constant resistance circle. This reactance can then be converted to a capacitance or inductance value using Equations (3-31) and (3-32).

$$C\ (series) = \frac{1}{\omega X_n} \qquad\qquad (3\text{-}31)$$

$$L \ (series) = \frac{X_n}{\omega} \tag{3-32}$$

where,

X = reactance measured along the arc length

ω = the design frequency, $2\pi f$

n = impedance normalizing value (50 Ω)

Example 3-6: Measure the amount of inductance required to move the impedance $Z = 25 + j25 \ \Omega$ from point A to point B at approximately $Z = 25 + j43 \ \Omega$ on the Smith Chart. Assume $f = 1000$ MHz.

Solution: The amount of reactance required in the inductor can be measured from the reactance lines that intersect the start point (A) and end point (B). As Figure 3-15 shows the reactance is: $43 - 25 = 18 \ \Omega$.

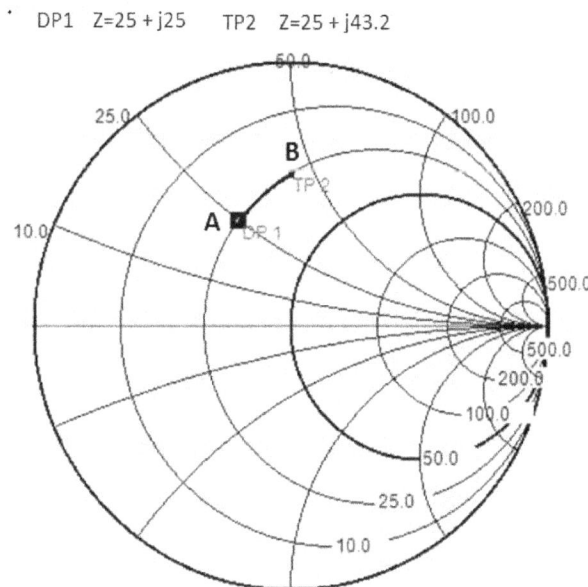

Figure 3-15 Moving point A to point B

Using a design frequency of 1000 MHz and Equation (3-30) the inductance is calculated to be 2.86 nH.

$$L(series) = \frac{18}{(2\pi)(1000)(10^6)} = 2.86 \quad nH$$

The load impedance presented in Figure 3-16 is equal to 25 + j25 Ω at 1000 MHz. Adding a series inductor to this load results in a new impedance equal to 25 + j43.2 Ω at 1000 MHz. This can be shown by simulating the input impedance of the schematic in Figure 3-16. Alternatively, one can use Smith V3.10 software and readily obtain the series inductance value by entering load impedance at 1000 MHz (point A) and adding a series inductor to reach point B. This procedure is captured in Figure 3-17, where moving from point A (DP1) to point B (TP2) is achieved by adding a series inductor. Note that a schematic is automatically generated in Smith v3.10 as a result of this action.

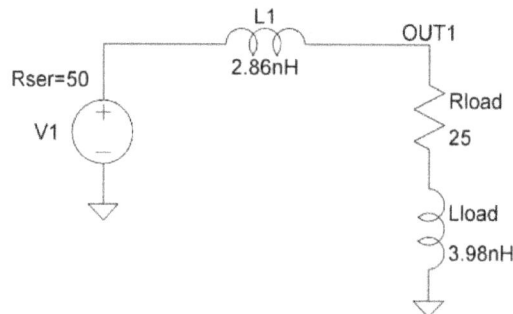

Figure 3-16 Series inductance added to a load

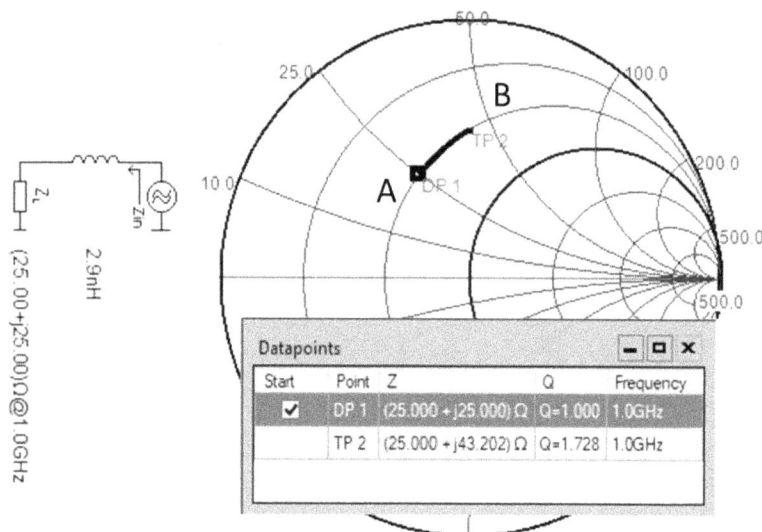

Figure 3-17 Moving point A to point B on Smith V3.10

3.9 Adding a Shunt Reactance to an Impedance

Adding a shunt element to an impedance point on the Smith Chart causes the resulting impedance to move along the constant conductance circle in which the impedance intersects. A shunt inductance will move the impedance in a counter-clockwise direction while a shunt capacitance will move the resulting impedance in a clockwise direction on the constant conductance circle. The susceptance that is added to the impedance by the shunt element can be read from the Smith Chart by finding the difference between the lines of susceptance that intersect the start point and end point on the constant conductance circle. This susceptance can then be converted to a capacitance or inductance value using Equations (3-25) and (3-26) [5].

$$L(shunt) = \frac{n}{\omega B} \tag{3-33}$$

$$C(shunt) = \frac{B}{\omega n} \tag{3-34}$$

Where,

B = susceptance measured along the arc length
ω = the design frequency $2\pi f$
n = impedance normalizing value (50 Ω)

Example 3-7: Continuing with the previous example calculate the value of the shunt capacitance required to move from point B to point C on the real impedance axis.

Solution: Add a shunt capacitance to move the impedance from point B to point C as shown in Figure 3-18. When adding a shunt element switch from the impedance grid to the admittance coordinates. The admittance follows the constant conductance circle in which the point lies. Measure the susceptance required to move to point C on the real impedance axis by the difference between the intersecting susceptance lines. The admittance at point B, $Y_B = 1/(Z_B) = 0.010 - j0.0173$, where $Z_B = 25 + j43.2$ Ohm. The susceptance as measured on the perimeter of the chart is $B = 0.0173$ mhos. Therefore, from Equation (3-34) the shunt capacitance value necessary to move point B to point C is (with $n = 1$):

$$C(shunt) = \frac{B}{\omega} = \frac{0.0173}{2.\pi.1000.10^6} = 2.75 \, pF$$

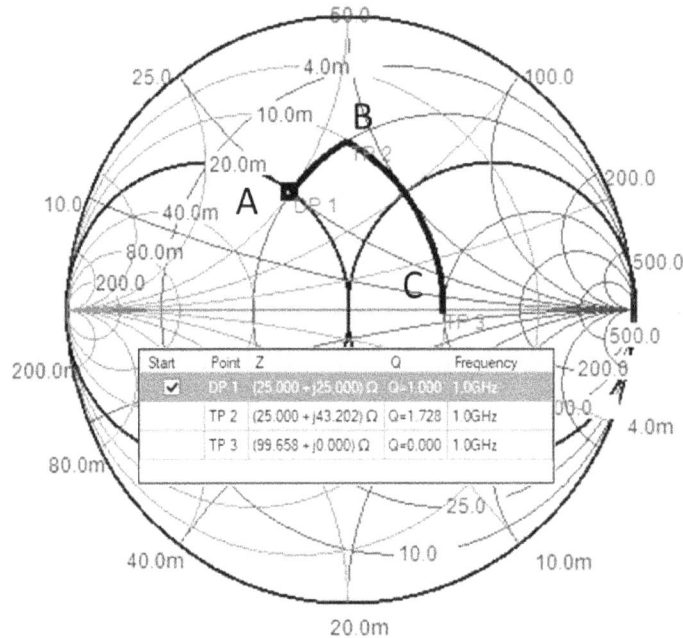

Figure 3-18 Moving point A / B to point B / C on impedance / admittance chart

Alternatively you can add a shunt capacitor to the circuit and make the capacitance value tunable. Start with a very low value of approximately 0.1 pF and increase the value of capacitance until the admittance is moved from point B to point C. This cab ne accomplished in LTspice by simulating input impedance (Figure 3-19).

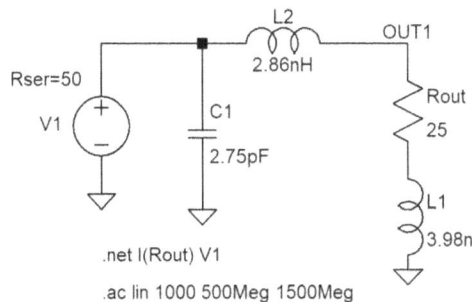

Figure 3-19 Adding shunt capacitance to the network

As shown in Figure 3-19, point C lies on the admittance circle equal to 10 mhos, and the impedance is equal to 100 Ω (= 1 / 0.010). Typical Smith Chart depicts normalized impedance and admittance values. Thus, the normalized admittance at point C is 0.5 (= 0.010*50). These techniques form the basis for performing impedance transformations using the Smith Chart. In this example a complex impedance of 25 + j25 Ω has been transformed to a pure resistance of 100 Ω.

3.10 VSWR Circles on the Smith Chart

Equation (2-29) demonstrated that the VSWR of a network is related to the magnitude of the reflection coefficient, independent of the angle. From plotting the reflection coefficient on the Smith Chart we know that the origin of the impedance vector is located in the center of the chart. This suggests that as the reflection coefficient vector rotates 360 degrees around the chart with a constant magnitude, the VSWR will remain constant. This locus of points around the center of the Smith Chart is known as the constant VSWR circle. In LTspice there is no direct way to plot constant VSWR circles, which are frequently used for Low Noise Amplifier design.

3.11 Adding a Transmission Line in Series with an Impedance

We have seen that adding a reactance in series with an impedance point causes the impedance to follow the constant resistance circles. Adding a transmission line of the same impedance as the Smith Chart's normalized impedance, in series with an impedance point causes the resulting impedance to follow the constant VSWR circle in which the impedance lies. The impedance moves in a clockwise direction on the constant VSWR circle.

Example 3-8: Calculate the electrical length of a series transmission line moving the 25+j25 Ω impedance from point A to point B on the Smith Chart, as shown in Figure 3-21. Then add transmission line to move the impedance from point B to point C, as shown in Figure 3-23. Assume frequency of operation is 1 GHz.

Solution: Add a series transmission line to the impedance and make the electrical length tunable. Increase the line length to move the impedance to point B. The electrical line length is 58.45 degrees. The impedance on the real axis (zero reactance) on the right hand side of the Smith Chart would represent a point of maximum voltage-minimum current along the transmission line.

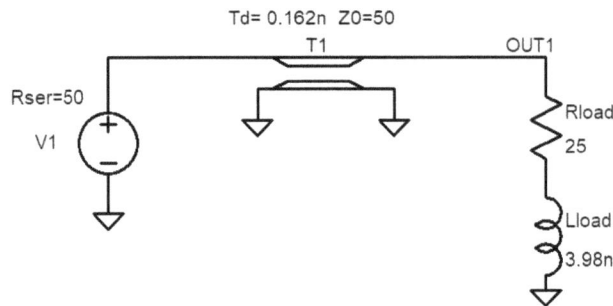

Figure 3-20 Series transmission line added to load

Transmission line lengths are sometimes referred to in terms of fractional wavelengths. Because one wavelength is equal to 360°, a 58.45° electrical length represents (58.45/360) 0.162λ. Continue to add electrical length to the transmission line to reach point C on the Smith Chart. The real impedance (zero reactance) on the left side of the horizontal axis on the Smith Chart represents a minimum voltage-maximum current point along the transmission line.

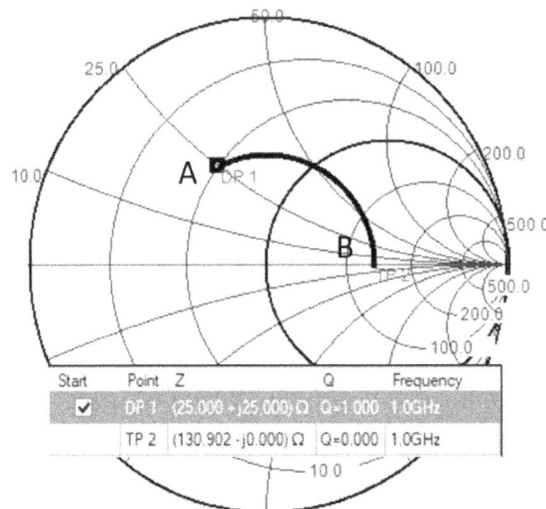

Figure 3-21 Series transmission line moves points A to B

The following schematic shows the series transmission line added to the impedance to bring it the minimum voltage point.

Td= 0.412n Z0=50

T1   OUT1

Rser=50

V1

Rload

25

Lload

3.98n

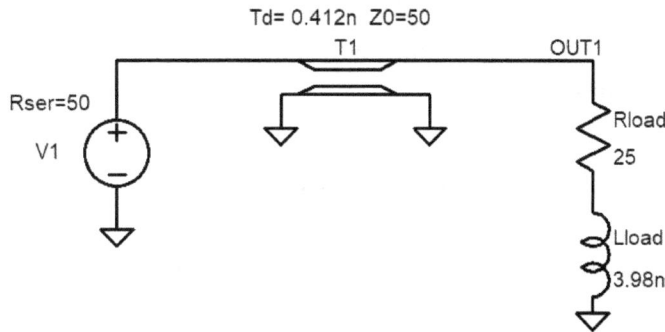

Figure 3-22 Series transmission line added to impedance

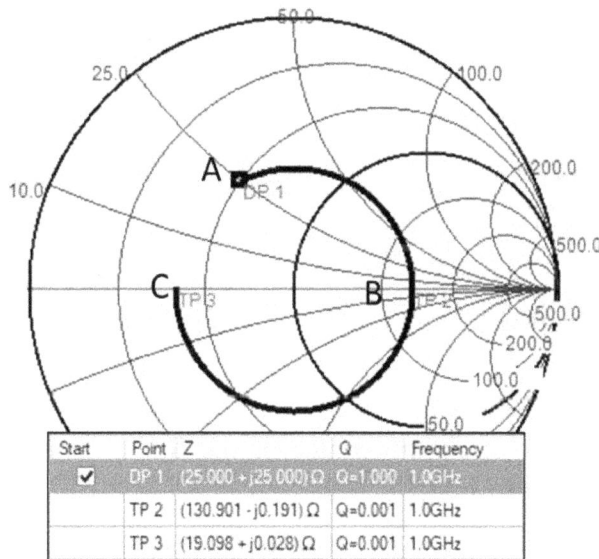

Start	Point	Z	Q	Frequency
☑	DP 1	(25.000 + j25.000) Ω	Q=1.000	1.0GHz
	TP 2	(130.901 - j0.191) Ω	Q=0.001	1.0GHz
	TP 3	(19.098 + j0.028) Ω	Q=0.001	1.0GHz

Figure 3-23 Series transmission line moves point B to C

Further increasing the length of the transmission line we find that we arrive back at point A at 180 degrees of electrical length. Therefore the electrical distance around the Smith Chart is $180°$ or $\lambda/2$ wavelength. The points of maximum voltage and minimum voltage will repeat every $\lambda/2$ wavelength.

3.12 Adding a Transmission Line in Parallel with an Impedance

In Chapter 2 section 2.11 we have seen that the open and short-circuited transmission lines could take on the equivalence of an inductor, capacitor, or series and parallel resonant circuits depending on the electrical length of the line. Therefore, the shunt transmission line will behave more like the lumped element movements on the Smith Chart.

3.13 Short Circuit Transmission Lines

At DC and low frequencies, a short circuited line is a very low inductance but this is not the case at higher RF and microwave frequencies. Figure 3-27 point A shows that a short circuit transmission line with 0 degree length (perfect short) appears at the short circuit point on the Smith chart. Assume $f = 1\text{GHz}$.

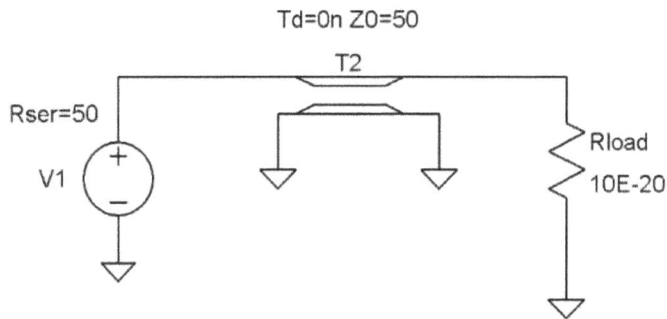

Td=0n Z0=50

T2

Rser=50

V1

Rload
10E-20

Figure 3-24 Short circuited transmission line 0 degree electrical lengths

If 45° of electrical length is added to the short circuit line, the impedance moves clockwise along the outer circumference of the Smith Chart to the position B shown at the top of Figure 3-27.

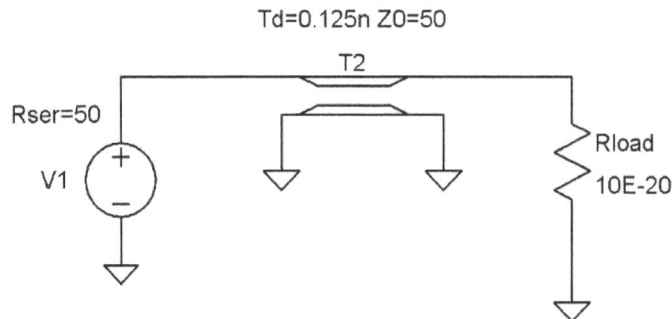

Td=0.125n Z0=50

T2

Rser=50

V1

Rload
10E-20

Figure 3-25 Short circuited transmission line 45 degree electrical lengths

As the line length is increased to 90° we see that the short circuit has been transformed to an open circuit at point C shown in Figure 3-27.

Td=0.25n Z0=50

T2

Rser=50

V1

Rload
10E-20

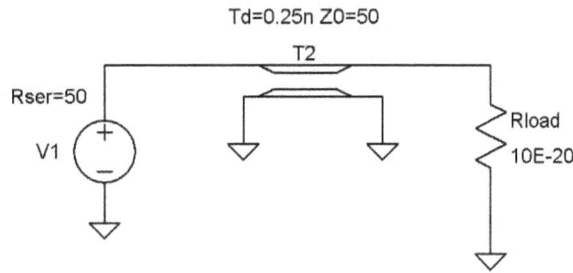

Figure 3-26 Short circuited transmission line 90 degree electrical lengths

At $180°$ line length the impedance will travel completely around the Smith Chart and appear as a short circuit again. Therefore depending on the line length the short circuit transmission line can be transformed into a shunt capacitor or shunt inductor.

The Impedance values at points A, B and C, utilizing Smith V3.10, are shown in Table 3-2.

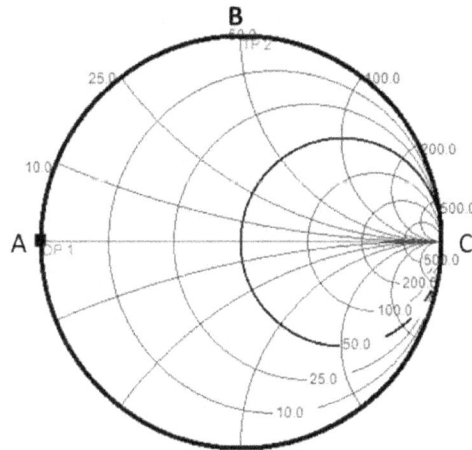

Figure 3-27 Short circuited transmission line 90 degree electrical lengths

Start	Point	Z	Q	Frequency
✓	DP 1	$(0.000 + j0.000)\,\Omega$	Q=99999.000	1.0GHz
	TP 2	$(0.000 + j50.000)\,\Omega$	Q=99999.000	1.0GHz
	TP 3	$(-\text{Infinity} + j\text{Infinity})\,\Omega$	Q=NaN	1.0GHz

Table 3-2 Data point values at points A, B, and C

Again, the quality factor, Q, is the ratio of reactive to resistive part of impedance (i.e., $Q = X / R$) at any point on the Smith chart. Point B represents impedance of an inductor. Therefore, its Q is infinite since impedance ($Z = R + jX$) is purely reactive ($R = 0$).

3.14 Open Circuit Transmission Lines

Figure 3-28 shows the characteristic of the open circuit transmission line. At 0° electrical line length it appears as a perfect open circuit, shown as point A in Figure 3-31.

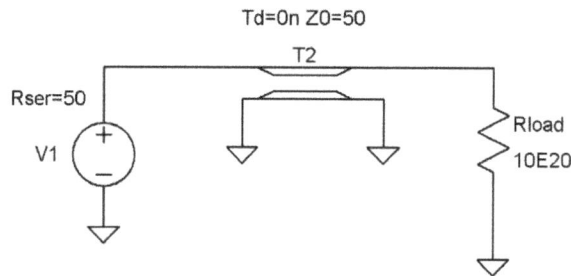

Table 3-28 Open circuited transmission line 0 degree electrical lengths

As the electrical length is increased, the impedance moves clockwise around the circumference to the 45° position at the bottom of the chart, shown as point B in Figure 3-31.

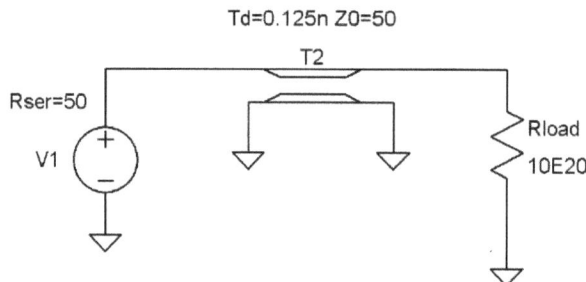

Figure 3-29 Open circuited transmission line 45 degree electrical lengths

At 90° electrical length in Figure 3-30, the open circuit now appears as a short circuit, shown as point C in Figure 3-31.

Td=0.25n Z0=50

T2

Rser=50

V1

Rload
10E20

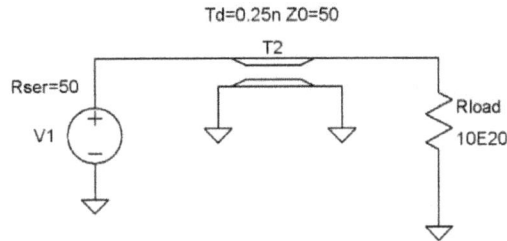

Figure 3-30 Open circuited transmission line 90 degree electrical lengths

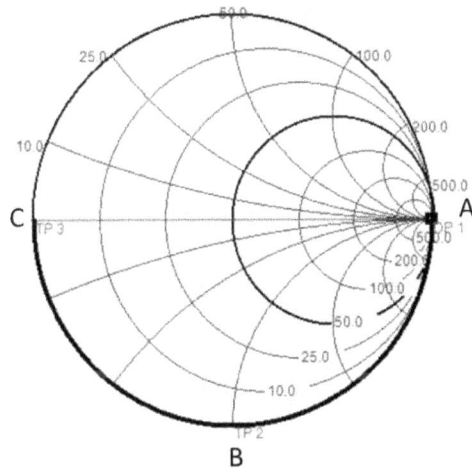

Figure 3-31 Open circuited transmission line 90 degree electrical lengths

The Impedance values at points A, B and C are shown in the following Table. This property of transforming open circuits to short circuits and vice versa is one that is used frequently throughout microwave circuit design.

Start	Point	Z	Q	Frequency
✔	DP 1	(999999950.526 + j0.000) Ω	Q=0.000	1.0GHz
	TP 2	(0.000 - j50.000) Ω	Q=99999.000	1.0GHz
	TP 3	(0.000 + j0.000) Ω	Q=0.000	1.0GHz

Table 3-3 Data point values at points A, B, and C

3.15 Open and Short Circuit Shunt Transmission Lines

For small fractional wavelength transmission lines the open circuit shunt transmission line acts as a shunt capacitor.

Example 3-9: Measure the electrical length of a shunt transmission line or the amount of shunt capacitance to move the impedance $Z = 25 + j25$ Ω from point A to point B the center of Smith Chart, as shown in Figure 3-34.

Solution: Plotting the impedance on the Smith Chart with the admittance circles shows that the impedance lies directly on the unit conductance circle. Therefore an open circuit shunt transmission line can move this impedance directly to 50 Ω. As the schematic of Figure 3-34 shows, a $45.38°$ electrical length of a 50 Ω shunt transmission line moves the impedance to the center of the Smith Chart.

The schematic in Figure 3-33 shows that this circuit is equivalent to a 3.2 pF capacitor in shunt with the load impedance.

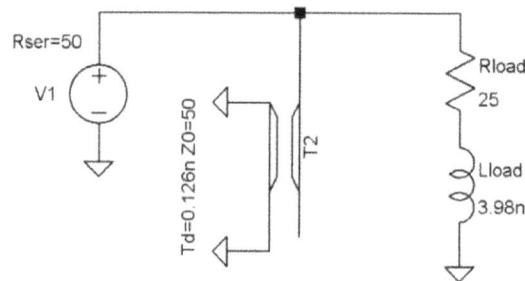

Figure 3-32 Open circuit transmission line at 45.38 degree lengths

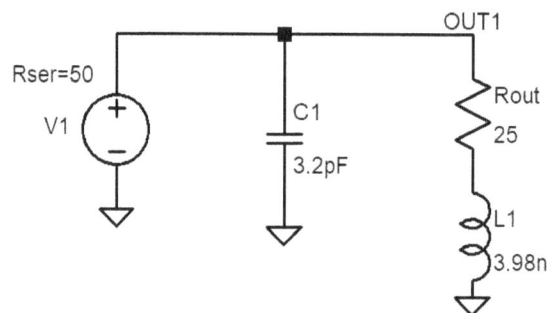

Figure 3-33 Equivalent to a 3.2 pF capacitor in shunt with the load

The Impedance values at points A and B are shown in Table 3-4. Note that a admittance chart is depicted in Figure 3-34. Furthermore, point B approximately represent center of Smith chart ($Z = 50 + j0$).

Start	Point	Z	Q	Frequency
☑	DP 1	(25.000 + j25.000) Ω	Q=1.000	1.0GHz
	TP 2	(49.999 - j0.265) Ω	Q=0.005	1.0GHz

Table 3-4 Data point values at points A and B

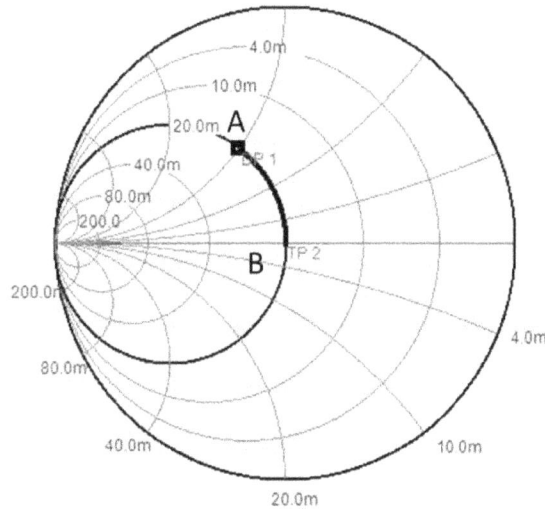

Figure 3-34 Moving from point A to point B utilizing Smith V3.10

For small fractional wavelength transmission lines the short circuit shunt transmission line acts as a shunt inductor. Consider the 4.3 –j14 Ω impedance as shown in Figure 3-36. This impedance lies on the unit conductance circle on the bottom half of the Smith Chart. A 50 Ω shunt transmission line added to the impedance moves along the constant conductance circle to the center of the Smith Chart.

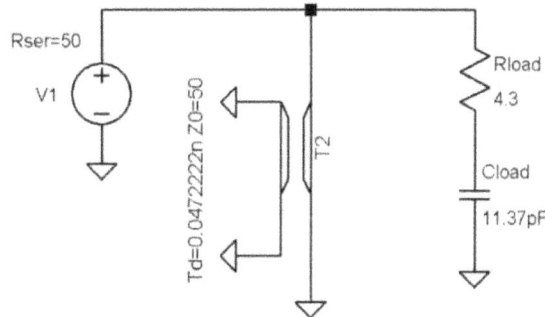

Figure 3-35 Short circuit shunt transmission line added to an impedance

Similarly a 2.4 nH shunt inductor has the same effect at a frequency of 1000 MHz. These movements form the basis for distributed network impedance matching which is covered in detail in Chapter 6.

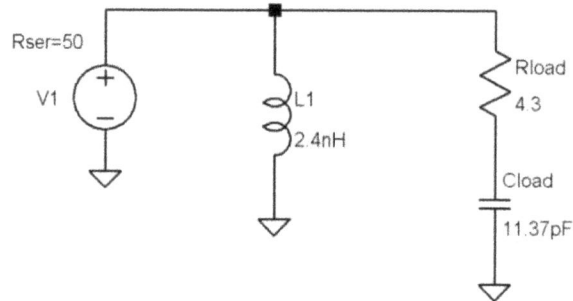

Figure 3-36 Shunt 2.4 nH inductor added to an impedance

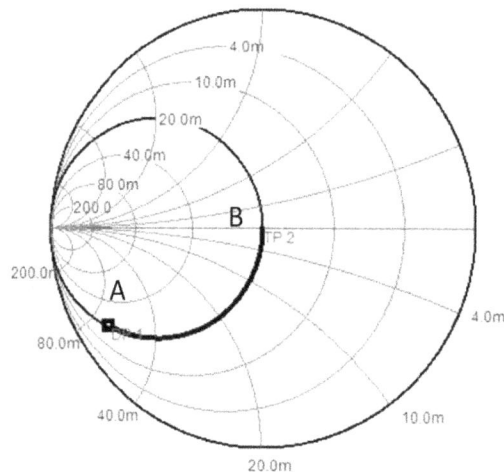

Figure 3-37 Moving from point A to point B Smith chart center

The impedance values at points A and B are shown in Table 3-5.

Start	Point	Z	Q	Frequency
✓	DP 1	(4.300 - j14.000) Ω	Q=3.256	1.0GHz
	TP 2	(49.881 + j0.000) Ω	Q=0.000	1.0GHz

Table 3-5 Data point values at points A and B

Problems

3-1.   Show that for a two port network having source impedance equal to Zs, the output reflection coefficient is given by

$$\Gamma_{OUT} = \frac{V_2^-}{V_2^+} = S_{22} + \frac{S_{12}S_{21}\Gamma_s}{(1-\Gamma_s S_{11})}.$$

3-2.   Determine the S parameters of Z in figure of example 3-1 when
       (a) $Z = 20 \ \Omega$.
       (b) $Z = j20 \ \Omega$.

3-3.   Determine the S parameters of Y in figure of example 3-2 for
       (a) A shunt resistor equal to $20 \ \Omega$.
       (b) A shunt capacitor of 2 pF at 1 GHz.

3-4.   The Π (Pi) attenuator is widely used in RF and microwave circuits to introduce voltage attenuation factor K similar to that of the T attenuator. The resistors R_1 and R_2 are calculated such that when the attenuator is loaded with the characteristic line impedance, it provides a matched load.

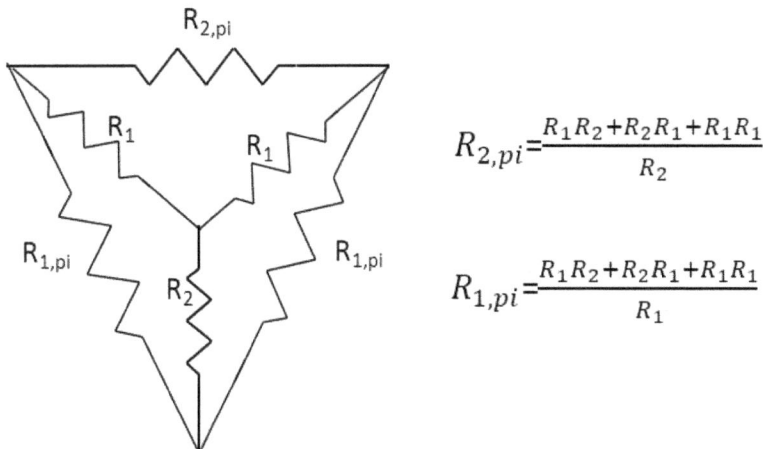

$$R_{2,pi} = \frac{R_1 R_2 + R_2 R_1 + R_1 R_1}{R_2}$$

$$R_{1,pi} = \frac{R_1 R_2 + R_2 R_1 + R_1 R_1}{R_1}$$

Figure P3-4 Illustration of T to Pi conversion.

(a) Use T- Π conversion procedure [6] above to determine R_1 and R_2 for a Pi attenuator in terms of attenuation factor K and characteristic line impedance. Show that the expressions for R_1 and R_2 in a Pi network reduce to

$$R_{1,pi} = \frac{1 + K}{1 - K} Z_0$$

$$R_{2,pi} = \frac{1 - K^2}{2K} Z_0.$$

Hint:

(b) Compute R_1 and R_2 for a 3-dB Pi attenuator in a 50Ω system.
(c) Compute R_1 and R_2 for a 3-dB Pi attenuator in a 75Ω system.
(d) Determine the S parameters in part (c).

3-5.   Place a 20 + j30 Ω impedance at point A on the Smith Chart. Add a series inductance to move the impedance along the constant resistance circle to point B having the impedance 20 + j50 Ω. Using a design frequency of 1000MHz, calculate the inductance that this reactance represents.

3-6.   Continuing with Problem 3-5 enable the admittance coordinates on the Smith Chart to add a shunt element. Add a shunt capacitance to move the impedance to the real axis. Measure the susceptance required to move to the real impedance axis by the difference between the intersecting susceptance lines.

3-7.   Use MATLAB to calculate the magnitude of reflection coefficient for a desired VSWR = 2.

3-8.   Create a constant VSWR circle for a VSWR=20. Comment on the VSWR value required to place the VSWR circle on the circumference of the Smith Chart.

3-9.   A load of 75 + j20 Ω is connected to a 50 Ω transmission line. Calculate the load admittance and the input impedance if the line is 0.2 wavelengths long.

3-10.   For the load impedance in Problem 3-9, determine the reflection coefficient and the transmission coefficient.

3-11.   For the load impedance in Problem 3-8, determine the normalized value of the load impedance if the impedance is normalized to 75 Ω.

3-12.   A series RLC load, R = 100 Ω, L = 20 nH, C = 20 pF is connected to a 50 Ω transmission line. Calculate the VSWR and reflection coefficient at the load at 100 MHz.

3-13.   Using the RLC load impedance of problem 3-12, determine the impedance with a series transmission line of characteristic impedance of 50 Ω and electrical length of 180 degrees.

3-14.   Determine the input impedance of a network that has a reflection coefficient of 0.5 at an angle of 90°.

3-15.   Determine the ABCD parameters of a lossy transmission line having length d, characteristic impedance Z_0, and complex propagation constant γ.

3-16.   Find the ABCD parameters of the ideal transformer in Figure P3-16 below. Assume that $(V_2/V_1) = (N_2/N_1)$ and $(I_2/I_1) = (N_1/N_2)$.

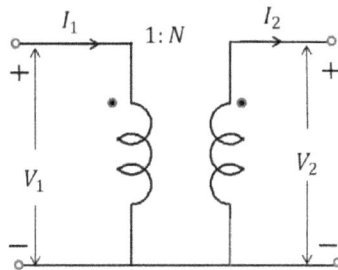

Figure P3-16. Schematic diagram of a transformer as a two-port network

3-17.   Find the scattering matrix of cascaded ideal transformers having 1:N and N:1 turn ratios. **Hint**: find the T matrix of the overall network first.

3-18.   Consider a finite length transmission line having characteristic impedance equal to Z_0, propagation constant β, and physical length equal to d. Show that its scattering matrix is given by

$$[S] = \begin{bmatrix} 0 & e^{-j\beta d} \\ e^{-j\beta d} & 0 \end{bmatrix}.$$

References and Further Readings

[1]   David M. Pozar, *Microwave Engineering*, Third Edition, John Wiley and Sons, Inc., 2005.

[2]   Robert E. Collin, *Foundations for Microwave*, Second Edition, McGraw-Hill, Inc., New York, 1996.

[3]   Guillermo Gonzales, *Microwave Transistor Amplifiers – Analysis and Design*, Second Edition, Prentice Hall Inc., Upper Saddle River, NJ.

[4]   Ali A. Behagi, *RF and Microwave Circuit Design*, A Design Approach Using (**ADS**), Techno Search, Ladera Ranch, CA 2015

[5]   Chris Bowick, *RF Circuit Design*, Second Edition, Newnes, Elsevier, 2008.

[6]   Charles K. Alexander and Matthew N. O. Sadiku, *Fundamentals of Electric Circuits*, Third Edition Boston, 2007

Chapter 4

Resonant Circuits and Filters

4.1 Introduction

Resonant circuits are used in many applications, such as filters, oscillators, tuners, tuned amplifiers, and microwave communication networks. The first half of this chapter examines resonant circuits. Lumped element resonant circuits and the lumped equivalent networks of mechanical and distributed resonators are considered. The analysis of basic lumped element series and parallel RLC resonant circuits is implemented software. The discussion turns to network parameters with an analysis of the Q factor measurement of transmission line resonators. Using the LTspice software a robust technique is demonstrated for the evaluation of Q factor from the measured S parameters of a resonant circuit. The second half of the chapter is an introduction to the vast subject of filter networks. The design of lumped element filters is introduced and followed by an introduction to distributed element filters.

4.2 Resonant Circuits

Near resonance, RF and microwave resonant circuits can be represented either as a lumped element series or parallel RLC network.

4.3 Series Resonant Circuits

In this section we analyze the behavior of the series resonant circuit.

Example 4.1: Consider the one port series resonator that is represented as a series RLC circuit of Figure 4-1. Analyze the circuit, with R = 10 Ω, L = 10 nH, and C = 10 pF.

Solution: Set up the schematic in as shown in Figure 4-1.

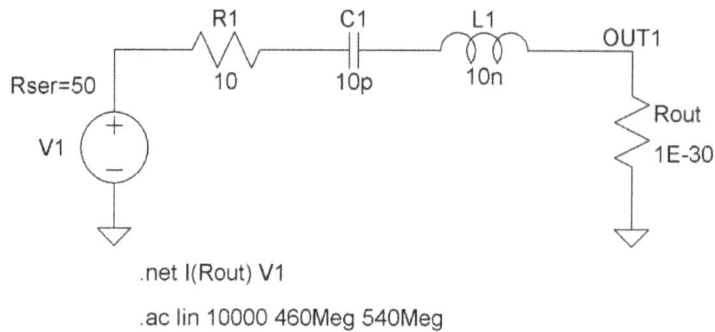

Figure 4-1 One-port series RLC resonator circuit

The plot of the resonator's input impedance in Figure 4-2 shows that the resonance frequency is about 503.3 MHz and the input impedance at resonance is 10 Ω, the value of the resistor in the network.

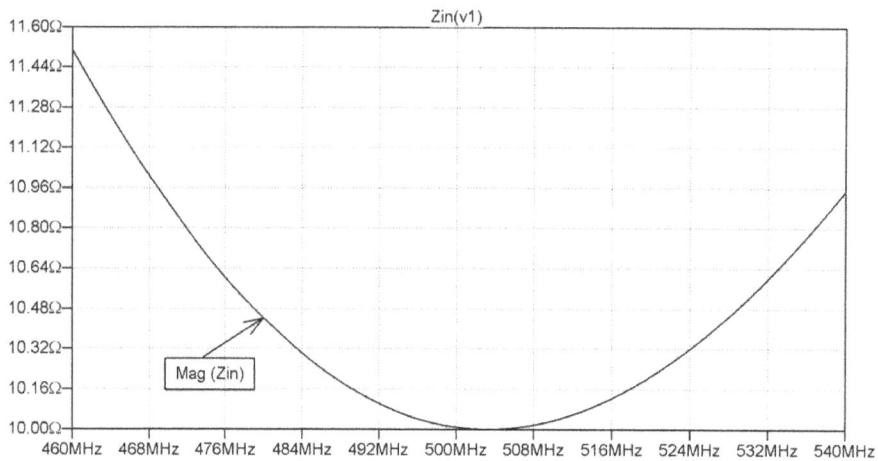

Figure 4-2 Input impedance plot showing the resonance frequency

The input impedance of the series RLC resonant circuit is given by,

$$Z_{in} = R + j\omega L - j\frac{1}{\omega C}$$

where, $\omega = 2\pi f$ is the angular frequency in radian per second.

If the AC current flowing in the series resonant circuit is I, then the complex power delivered to the resonator is

$$P_{in} = \frac{|I|^2}{2} Z_{in} = \frac{|I|^2}{2}\left(R + j\omega L - j\frac{1}{\omega C}\right) \qquad (4\text{-}1)$$

At resonance the reactive power of the inductor is equal to the reactive power of the capacitor. Therefore, the power delivered to the resonator is equal to the power dissipated in the resistor

$$P_{in} = \frac{|I|^2 R}{2} \qquad (4\text{-}2)$$

4.4 Parallel Resonant Circuits

Example 4.2: Analyze a rearrangement of the RLC components of Figure 4-1 into the parallel configuration of Figure 4-3. The schematic of Figure 4-3 represents the lumped element representation of the parallel resonant circuit.

Solution: The one port parallel resonant circuit is shown in Figure 4-3.

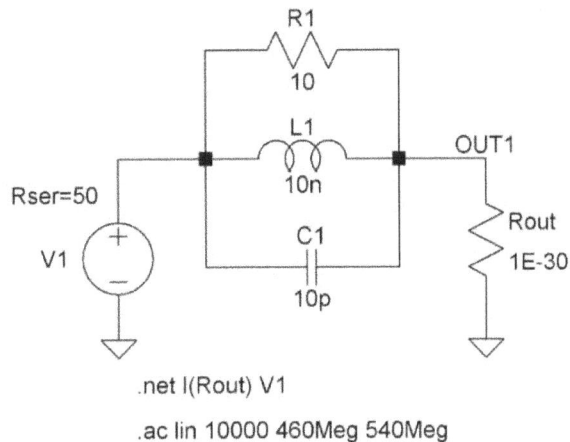

Figure 4-3 One-port parallel RLC resonant circuit

Simulate the schematic and display the input impedance in a rectangular plot. The plot of the magnitude of the input impedance shows that the resonance frequency is still 503.3 MHz where the input impedance is R = 10 Ω. Again this shows that the impedance of the inductor cancels the impedance of the capacitor at resonance. In other words, the reactance, X_L, is equal to the reactance, X_C, at the resonance frequency.

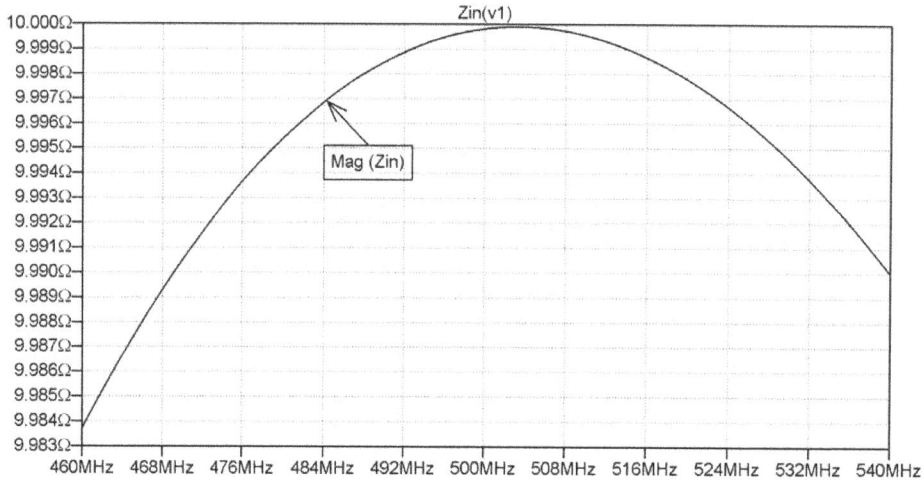

Figure 4-4 Input impedance of parallel RLC resonant circuit

The input admittance of the parallel resonant circuit is given by:

$$Y_{in} = \frac{1}{R} + j\omega C - j\frac{1}{\omega L}$$

If the AC voltage across the parallel resonant circuit is V, then the complex power delivered to the resonator is

$$P_{in} = \frac{|V|^2}{2}\,Y_{in} = \frac{|V|^2}{2}\left(\frac{1}{R} + j\omega C - j\frac{1}{\omega L}\right) \tag{4-3}$$

At resonance the reactive power of the inductor is equal to the reactive power of the capacitor. Therefore, the power delivered to the resonator is equal to the power dissipated in the resistor.

$$P_{in} = \frac{|V|^2}{2R} \qquad (4\text{-}4)$$

The resonance frequency for the parallel resonant circuit as well as the series resonant circuit is obtained by setting $\omega_0 C = \dfrac{1}{\omega_0 L}$ or:

$$\omega_o = 2\pi f_o = \frac{1}{\sqrt{L\,C}} \qquad (4\text{-}5)$$

where, ω_0 is the angular frequency in radian per second and f_0 is equal to the frequency in Hertz.

4.5 Resonant Circuit Loss

In Figure 4-1 and 4-3 the resistor R1 represents the loss in the resonator. It includes the losses in the capacitor as well as the inductor. The Q factor can be shown to be a ratio of the energy stored in the inductor and capacitor to the power dissipated in the resistor as a function of frequency [6]. For the series resonant circuit of Figure 4-1 the unloaded Q factor is defined by:

$$Q_u = \frac{X}{R} = \frac{\omega_o L}{R} = \frac{1}{\omega_o RC} \qquad (4\text{-}6)$$

The unloaded Q factor of the parallel resonant circuit in Figure 4-3 is simply the inverse of the series resonant circuit.

$$Q_u = \frac{R}{X} = \frac{R}{\omega_o L} = \omega_o RC \qquad (4\text{-}7)$$

We can clearly see that as the resistance increases in the series resonant circuit, the Q factor decreases. Conversely as the resistance increases in the parallel resonant circuit, the Q factor increases. The Q factor is a measure of loss in the resonant circuit. Thus a higher Q corresponds to lower loss and a lower Q corresponds to a higher loss. It is usually desirable to achieve high Q factors in a resonator as it will lead to lower losses in filter applications or lower phase noise in oscillators. Note that the resonator Q of Equation (4-6) and (4-7) is defined as Q_u, the unloaded Q of the resonator. This means that the resonator is not connected to any source or load impedance and as such is unloaded. The measurement of Q_u requires that the resonator be attached (coupled) to a signal source or load of some finite impedance. Equations (4-6) and (4-7) would then have to be modified to include the source and load resistance. We might also surmise that any reactance associated with the source or load impedance may alter the resonant frequency of the resonator. This leads to two additional definitions of Q factor that the engineer must consider: the loaded Q and external Q.

4.6 Loaded Q and External Q

Example 4.3: Analyze the parallel resonator that is attached to a 50 Ω source and load as shown in Figure 4-5.

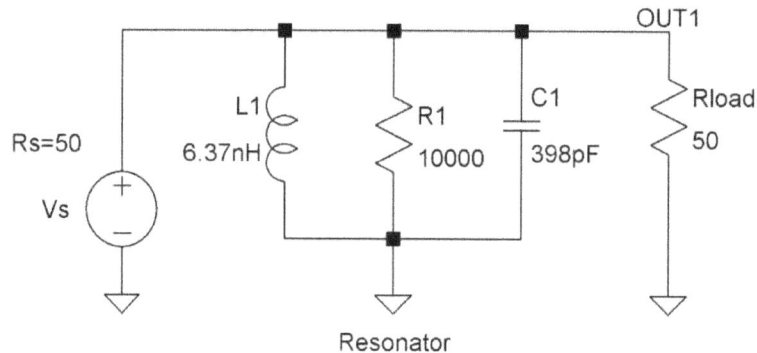

Figure 4-5 Parallel resonator with source and load impedance attached

Solution: Using Equation (4-7) to define the Q factor for the circuit requires that we include the source and load resistance which is 'loading' the resonator. This leads to the definition of the loaded Q, Q_L, for the parallel resonator as defined by Equation (4-8).

$$Q_L = \frac{R_S + R + R_L}{\omega_o L} \qquad (4\text{-}8)$$

Conversely we can define a Q factor in terms of only the external source and load resistance. This leads to the definition of the external Q, Q_E.

$$Q_E = \frac{R_S + R_L}{\omega_o L} \qquad (4\text{-}9)$$

The three Q factors are related by the inverse relationship of Equation (4-10).

$$\frac{1}{Q_L} = \frac{1}{Q_E} + \frac{1}{Q_U} \qquad (4\text{-}10)$$

At RF and microwave frequencies it is difficult to directly measure the Q_u of a resonator. We may be able to calculate the Q factor based on the physical properties of the individual inductors and capacitors as we seen in chapter 1. This is usually quite difficult and the Q factor is typically measured using a Vector Network Analyzer, VNA. Therefore, the measured Q factor is usually the loaded Q, Q_L. External Q is often used with oscillator circuits that are generating a signal. In this case the oscillator's load impedance is varied so that the external Q can be measured. The loaded Q of the network is then related to the fractional bandwidth by Equation (4-11).

$$Q_L = \frac{\sqrt{f_l f_h}}{BW_{-3dB}} \qquad (4\text{-}11)$$

where, BW (BW = $f_h - f_l$) is the -3 dB bandwidth in Hz and f_l, f_h are the lower and upper frequencies at -3dB points.

4.7 Lumped Element Parallel Resonator Design

Example 4.4: In this example we design a lumped element parallel resonator at a frequency of 100 MHz. The resonator is intended to operate between a source resistance of 100 Ω and a load resistance of 400 Ω.

Solution: Best accuracy would be obtained by using S parameter files or Modelithics models for the inductor and capacitor. However a good first order model can be obtained by using the inductor and capacitor models that include the component Q factor. These models save us the work of calculating the equivalent resistive part of the inductor and capacitor model. Use the Q factors shown in the schematic of Figure 4-6.

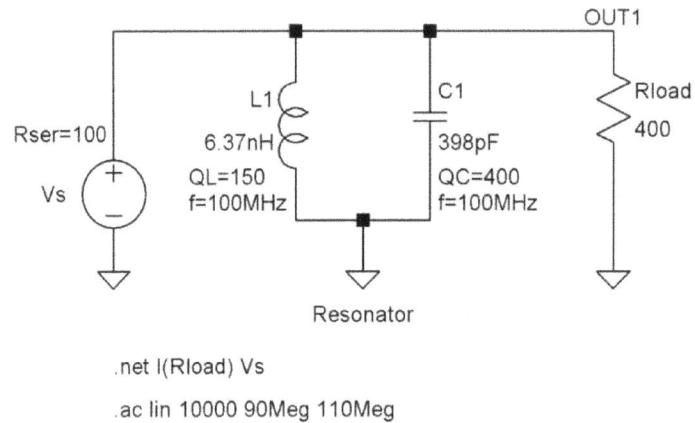

Figure 4-6 Parallel resonator using inductor and capacitor with assigned Q_u values

Simulate the schematic and display the insertion loss, S_{21}, in a rectangular plot.

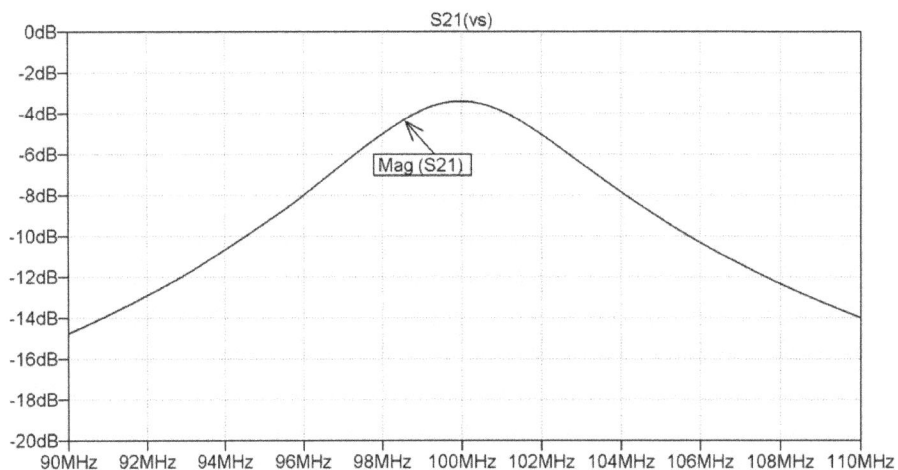

Figure 4-7 Insertion loss of Parallel resonator using inductor and capacitor

The -3 dB bandwidth can be determined by inserting cursors at -3 dB points. To place cursors 1 and 2 manually at -3 dB points, first double-click on "S21(vs)" in the waveform window. Read the peak value of S21 in dB and record the corresponding frequency. Next, move cursor 1 and cursor 2 to frequencies to the and right of center frequency where S21(vs) drops by 3 dB. Record the -3 dB points f_l = 97.05 MHz and f_h = 103 MHz. The loaded Q, Q_L can be calculated using Equation (4-11).

$$Q_L = \frac{\sqrt{f_l f_h}}{BW} = \frac{99.98\,MHz}{5.95\,MHz} = 16.80$$

The designer must use caution when sweeping resonant circuits. Particularly high Q band pass networks require a large number of discrete frequency steps in order to achieve the necessary resolution required to accurately measure the 3 dB bandwidth. In this example the Linear Analysis is set up to sweep the circuit from 90 MHz to 110 MHz using 10000 points.

4.8 Effect of Load Resistance on Bandwidth and QL

In RF circuits and systems the impedances encountered are often quite low, ranging from 1 Ω to 50 Ω. It may not be practical to have a source impedance of 100 Ω and a load impedance of 400 Ω.

Example 4.5: Using the parallel LC example, change the load resistance from 400 Ω to 50 Ω and re-examine the circuit's 3 dB bandwidth and QL.

Solution: Change the load resistance from 400 Ω to 50 Ω in Figure 7 and simulate the schematic.

Simulate the schematic and display the insertion loss, S_{21}, in a rectangular plot. The 3 dB bandwidth is now 12.87 MHz resulting in a loaded Q factor of 7.766.

$$Q_L = \frac{\sqrt{f_l f_h}}{BW} = \frac{99.958\,MHz}{12.87\,MHz} = 7.766$$

The loaded Q factor has decreased by nearly half of the original value. We have increased the bandwidth or de-Q'd the resonator. This can also be thought of as tighter coupling of the resonator to the load.

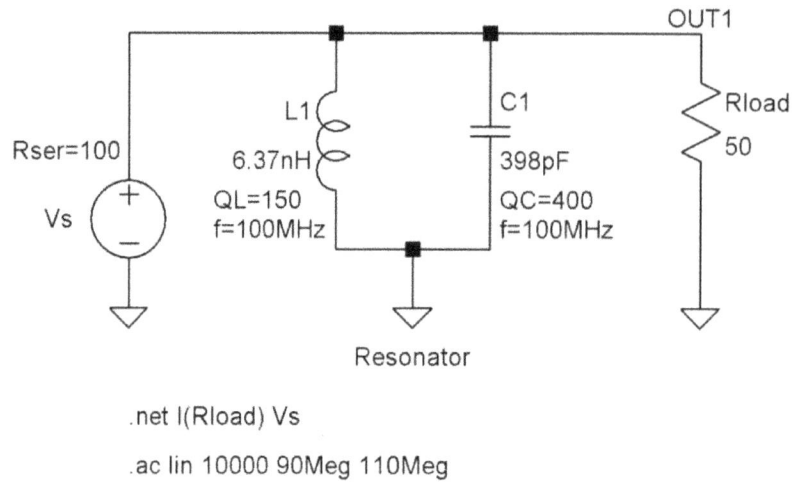

.net I(Rload) Vs

.ac lin 10000 90Meg 110Meg

Figure 4-8 A parallel resonance circuit

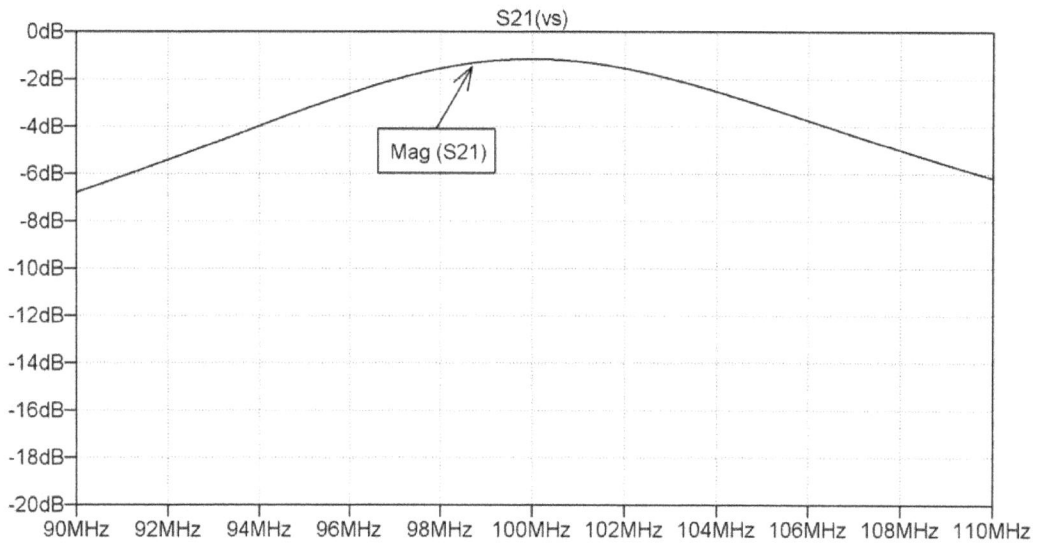

Figure 4-9 Insertion loss of parallel resonance circuit

4.9 Lumped Element Resonator Decoupling

To maintain the high Q of the resonator when attached to a load such as 50 Ω, it is necessary to transform the low impedance to high impedance presented to the load. The 50 Ω impedance can be transformed to the higher impedance of the parallel resonator thereby resulting in less loading of the resonator impedance. This is referred to as loosely coupling the resonator to the load. The tapped-capacitor and tapped-inductor networks can be used to accomplish this Q transformation in lumped element circuits.

4.10 Tapped Capacitor Resonator Design

Example 4.6: Consider rearranging the parallel LC network of Figure 4-8 with the tapped capacitor network shown in Figure 4-10. Re-examine the circuit's 3 dB bandwidth and QL.

Solution: The new capacitor values for C1 and C2 can be found by the simultaneous solution of the following equations.

$$C_T = \frac{C1 \cdot C2}{C1 + C2} \qquad\qquad (4\text{-}12)$$

$$R_{L1} = R_L \left(1 + \frac{C1}{C2}\right)^2 \qquad\qquad (4\text{-}13)$$

R_{L1} is the higher, transformed, load resistance. In this example substitute R_{L1} = 400 Ω, the original load resistance value. C_T is simply the original capacitance of 398 pf. The capacitor values are found to be: C1 = 1126.23 pF and C2 = 616.1 pF. The new resonator circuit is shown in Figure 4-10.

Sweeping the circuit we see that the response in Figure 4-11 has returned to the original performance of Figure 4-7.

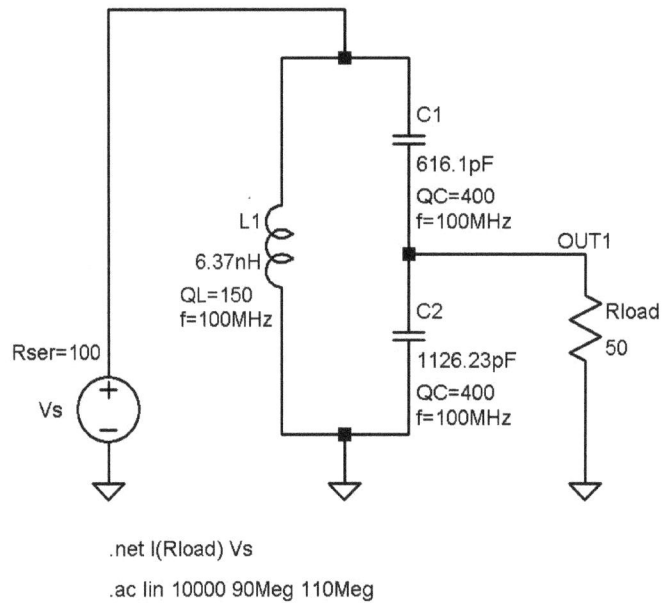

.net I(Rload) Vs

.ac lin 10000 90Meg 110Meg

Figure 4-10 Parallel LC resonator using tapped capacitor

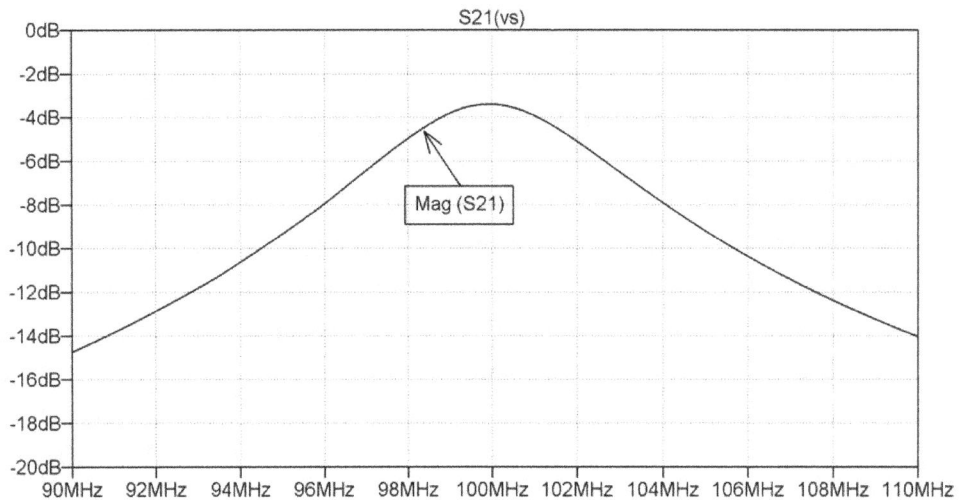

Figure 4-11 Response of parallel LC resonator using a tapped capacitor

The 3 dB bandwidth has returned to 5.89 MHz making the Q_L equal to:

$$Q_L = \frac{\sqrt{f_l f_h}}{BW} = \frac{99.958\,MHz}{5.89\,MHz} = 16.97$$

The 50 Ω load resistor has been successfully decoupled from the resonator. The tapped capacitor and inductor resonators are popular methods of decoupling RF and lower microwave frequency resonators. It is frequently seen in RF oscillator topologies such as the Colpitts oscillator in the VHF frequency range.

4.11 Tapped Inductor Resonator Design

Example 4.7: Similarly design a tapped inductor network to decouple the 50 Ω source impedance from loading the resonator.

Solution: Replace the 100 Ω source impedance with a 50 Ω source and use a tapped inductor network to transform the new 50 Ω source to 100 Ω. Modify the circuit to split the 6.37 nH inductor, L_T, into two series inductors, L1 and L2. The inductor values can then be calculated by solving the following equation set simultaneously. R_{S1} is the higher, transformed, source resistance. In this example substitute R_{S1} = 100 Ω,

$$Rs1 = Rs\left(\frac{L_T}{L_1}\right)^2 \qquad\qquad (4\text{-}14)$$

$$L_T = L_1 + L_2 \qquad\qquad (4\text{-}15)$$

Solving the equation set results in values of L1=4.5 nH and L2=1.87 nH. The resulting schematic and response is shown in Figures 4-12 and 4-13.

The new response is identical to the plot of Figure 4-11. Therefore we now have a source and load resistance of 50 Ω and have not reduced the Q of the resonator from what we had with the original source resistance of 100 Ω and a load resistance of 400 Ω.

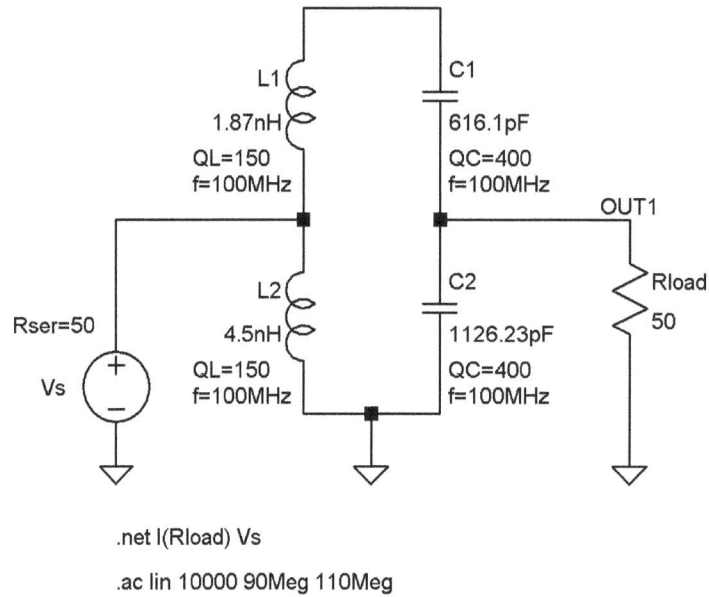

.net I(Rload) Vs

.ac lin 10000 90Meg 110Meg

Figure 4-12 Tapped-inductor parallel resonant circuit

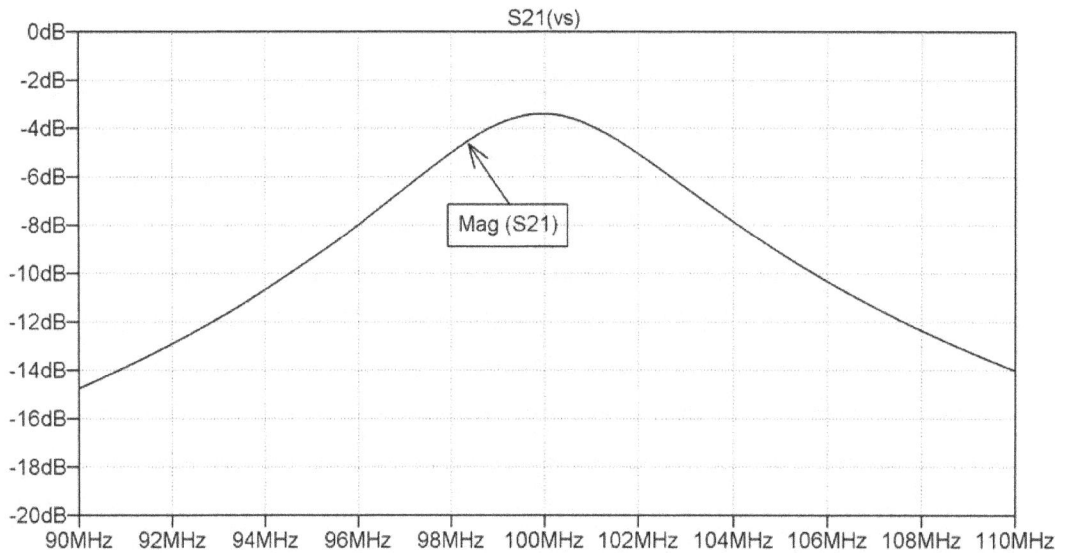

Figure 4-13 Response of the parallel resonant circuit

4.12 Transmission Line Resonators

From Figure 2-11 we have seen that a quarter-wave short-circuited transmission line results in a parallel resonant circuit. Similarly Figure 2-15 showed that a half-wave open circuited transmission line results in a parallel resonant circuit. Such parallel resonant circuits are often used as one port resonators. Near the resonant frequency, the one port resonator behaves as a parallel RLC network as shown in Figure 4-3. As the frequency moves further from resonance the equivalent network becomes more complex typically involving multiple parallel RLC networks. One port resonators are coupled to one another to form filter networks or directly to a transistor to form a microwave oscillator. Knowing the losses due to the physical and electrical parameters of the transmission line, one can calculate the Q_u of the transmission line resonator. The microstrip resonator Q_u is comprised of losses due to the conductor metal, the substrate dielectric, and radiation losses. The Q_u is often dominated by the conductor Q. Unfortunately it can be quite difficult to accurately determine the conductor losses in a microstrip resonator. T. C. Edwards has developed a set of simplified expressions for the conductor losses [4]. Equation (4-16) is an approximation of the conductor losses that treats the transmission line as a perfectly smooth surface.

$$\alpha_c = 0.072 \frac{\sqrt{f}}{W_e Z_o} \lambda_g \quad \text{dB/inch} \qquad (4\text{-}16)$$

where,

f = the frequency in GHz

W_e = the effective conductor width (inches)

Z_o = the characteristic impedance of the line

α_c = Conductor loss in dB/inch

λ_g = wavelength in dielectric in inches

The corresponding Q factor related to the conductor is then given by:

$$Q_c = \frac{27.3\sqrt{\varepsilon_{\text{eff}}}}{\alpha_c \lambda_o} \qquad (4\text{-}17)$$

4.13 Microstrip Resonator Example

Example 4.8: Consider a 5 GHz half wavelength open circuit microstrip resonator. The resonator is realized with a 50 Ω microstrip line on Roger's RO3003 dielectric. Calculate the unloaded Q factor of the resonator. The substrate parameters are defined as:

Dielectric constant	$\varepsilon_r = 3$
Substrate height	$h = 0.030$ in.
Conductor thickness	$t = 0.0026$ in.
Line Impedance	$Z_o = 50$ Ω
Conductor width	$w = 0.075$ in.
Loss tangent	$\tan\delta = 0.0013$

Solution: Using the simplified expression of Equation (4-16) for a smooth microstrip line the loss and conductor Q factor is calculated as:

$$\alpha_c = 0.072 \frac{\sqrt{5}}{0.077(50)}(1.52) = 0.063 \; dB/inch$$

$$Q_c = \frac{27.3\sqrt{2.41}}{(0.063)(2.36)} = 283.1$$

The dielectric loss and Q factor are then calculated from:

$$\alpha_d = 27.3 \frac{(3)(2.41 - 1)(0.0013)}{\sqrt{2.41}\,(3 - 1)(2.36)} = 0.021 \; dB/inch$$

$$Q_d = 27.3 \frac{\sqrt{\varepsilon_{eff}}}{\alpha_d \lambda_0} = 27.3 \frac{\sqrt{2.41}}{(0.021)\,(2.36)} = 875.2, \text{ where}$$

α_d = loss of dielectric material dB/inch
Q_d = quality factor for the dielectric material.

For simplicity the radiation Q factor will be omitted. We will model the resonator using the linear simulator. Linear simulators often do not model the radiation effects of the microstrip line. Therefore the overall unloaded Q factors then becomes.

$$\frac{1}{Q_u} = \frac{1}{283.1} + \frac{1}{875.2} = \frac{1}{213.9}$$

$$Q_u = 213.9$$

4.14 Microstrip Resonator Model

The 5 GHz half wave open circuit microstrip resonator is modeled in Fig 4-14. Note that the source and load impedance has been increased to 5000 Ω to avoid loading the impedance of the parallel resonant circuit.

Example 4-9: The 5 GHz half wave open-circuited microstrip resonator is shown in Figure 4-14. Simulate the resonator and measure its loaded Q factor.

Solution: Schematic of the microstrip half wave resonator is shown in Figure 4-14.

Simulate the schematic and display S$_{21}$ in a rectangular plot, as shown in Figure 4-15.

Using the -3 dB data points in the Equation (4-10), the loaded Q factor is:

$$Q_L = \frac{\sqrt{(4968)(5031)}}{5031 - 4967} = 79.35$$

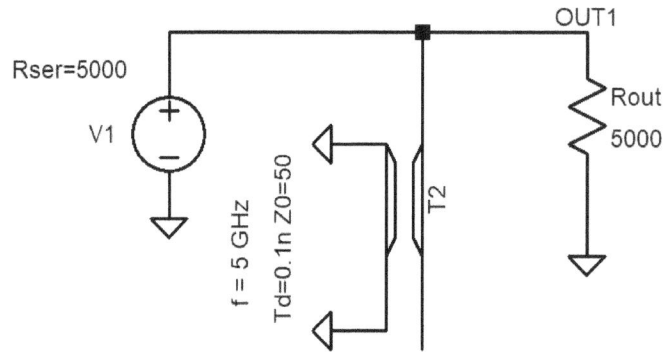

.net I(Rout) V1

.ac lin 10000 4500Meg 5400Meg

Figure 4-14 Half-wave open ended microstrip resonator

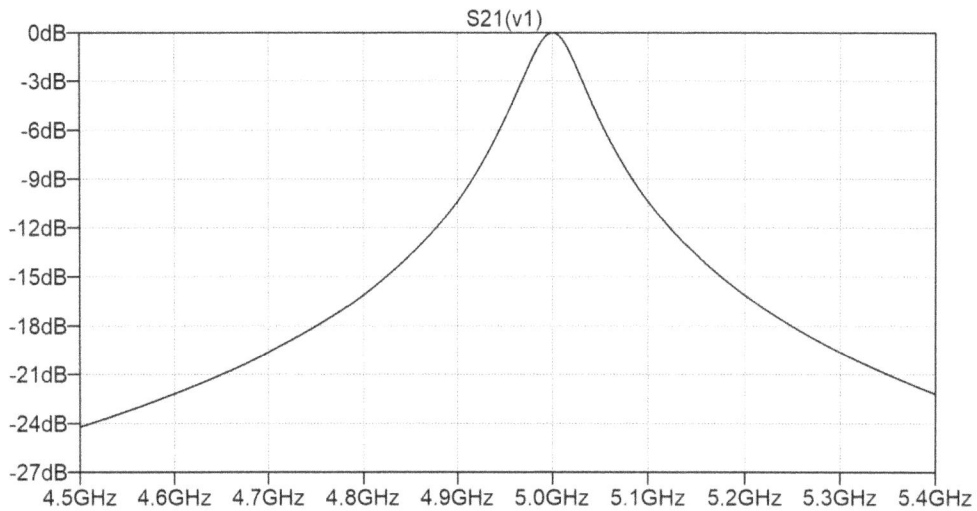

Figure 4-15 Response of the half-wave open circuit microstrip resonator

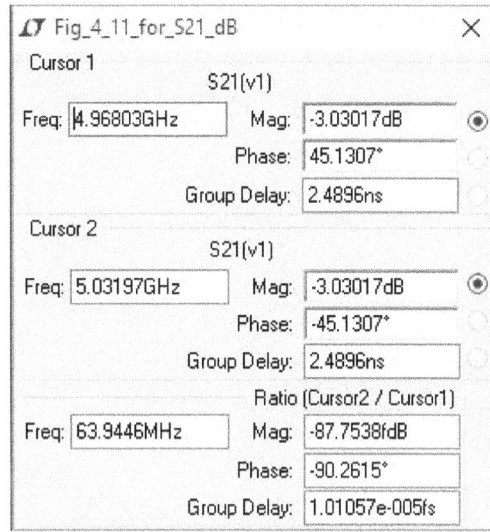

Figure 4-16: Data points at −3dB

4.15 Filter Design at RF and Microwave Frequency

In Section 4.8 we have seen that it is possible to change the shape of the frequency response of a parallel resonant circuit by choosing different source and load impedance values. Likewise multiple resonators can be coupled to one another and to the source and load to achieve various frequency shaping responses. These frequency shaped networks are referred to as filters.

4.16 Filter Topology

The subject of filter design is a complex topic and the subject of many dedicated texts [1–2]. This section is intended to serve as a fundamental primer to this vast topic. It is also intended to set a foundation for successful filter design using the LTspice software. The four most popular filter types are: Low Pass, High Pass, Band Pass, and Band Stop. The basic transmission response of the filter types is shown in Figure 4-17. The filters allow RF energy to pass through their designed pass band. RF energy that is present outside of the pass band is reflected back toward the source and not transmitted to the load. The amount of energy present at the load is defined by the S21 response. The amount of energy reflected back to the source is defined by the S11 response.

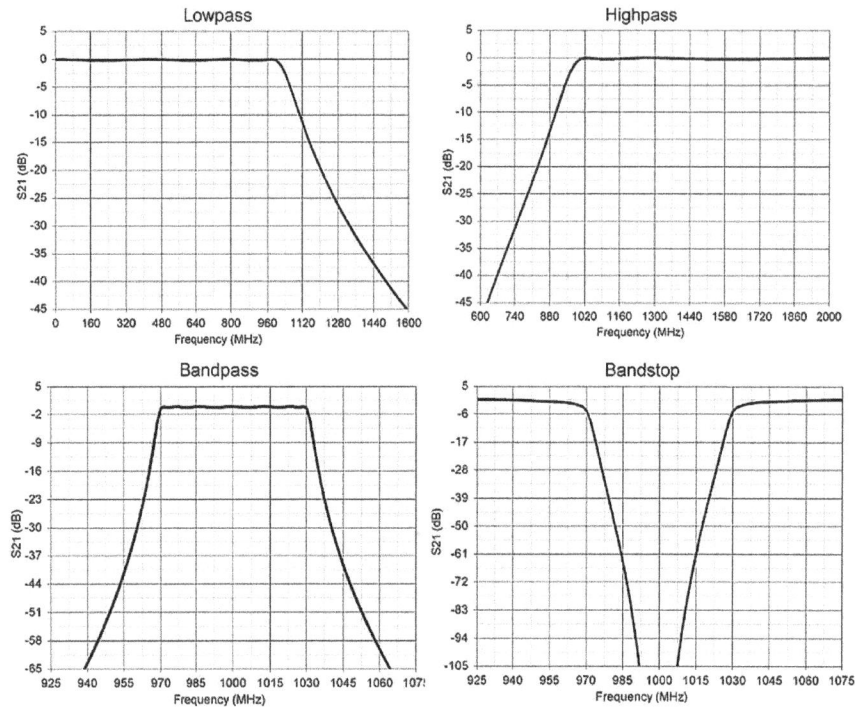

Figure 4-17 Filter transmission versus frequency

4.17 Filter Order

The design process for all of the major filter types is based on determination of the filter pass band, and the attenuation in the reject band. The attenuation in the reject band that is required by a filter largely determines the slope needed in the transmission frequency response. The slope of the filter's response is related to the order of the filter. The steeper the slope or 'skirt' of the filter; the higher is the order. The term order comes from the mathematical transfer function that describes a particular filter. The highest power of s in the denominator of the filter's Laplace transfer function is the order of the filter. For the simple low pass and high pass filters presented in this chapter the filter order is the same as the number of elements in the filter. However this is not the case for general filter networks. In more complex types of filters as well as bandpass and bandstop filters the filter order will not be equal to the number of elements in the filter. In general the filter order is the total of the number of transmission zeros at frequencies:

$$F = 0 \text{ (DC)}$$
$$F = \infty$$
$$0 < F < \infty \text{ (specific frequencies between DC and } \infty)$$

Transmission zeros block the transfer of energy from the source to the load. In fact the order of a filter network can be solved visually by adding up the number of transmission zeros that satisfy the above criteria. Figure 4-18 shows the relationship between the filter order and slope of the response for a Low Pass filter. Each filter of Figure 4-18 has the same cutoff frequency of 1000 MHz. The third order filter has an attenuation of about 16 dB at a rejection frequency of 2000 MHz. The fifth order filter shows an attenuation of 39 dB and the seventh order filter has more than 61 dB attenuation at 2000 MHz. It is therefore clear that the order of the filter is one of the first criteria to be determined in the filter design. It is dependent on the cutoff frequency of the pass band and the amount of attenuation desired at the rejection frequency.

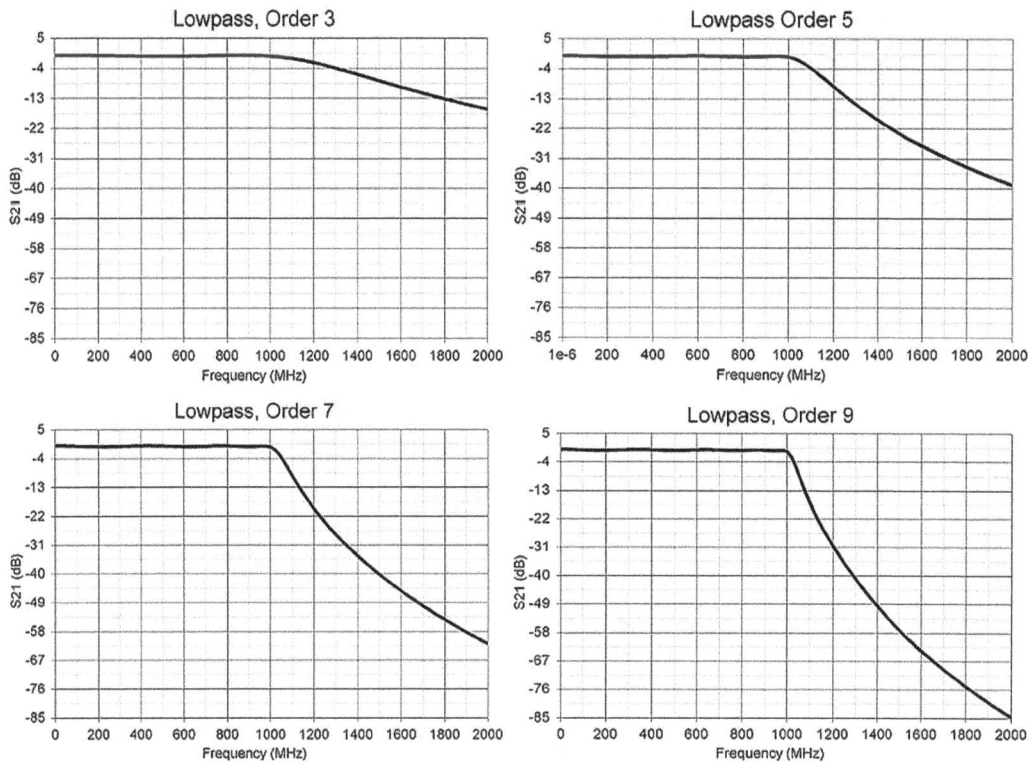

Figure 4-18 Relationship between filter order and the slope of S_{21}

4.18 Filter Type

The shape of the filter passband and attenuation skirt can take on different shape relationships based on the coupling among the various reactive elements in the filter. Over the years several polynomial expressions have been developed for these shape relationships. Named after their inventors, some of the more popular passive filter types include: Bessel, Butterworth, Chebyshev, and Cauer. Figure 4-19 shows the general shape relationship among these filter types for a given seventh order filter. The Bessel filter type is a low Q filter and does not exhibit a steep roll off compared to its counterparts. The benefit of the Bessel filter is its linear phase or flat group delay response. This means that the Bessel filter can pass wideband signals while introducing little distortion. The Butterworth is a medium Q filter that has the flattest pass band of the group. The Chebyshev response is a higher Q filter and has a noticeably steeper skirt moving toward the reject band.

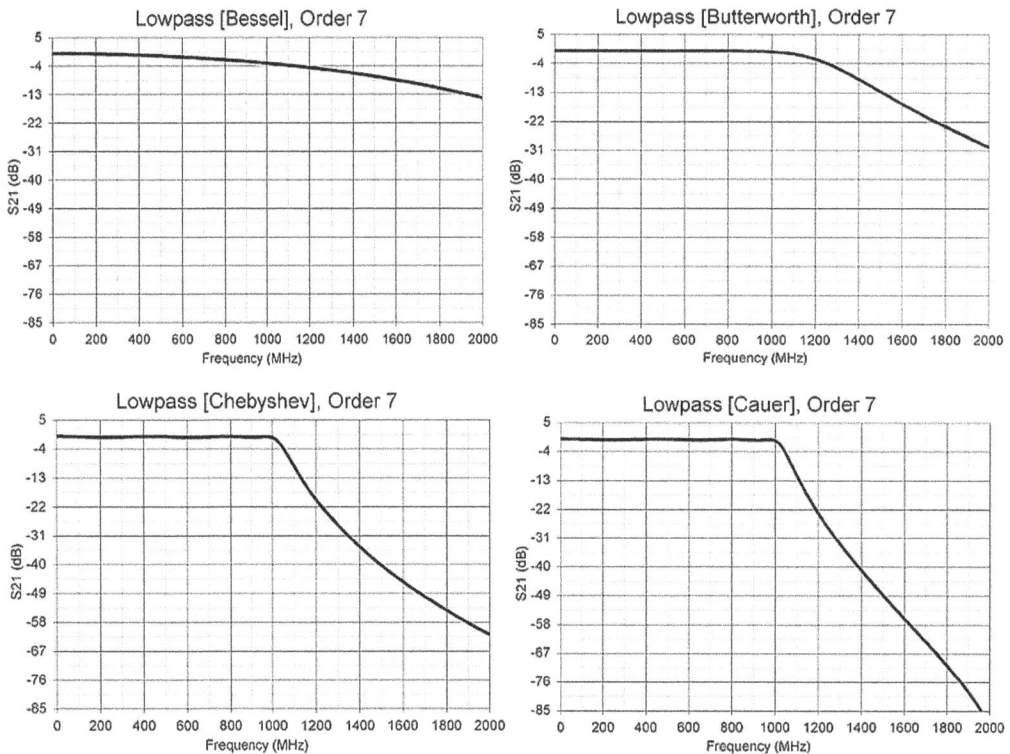

Figure 4-19 Shape of Bessel, Butterworth, Chebyshev, and Cauer filters

As a result it exhibits more transmission ripple in the pass band. The Cauer filter has the steepest slope of all of the four filter types. The Cauer filter is also known as an elliptic filter. Odd order Chebyshev and Cauer filters can be designed to have an equal source and load impedance. The even order Chebyshev and Cauer filters will have a different output impedance from the specified input impedance. Another interesting characteristic of the Cauer filter is that it has the same ripple in the rejection band as it has in the pass band. The Butterworth, Chebyshev and Cauer filters differ from the Bessel filter in their phase response. The phase response is very nonlinear across the pass band. This nonlinearity of the phase creates a varying group delay. The group delay introduces varying time delays to wideband signals which, in turn, can cause distortion to the signal. Group delay is simply the derivative, or slope, of the transmission phase and defined by Equation (4-34). Figure 4-20 shows the respective set of filter transmission characteristics with their corresponding group delay. Note the relative values of the group delay on the right hand axis.

$$\tau_g = -\frac{d\phi}{d\omega}$$

(4-18)

Where, ϕ is the phase shift in radians and ω is the frequency in radians per second.

From the group delay plots of Figure 4-20 it is evident that the group delay peaks near the corner frequency of the filter response. The sharper cutoff characteristic results in greater group delay at the band edge.

4.19 Filter Return Loss and Passband Ripple

The Bessel and Butterworth filters have a smooth transition between their cutoff frequency and rejection frequency. The forward transmission, S_{21}, is very flat vs. frequency. The Chebyshev and Cauer filters have a more abrupt transition between their cutoff and rejection frequencies. This makes these filter types very popular for many filter applications encountered in RF and microwave engineering. It is important to note however that the steeper filter skirt results as a certain amount of impedance mismatch between the source and load impedance.

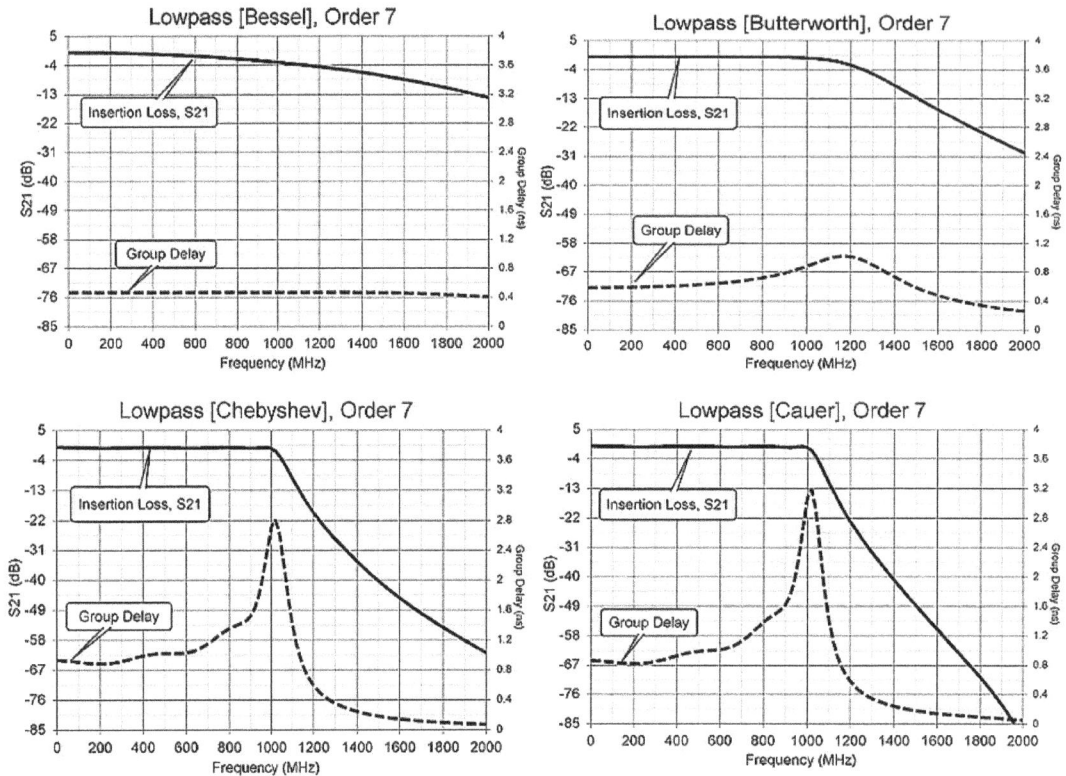

Figure 4-20 Group delay characteristic for various lowpass filter types

The Chebyshev and Cauer filter types have ripple in the forward transmission path, S_{21}. The amount of ripple is caused by the degree of mismatch between the source and load impedance and thus the resulting return loss that is realized by these filter types. For a given Chebyshev or Cauer filter order, the roll off of the filter response is also steeper for greater values of passband ripple. The cutoff frequency of the filters that have passband ripple is then defined as the passband ripple value. For all-pole filters such as the Butterworth, the cutoff frequency is typically defined as the 3 dB rejection point. Figure 4-21 shows the passband ripple of a fifth order low pass filter for ripple values of 0.01, 0.1, 0.25, and 0.5 dB. Note that the ripple shown is produced by ideal circuit elements. In practice the finite unloaded Q or losses in the inductors and capacitors will tend to smooth out this ripple.

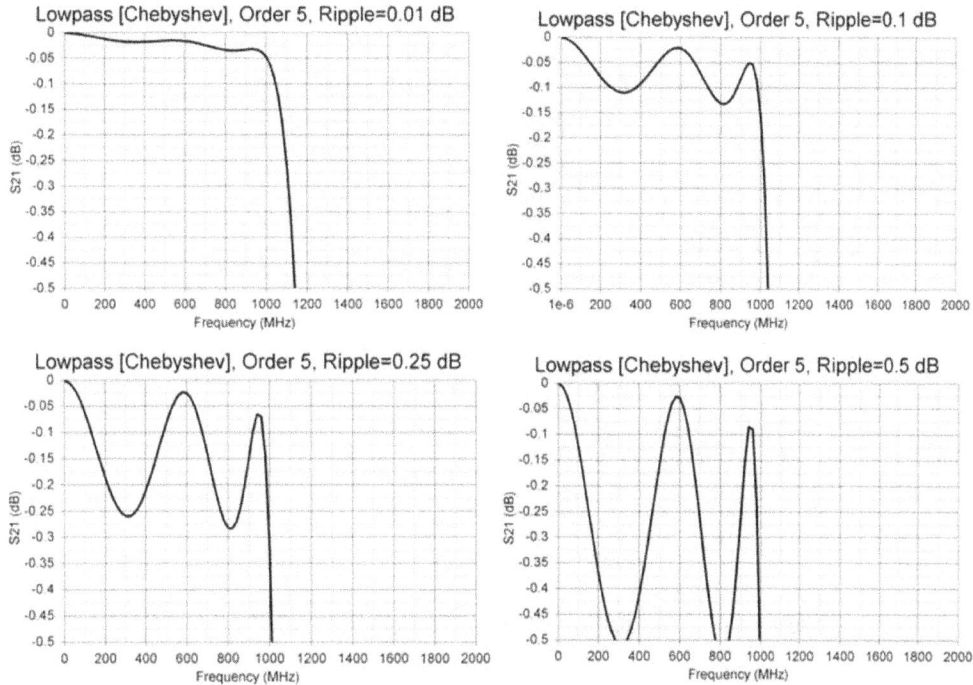

Figure 4-21 Passband ripples in lowpass Chebyshev filter

In Chapter 2 the relationship for mismatch loss between a source and load was presented. For the Chebyshev and Cauer filters this mismatch loss is the passband ripple.

Figure 4-22 shows the same filters from Figure 4-21 with the return loss plotted along with the insertion loss, S_{21}. We can see that for a given filter order, there is a tradeoff between filter rejection and the amount of ripple, or return loss, that can be tolerated in the passband. In most RF and microwave filter designs the 0.01 and 0.1 dB ripple values tend to be more popular. This is due to the tradeoff between good impedance match and reasonable filter skirt slope. Table 4-4 shows the calculation of return loss and the ripple for different values of VSWR. Figure 4-22 shows good correlation of the worst case return loss with that calculated in Table 4-4. When tuning filters using modern network analyzers it is sometimes easier to see the larger changes in the return loss as opposed to the fine grain ripple as shown in Figure 4-21. For this reason it is common to tune the forward transmission of the filter by observing the level and response of the filter's return loss. Return loss is a very sensitive indicator of the filter's alignment and performance.

VSWR	Return Loss (dB)	Ripple (dB)
1.1	26.444	0.01
1.239	19.433	0.05
1.3	17.692	0.075
1.355	16.435	0.1
1.405	15.473	0.125
1.452	14.688	0.15
1.538	13.474	0.2
1.62	12.518	0.25
1.984	9.636	0.5

Table 4-4 Calculation of filter ripple versus VSWR

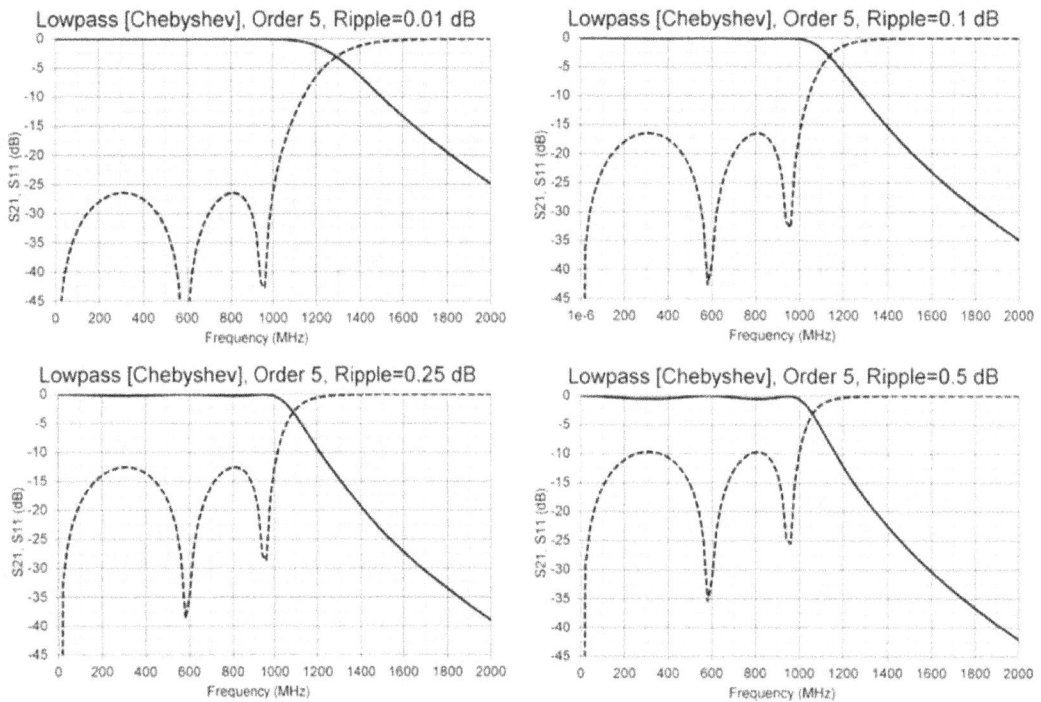

Figure 4-22 Lowpass Chebyshev filter rejection and return loss

4.20 Lumped Element Filter Analysis

Classical filter design is based on extracting a prototype frequency-normalized model from a myriad of tables for every filter type and order [1]. Although these tables are built into many filter synthesis software applications that are readily available, we will analyze two practical filter examples using LTspice, one low pass and one high pass filter.

4.21 Low Pass Filter Example

Example 4.10: As a practical filter, consider a low pass (LP) filter with the following specifications:
- Chebyshev response with 0.1 dB pass band ripple.
- A passband cutoff frequency at 160 MHz
- It has at least -40 dB rejection at 435 MHz.

Construct and simulate this filter showing its input return loss (S11) and insertion loss (S21).

Solution: The low pass filter with the above specifications is shown in Fig. 4-23.

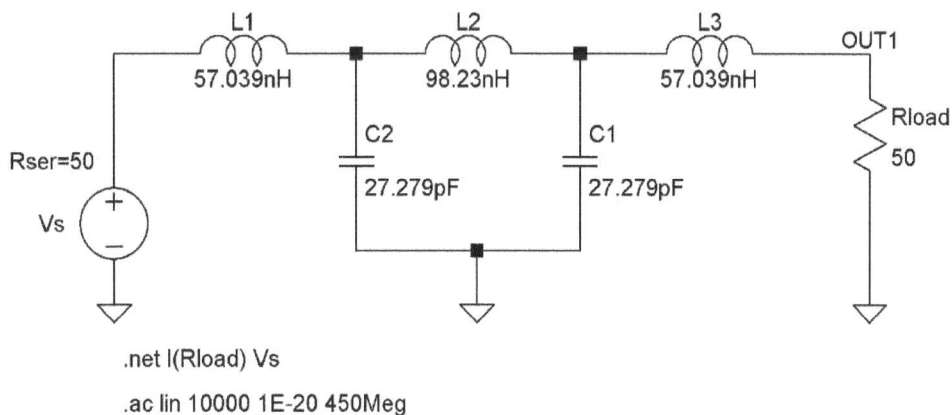

Figure 4-23 Schematic diagram of the LP filter for simulation of S parameters

The response of the low pass filter is shown in Figure 4-24.

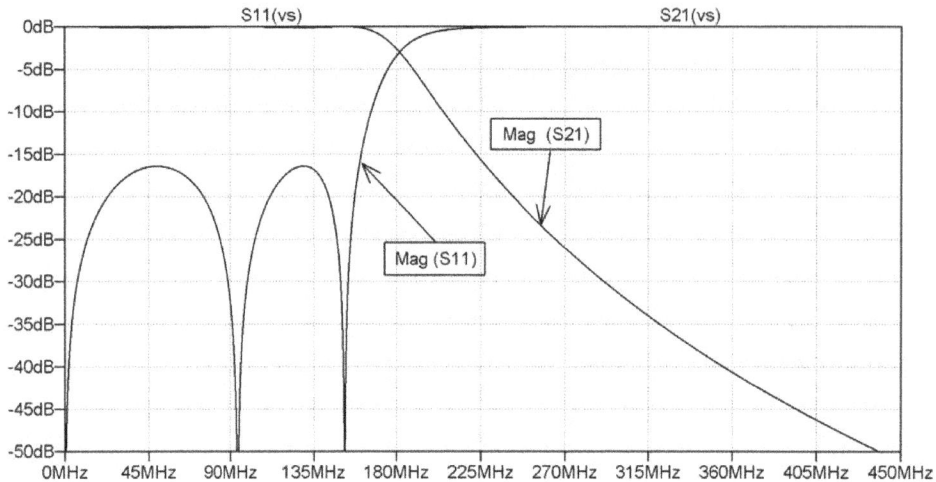

Figure 4-24 Response of the lowpass filter

4.22 High Pass Filter Design Example

Example 4.11: Analyze a high pass filter in the 420 MHz to 450 MHz range with the following specifications:

- The pass band cutoff frequency is 420 MHz.
- The filter has a Chebyshev response with 0.1 dB pass band ripple.
- The reject requirement is at least -60 dB at 146 MHz.

Solution: A high pass filter of order.7 is shown in Figures 4-25. The corresponding input return loss and its insertion loss plots are shown in Figure 4-26.

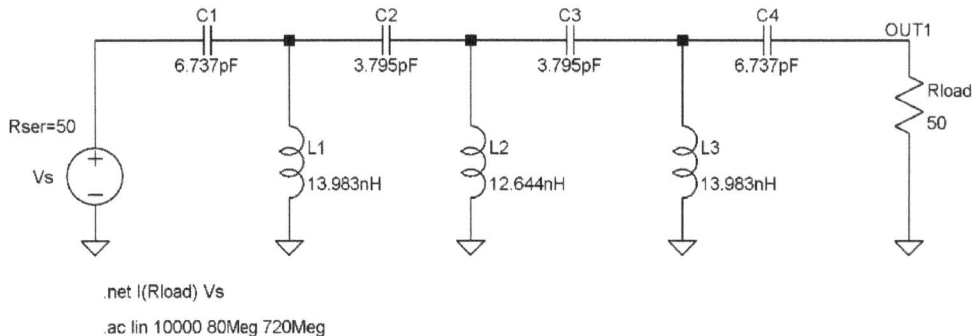

Figure 4-25 Synthesis model of the high pass filter

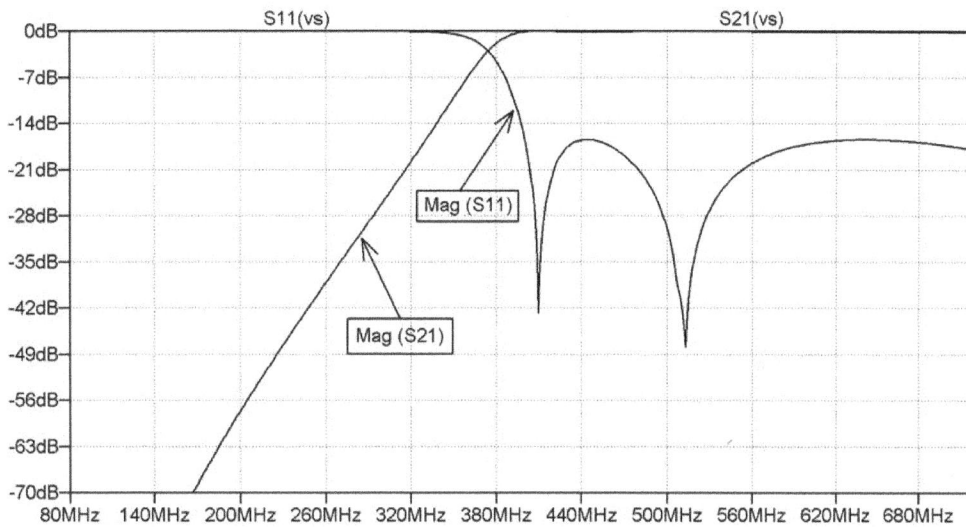

Figure 4-26 Response of the high pass filter

4.23 Distributed Filter Design

In the microwave frequency region filters can be designed using distributed transmission lines. Series inductors and shunt capacitors can be realized with microstrip transmission lines. In the next section we will explore the conversion of a lumped element low pass filter to a design that is realized entirely in microstrip.

Example 4.12: Consider the lumped element 2 GHz low pass filter schematic and response shown in Figures 4-27 and 4-28. This low pass filter has a 3 dB bandwidth of approximately 2485 MHz. Use the microstrip equivalent models of series inductance and shunt capacitance to realize the filter in microstrip. The microstrip substrate is Rogers's 6010 material with a 0.025 dielectric thickness.

Solution: The lumped element low pass filter is shown in Figure 4-27 [1].

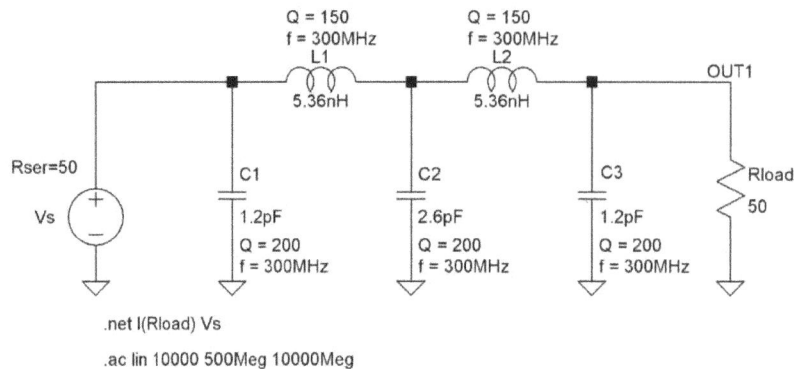

Figure 4-27 Lumped element 2.2 GHz low pass filter

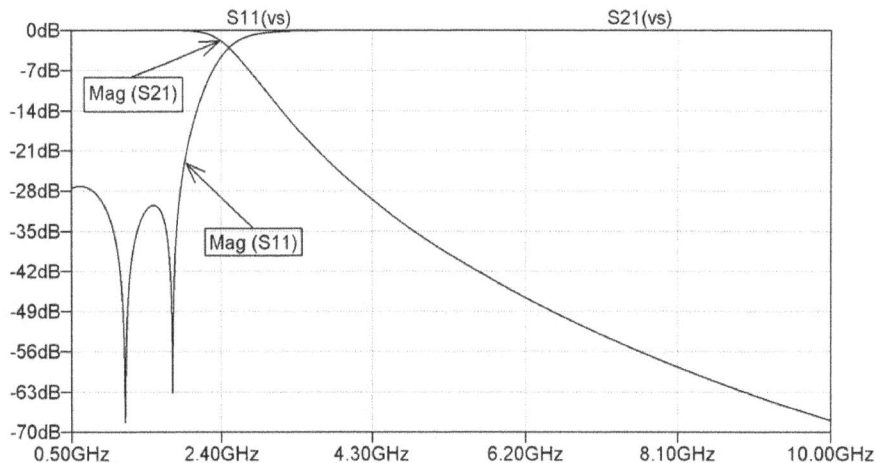

Figure 4-28 Response of the 2.2 GHz low pass filter response

4.24 Lumped to Distributed Element Conversion

Example 4-13: Convert the lumped element filter to distributed element filter.

Solution: The series inductors will be realized as 80 Ω transmission lines of sufficient length to act as a 5.36 nH inductor. The APCAD program is used to calculate the microstrip line width for an 80 Ω transmission line on the Rogers 0.025 in. RO3010 material. As Figure 4-30 shows the 80 Ω transmission line has a line width of 6.26 mils with an effective dielectric constant $\varepsilon_{\text{eff}} = 6.08$. To realize the required inductance value a specific length of 80 Ω transmission line is required. The length of line required for a 5.36 nH inductor is then found to be 321 mils. The shunt capacitors will be realized as 20 Ω transmission lines. The 20 Ω line width is calculated to be 102.8 mils with an effective dielectric constant ε_{eff} = 8.00. The line length for the 1.2 pF capacitors is then found to be 100 mils.

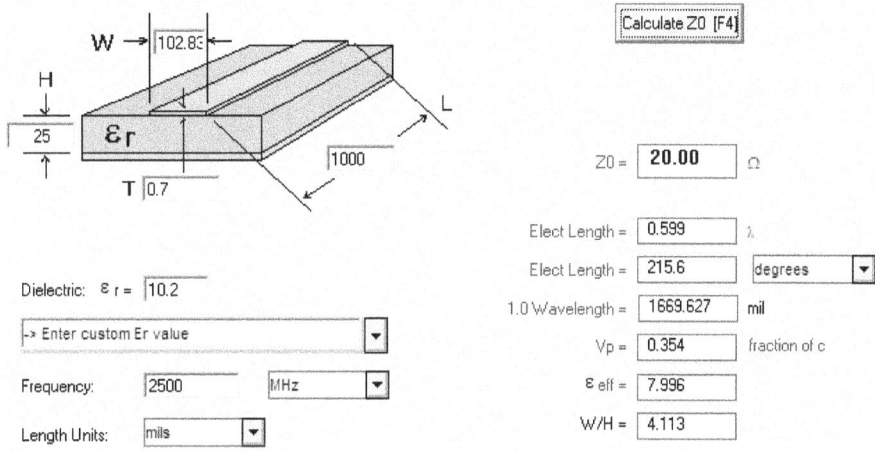

Figure 4-29 Calculation of 20 Ω microstrip lines (W = 102.833 mils)

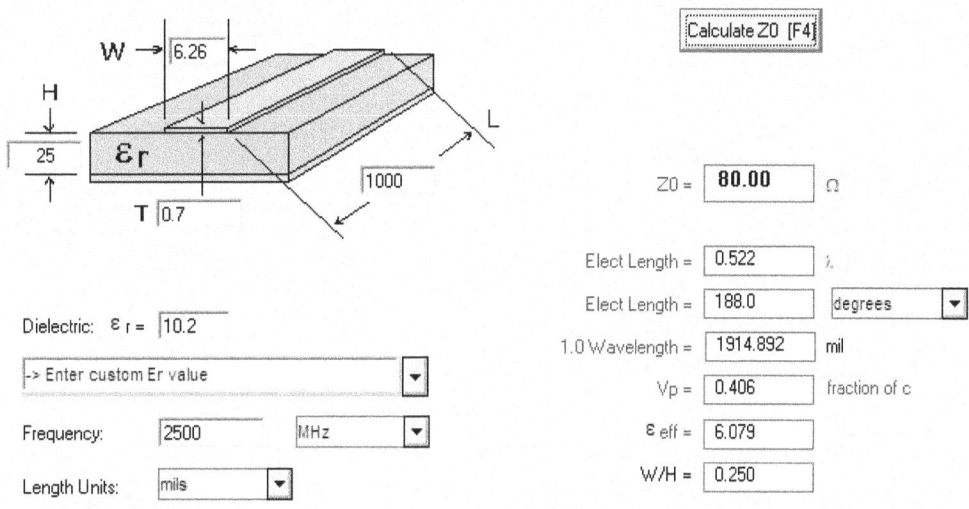

Figure 4-30 TLINE calculation of 80 Ω microstrip lines (W = 6.26 mils)

Adding a short 50 Ω section to the input and output, the initial low pass filter schematic and response is shown in Figures 4-31 and 4-33. Layout representation of the schematic in Figure 4-31 is depicted in Figure 4-32.

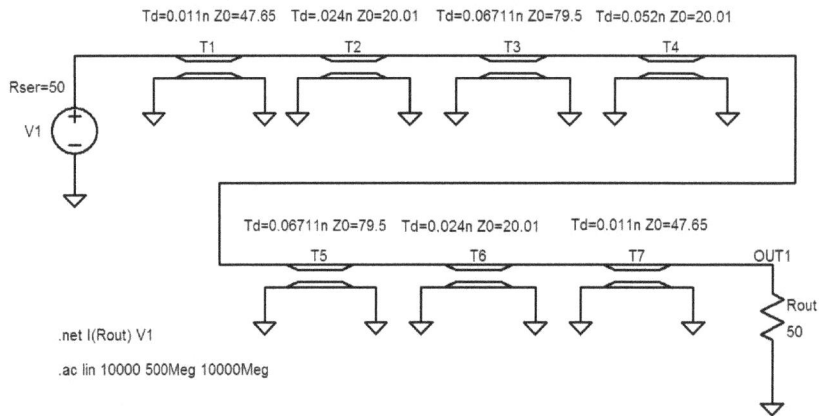

Figure 4-31 I Schematic of the low pass filter

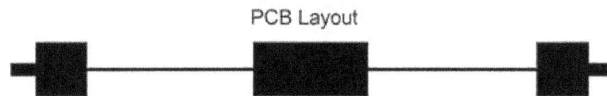

Figure 4-32 PCB layout of the low pass filter

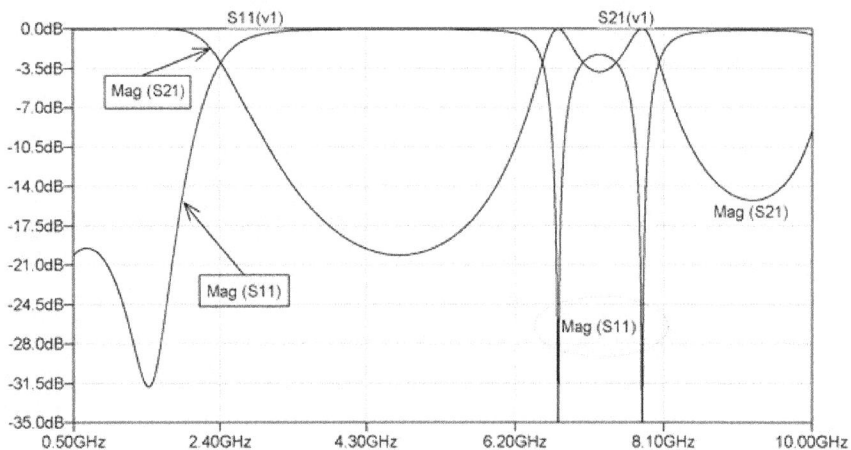

Figure 4-33 Response of the low pass filter

Problems

4-1.   Consider the one port resonator that is represented as a series RLC circuit
        as shown. Analyze the circuit, with R = 2 Ω, L = 5 nH, and C = 5 pF. Plot
        the magnitude of the resonator input impedance and measure the
        resonance frequency.

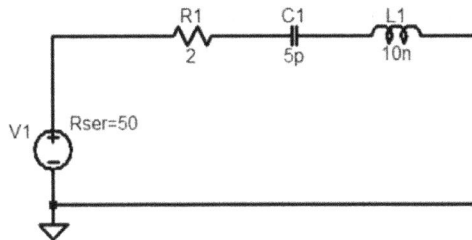

Figure P4-1 A series RLC circuit

4-2.   Consider the one port resonator that is represented as a parallel RLC
        circuit as shown. Analyze the circuit, with R = 500 Ω, L = 50 nH, and C =
        50 pF. Plot the magnitude of the resonator input impedance and measure
        the resonance frequency.

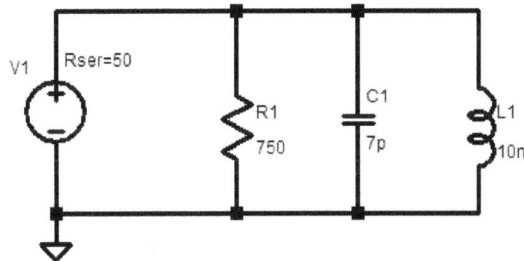

Figure P4-2 A parallel RLC circuit

4-3.   Design a 75 Ω transmission line of sufficient length to act as a 10 nH
        inductor. Calculate the microstrip line width on the Rogers 0.025 inch
        RO3010 material.

4-4.   Use APCAD to determine the width of a 50-ohm microstrip line having 30
        mils height and using FR4 material. Assume frequency of 1 GHz.
        Determine the electrical length if the physical length is 1 inch.

References and Further Readings

[1]   Ali A. Behagi, *RF and Microwave Circuit Design*, A Design Approach Using (**ADS**), Techno Search, Ladera Ranch, CA 2015

[2]   Ali Behagi and Manou Ghanevati, *Fundamentals of RF and Microwave Circuit Design,* Ladera Ranch, CA 2017

[3]   Handbook of Filter Synthesis, Analtol I. Zverev, Wiley 1967

[4]   *RF Circuit Design*, Second Edition, Christopher Bowick, Elsevier 2008

[5]   Foundations for Microstrip Circuit Design, T.C. Edwards, John Wiley & Sons, New York, 1981

[6]   *Q Factor*, Darko Kajfez, Vector Fields, Oxford Mississippi, 1994

[7]   David M. Pozar, *Microwave Engineering*, Second Edition, John Wiley and Sons, Inc. 1998

[8]   Q Factor Measurement with Network Analyzer, Darko Kajfez and Eugene Hwan, IEEE Transactions on Microwave Theory and Techniques, Vol. MTT-32, No. 7, July 1984.

[9]   *High Frequency Techniques*, Joseph F. White, John Wiley & Sons, Inc., 2005

[10]   *Soft Substrates Conquer Hard Designs*, James D. Woermbke, Microwaves, January 1982.

[11]   David M. Pozar, *Microwave Engineering*, Third Edition, John Wiley and Sons, 2005.

Chapter 5

Power Transfer and Impedance Matching

5.1 Introduction

Impedance matching is an integral part of RF and microwave circuit design and is used mainly for the efficient transfer of power from a source to the load. For example, in microwave amplifier design, the need for impedance matching arises when the amplifier must be properly terminated at both terminals in order to deliver maximum power from the source to the load. In narrowband applications impedance matching can be achieved, at a single frequency, with a two-element lossless network, known as L-network or L-section. In this chapter the basics of power transfer and the conditions for maximum power transfer are presented. The mathematical equations for the design of discrete L-networks are derived using MATLAB compatible expressions. This builds a solid foundation for the engineer to handle more complicated matching problems including broadband matching applications. In this chapter the analytical techniques for the design of matching arbitrary source and load impedances are developed. The analytical techniques are then used, in several examples, to design discrete narrowband and broadband impedance matching networks. For verification of the design the LTspice software is used throughout.

5.2 Power Transfer Basics

At low frequencies, where the electrical wavelengths of the signals are much longer than the physical dimensions of the wires and lumped components, the phase of the voltage and current waveforms do not change significantly along the length of wires or components. Therefore, power is transmitted from the source to the load with little loss. At RF and microwave frequencies, where the signal wavelengths are equal to or shorter than the physical dimension of the network components, the amplitude and phase of the voltage and current waveforms change significantly as they travel from source to the load. In this case power is reflected at discontinuities and the maximum power will not reach the load. To achieve maximum power transfer we need to eliminate reflections at discontinuities by inserting proper impedance matching networks.

5.3 Maximum Power Transfer Conditions

For the network of Figure 5-1 where a voltage source, V_S, and the series impedance $Z_S = R_S + jX_S$ are connected to a network, having the input impedance $Z_{IN} = R_{IN} + jX_{IN}$, the power transferred to the network is given by Equation (5-1).

$$P_{Network} = \frac{\text{Re}\left[V_{IN} I_{IN}^*\right]}{2} \qquad (5\text{-}1)$$

In Equation (5-1) Re denotes the real part and the symbol * denotes the conjugate value.

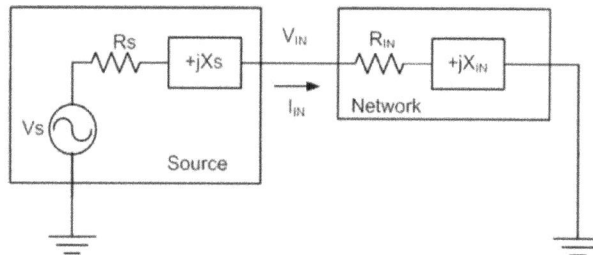

Figure 5-1 Voltage source connected to complex load impedance
(Courtesy of BT Microwave LLC)

In Figure 5-1, we assume V_S is sinusoidal steady state voltage source, R_S and R_{IN} are positive, and X_S and X_{IN} are real numbers. The input voltage and current to the network can be related to the source voltage as:

$$V_{IN} = \frac{V_S (R_{IN} + jX_{IN})}{(R_{IN} + R_S) + j(X_{IN} + X_S)}$$

And,

$$I_{IN}^* = \frac{V_S}{(R_{IN} + R_S) - j(X_{IN} + X_S)}$$

Therefore, multiplying V_{IN} by I_{IN}^*, Equation (5-1) can be written as:

$$P = \frac{\left(\dfrac{V_S^2 R_{IN}}{2}\right)}{(R_{IN} + R_S)^2 + (X_{IN} + X_S)^2} \tag{5-2}$$

To maximize the power transfer, we differentiate Equation (5-2) with respect to R_{IN} and X_{IN} and set them equal to zero,

$$\frac{V_S^2}{2}\left[(R_{IN} + R_S)^2 + (X_{IN} + X_S)^2\right] - \frac{V_S^2 R_{IN}}{2}\left[2R_{IN} + 2R_S\right] = 0$$

$$-\frac{V_S^2 R_{IN}}{2}\left(2X_{IN} + 2X_S\right) = 0$$

A simultaneous solution of these two equations for R_{IN} and X_{IN}, le to:

$$R_{IN} = R_S \tag{5-3}$$

and

$$X_{IN} = -X_S \tag{5-4}$$

Therefore, the maximum power transfer condition is that the load impedance be equal to the conjugate of the source impedance given in Equation (5-5).

$$Z_{IN} = Z_S^* \tag{5-5}$$

This maximum power transfer condition between the source and load impedance can be divided into three cases.
Case 1:

If the source impedance is purely resistive the load impedance must also be purely resistive and equal to source resistance. In this case:

$$Z_{IN} = R_{IN} = R_S$$

Case 2:
If the source impedance is a resistor in series with a capacitor, $Z_S = R_S - jX_S$, the load impedance must be a resistor in series with an inductor such that:

$$Z_{IN} = R_S + jX_S$$

Case 3:

If the source impedance is a resistor in series with an inductor, $Z_S = R_S + jX_S$, the load impedance must be a resistor in series with a capacitor such that:

$$Z_{IN} = R_S - jX_S$$

5.4 Maximum Power Transfer with Resistive Source and Load Impedance

This section explores in detail the three cases for maximum power transfer along with solutions using the LTspice software.

Example 5.1: Prove the maximum power transfer condition where the source and load impedance is purely resistive as shown in Figure 5-2.

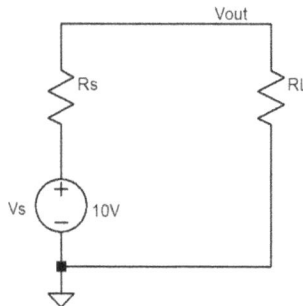

Figure 5-2 Network with purely resistive source and load impedance

Solution: When the source impedance is purely resistive and the load resistance is R_L, the maximum power is transferred to the load when $R_L = R_S$. Using the resistor voltage-divider principle we can determine the output voltage, V_{out} for the following three possible cases:

Case I: $R_L = R_S$
Case II: $R_L < R_S$
Case III: $R_L > R_S$

Case I: If the input voltage is 10 VDC and R_L = R_S = 50 Ω, the output voltage is 5 Volts and the output power is 0.5 Watts. This is the maximum power that can be transferred.

$$V_{out} = V_S \frac{R_L}{(R_S + R_L)} = 10 \frac{50}{(50 + 50)} = 5 \; volts$$

$$P_L = \frac{V_{out}^2}{R_L} = \frac{5^2}{50} = 0.5 \; Watts$$

Case II: If R_L = 25 Ω and R_S = 50 Ω, the output voltage is 3.333 Volts and the output power is 0.444 Watts.

$$V_{out} = V_S \frac{R_L}{(R_S + R_L)} = 10 \frac{25}{(50 + 25)} = 3.333 \; volts$$

$$P_L = \frac{V_{out}^2}{R_L} = \frac{3.333^2}{25} = 0.444 \; Watts$$

Case III: If R_L = 100 Ω and R_S = 50 Ω, the output voltage is 6.666 Volts and the output power is 0.444 Watts.

$$V_{out} = V_S \frac{R_L}{(R_S + R_L)} = 10 \frac{100}{(50 + 100)} = 6.666 \; volts$$

$$P_L = \frac{V_{out}^2}{R_L} = \frac{6.666^2}{100} = 0.444 \; Watts$$

Notice that the power output in Case I is greater than either Cases II or III. Therefore, the maximum power transfer is achieved only when R_L = R_S.

Example 5.2: Use the linear simulation techniques to demonstrate the maximum power transfer condition in a purely resistive system.

Solution: This is a very simple schematic where the voltage source is directly connected to 50 Ohm load, as shown in Figure 5-3. The input port is an RF signal source with 50 Ω source impedance and the output port is simply a 50 Ω resistive load. Simulate the schematic from 450 to 550 MHz and add a rectangular graph with the insertion loss, S_{21}, in dB. Figure 5- shows the plot indicating that $S_{21} = 0$ dB. Because there is zero insertion loss between the source and load, there is maximum power transfer.

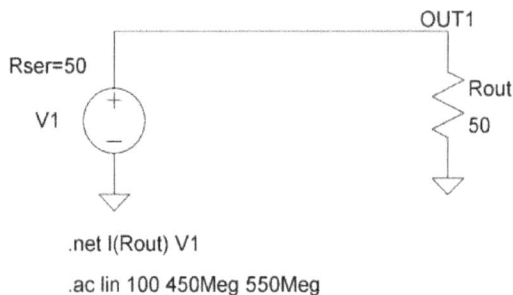

.net I(Rout) V1

.ac lin 100 450Meg 550Meg

Figure 5-3 Case I power transfer with $R_L = R_S$

Simulate the schematic and display the insertion loss, S_{21}, in a rectangular plot as shown in Figure 5-4.

Notice that when the load impedance is equal to the source impedance the insertion loss is 0 dB indicating maximum power transfer.

Change the load impedance in Figure 5-3 to 25 Ω as shown in Figure 5-5.

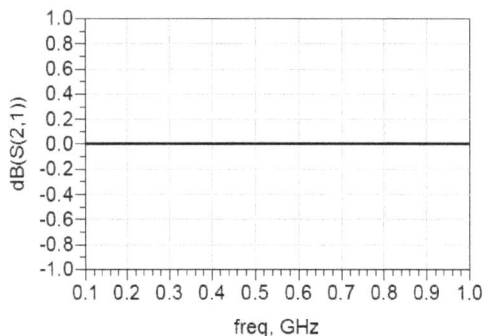

Figure 5-4 Insertion Loss with $R_L = R_S$

Simulate the schematic and display the insertion loss, S_{21}, in a rectangular plot as shown in Figure 5-6.

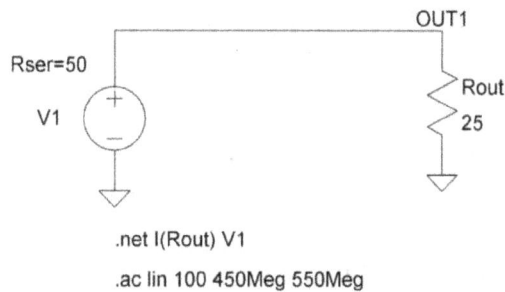

.net I(Rout) V1

.ac lin 100 450Meg 550Meg

Figure 5-5 Case II power transfer with $R_L < R_S$

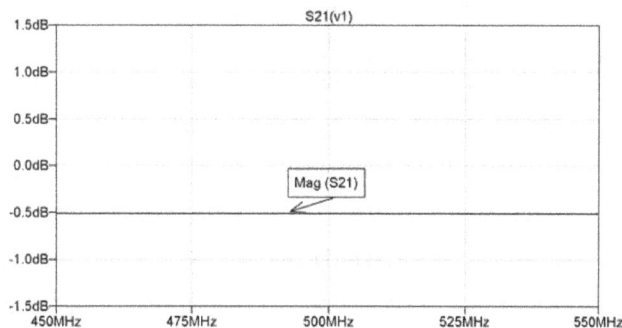

Figure 5-6 Insertion Loss with $R_L < R_S$

Notice that S_{21} = -0.512 dB indicating insertion loss. To calculate the amount of power loss, we know that the maximum power transfer is 0.5 Watts. To determine the amount of 0.512 dB power loss we first convert the 0.5 Watts to dBm which is the power in dB relative to 1 mW. Utilizing the conversion equation,

$$10 \log (\text{power in mW}) = \text{power in dBm}.$$

$$10 \log (500 \text{ mW}) = 26.98 \text{ dBm}$$

The loss of 0.512 dB is subtracted from the 26.98 dBm to result in (26.98 dBm – 0.512 dB) = 26.47 dBm. Then converting from dBm back to mW we get 444 mW as calculated in Example 5.2-1.

$$10^{\left(\frac{26.47}{10}\right)} = 444 \; mW \quad or \quad 0.444 \; Watts$$

5.5 Maximum Power Transfer with Complex Load Impedance

According to Equation 5-5 the maximum power transfer occurs when $Z_S = Z_L*$. Therefore, if $Z_L = R_L - jX_L$, then for maximum power transfer we must have $Z_S = R_L + jX_L$.

Example 5.3: If the load is 50 Ω in series with a 15 pF series capacitance, find the source impedance to have maximum power transfer at 500 MHz.

Solution: We can find a source inductance that cancels the reactance of the load at this frequency. The reactance of the 15 pF capacitor is:

$$X_C = \frac{1}{2\pi f C} = 21.231 \quad \Omega$$

Therefore,

$$Z_L = 50 - j21.231 \quad \Omega$$

For maximum power transfer the source impedance must be:

$$X_S = 50 + j21.231 \quad \Omega$$

At 500 MHz the value of the source series inductor is:

$$L = \frac{21.231}{2\pi f} = 6.76 \quad nH$$

To demonstrate the maximum power transfer, create a schematic as shown in Figure 5-7. Place the specified 15 pF capacitance in series with the load and the 6.76 nH inductance in series with the source. Plot the insertion loss and VSWR on a rectangular graph as shown in Figures 5-8 and 5-9. Note that unlike the purely resistive source and load case, a complex conjugate match occurs at a single frequency. A perfect 1:1 VSWR is achieved at the conjugate match frequency of 500 MHz.

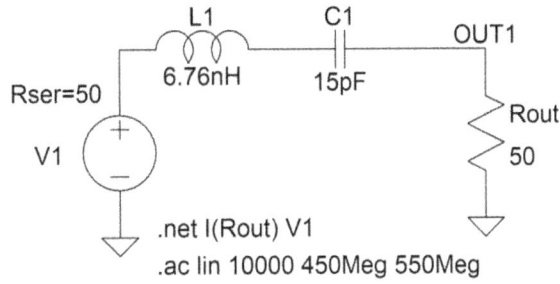

Figure 5-7 Maximum power transfer with complex impedance

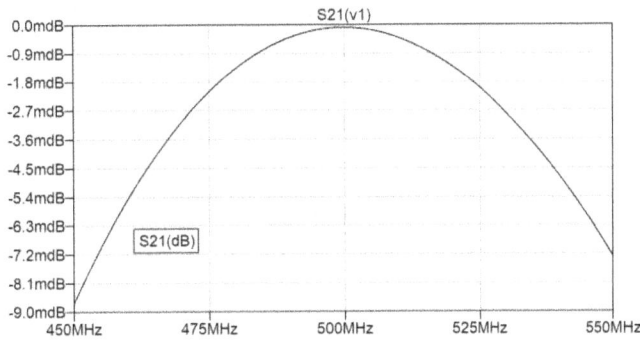

Figure 5-8 Maximum power transfer response

The Voltage Standing Wave Ratio is shown in Figure 5-9.

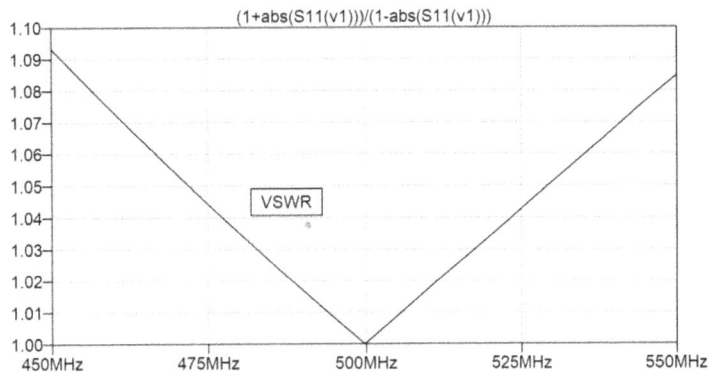

Figure 5-9 Perfect VSWR at 500 MHz

5.6 Analytical Design of Impedance Matching Networks

One of the important tasks in RF and microwave engineering is the determination of how an arbitrary complex load impedance, $Z_L = R_L + jX_L$, is analytically matched to any complex source impedance, $Z_S = R_S + jX_S$, as shown in Figure 5-10. This problem arises mainly in the design of inter-stage matching networks between active devices or between an antenna and a transmitter.

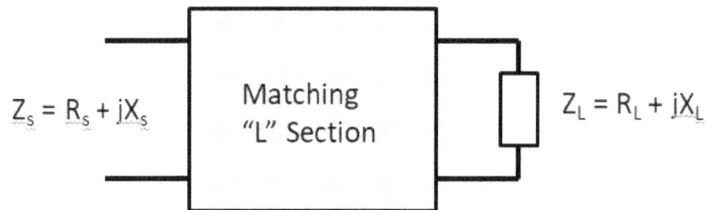

Figure 5-10 General impedance matching with an L-network

In Figure 5-10 the complex load impedance, $Z_L = R_L + j X_L$, is to be matched to the source impedance $Z_S = R_S + jX_S$. The only condition for impedance matching is that both R_S and R_L must be nonnegative while X_S and X_L could take any real value. In section 5.2 it was shown that maximum power would be transferred from the source to the load when the load impedance is the conjugate of the source impedance. For L-network matching there are two configurations that can match arbitrary load impedances to arbitrary source impedances. In the first configuration the first element adjacent to the load is a series element as shown in Figure 5-11.

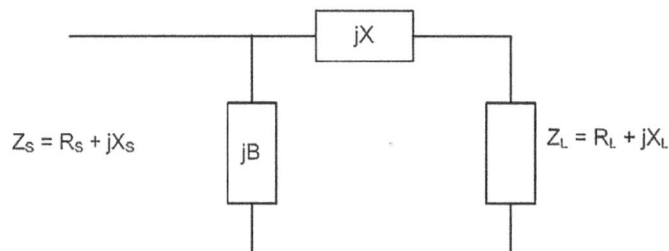

Figure 5-11 First impedance matching network configuration

In the second configuration the first element adjacent to the load is a shunt element as shown in Figure 5-12.

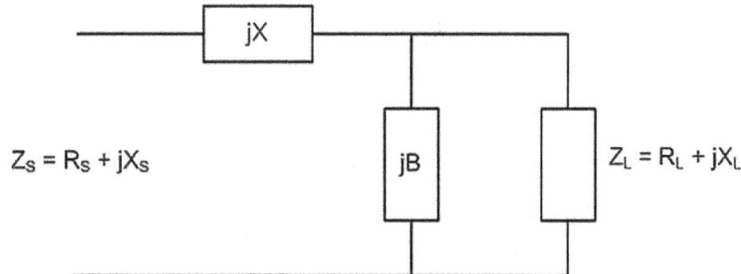

Figure 5-12 Second impedance matching network configuration

5.7 Matching a Complex Load to Complex Source Impedance

To match a complex load to any complex source impedance with a single L-network, either the first or the second configuration may be used. The choice of configurations depends on the conditions that source and load impedances dictate. Applying Equation $Z_S^* = Z_{IN}$ to the first matching configuration in Figure 5-11 we get:

$$R_S - jX_S = \cfrac{1}{jB + \left(\cfrac{1}{jX + R_L + jX_L}\right)} \quad (5\text{-}6)$$

By separating the real and imaginary parts of Equation (5-6) we obtain two solutions for B and X as follows:

$$B_1 = \frac{R_L X_S + \sqrt{R_L R_S (R_S^2 + X_S^2 - R_L R_S)}}{R_L (R_S^2 + X_S^2)} \quad (5\text{-}7)$$

$$X_1 = \frac{R_L X_S - R_S X_L}{R_S} + \frac{R_S - R_L}{B_1 R_S} \quad (5\text{-}8)$$

And

$$B_2 = \frac{R_L X_S - \sqrt{R_L R_S (R_S{}^2 + X_S{}^2 - R_L R_S)}}{R_L (R_S{}^2 + X_S{}^2)} \qquad (5\text{-}9)$$

$$X_2 = \frac{R_L X_S - R_S X_L}{R_S} + \frac{R_S - R_L}{B_2 R_S} \qquad (5\text{-}10)$$

Similarly, applying the same procedure to the second configuration we have:

$$R_S - jX_S = jX + \cfrac{1}{jB + \left(\cfrac{1}{R_L + jX_L}\right)} \qquad (5\text{-}11)$$

Separating the real and imaginary parts of Equation (5-11), we also get two sets of solutions for B and X:

$$B_3 = \frac{R_S X_L + \sqrt{R_L R_S (R_L{}^2 + X_L{}^2 - R_L R_S)}}{R_S (R_L{}^2 + X_L{}^2)} \qquad (5\text{-}12)$$

$$X_3 = \frac{R_S X_L - R_L X_S}{R_L} + \frac{R_L - R_S}{B_3 R_L} \qquad (5\text{-}13)$$

And

$$B_4 = \frac{R_S X_L - \sqrt{R_L R_S (R_L{}^2 + X_L{}^2 - R_L R_S)}}{R_S (R_L{}^2 + X_L{}^2)} \qquad (5\text{-}14)$$

$$X_4 = \frac{R_S X_L - R_L X_S}{R_L} + \frac{R_L - R_S}{B_4 R_L} \qquad (5\text{-}15)$$

Conditions for the validity of solutions are that the arguments of the square roots in Equations (5-7), (5-9), (5-12) and (5-14) be positive or zero.

If $R_S{}^2 + X_S{}^2 - R_L R_S > 0$ and $R_L{}^2 + X_L{}^2 - R_L R_S < 0,$

The two solutions obtained from Equations (5-7) through (5-10) are the only valid solutions.

If $R_S^2 + X_S^2 - R_L R_S < 0$ and $R_L^2 + X_L^2 - R_L R_S > 0$,

The two solutions obtained from Equations (5-12) through (5-15) are the only valid solutions.

If $R_S^2 + X_S^2 - R_L R_S > 0$ and $R_L^2 + X_L^2 - R_L R_S > 0$,

All four solutions obtained from Equations (5-7) through (5-10) and Equations (5-12) through (5-15) are valid.

Once the real values for B and X are calculated, the values of the matching elements are obtained from the following equations:

If B is positive, the matching element is a capacitor given by:

$$C = \frac{B}{2\pi f} \qquad (5\text{-}16)$$

If B is negative, the matching element is an inductor given by:

$$L = -\frac{1}{2\pi f B} \qquad (5\text{-}17)$$

If X is positive, the matching element is an inductor given by:

$$L = \frac{X}{2\pi f} \qquad (5\text{-}18)$$

If X is negative, the matching element is a capacitor given by:

$$C = -\frac{1}{2\pi f X} \qquad (5\text{-}19)$$

In the above equations, if the frequency is in Hz, capacitors and inductors are in Farad and Henry, respectively.

We utilize Equations (5-7) through (5-15) to design the matching L-networks.

Example 5-4: Analytically design L-networks that matches a complex load impedance, $Z_L = 10 - j15\ \Omega$, to a complex source impedance, $Z_S = 15 - j20\ \Omega$, at 2 GHz.

Note: Since $R_S^2 + X_S^2 - R_L R_S = 475 > 0$, and $R_L^2 + X_L^2 - R_L R_S = 175 > 0$ this Example has four solutions.

First Solution: We utilize Equations (5-7) and (5-8) in MATLAB Script to calculate the element values of the matching L-network. The procedure follows.

1. Enter design parameters

RS = 15; XS = -20; RL = 10; XL = -15; f = 2e9

2. Use Equations (5-7) and (5-8) to calculate the matching element values of the first L-network.

B1= ((RL*XS) + sqrt(RS*RL*(RS^2+XS^2-RS*RL)))/(RL*(RS^2+XS^2))
X1= (RL*XS-RS*XL)/RS + (RS-RL)/(B1*RS)
L1= X1/(2*pi*f)
C2= B1/(2*pi*f)

The calculation results show that B1 = 0.011 and X1 = 32.795, therefore, the series element is an inductor, L1 = 2.61 nH, and the shunt element is a capacitor, C2 = 0.852 pF.

To verify the solution, create a new schematic in and add the lumped RLC elements as shown in Figure 5-13.

Simulate the schematic and display the magnitude of the input reflection coefficient, Gamma, in a rectangular plot as shown in Figure 5-14.

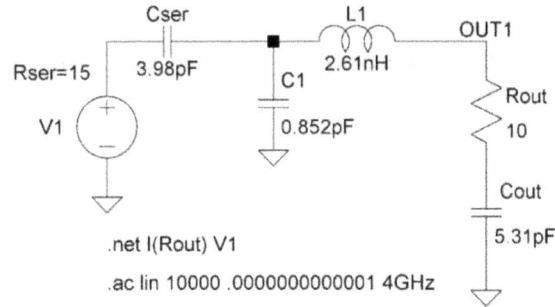

Figure 5-13 Schematic of the first matching L-network

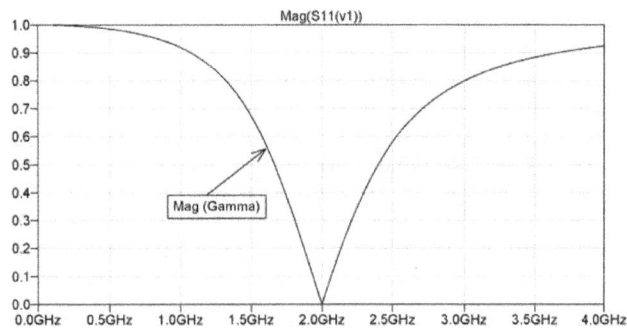

(a)

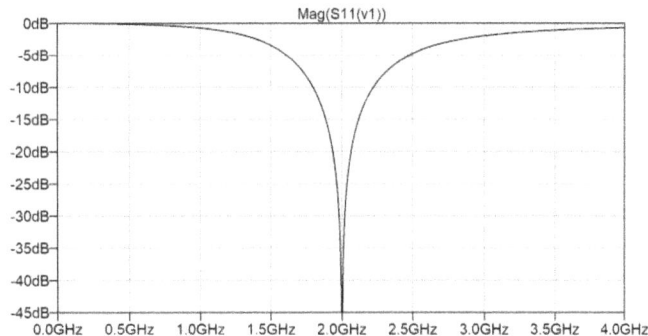

(b)

Figure 5-14 Magnitude of the input reflection coefficient in linear unity scale (a) and (b) in dB (i.e., S_{11}dB)

The magnitude of Gamma in Figure 5-14 shows that the matching L-network has a narrow bandwidth at 0.1 and perfectly matches the load impedance to the source impedance at 2 GHz.

Example 5.5: Design the second L-network that matches the complex load impedance, $Z_L = 10 - j15\ \Omega$, to a complex source impedance, $Z_S = 15 - j20\ \Omega$, at 2 GHz.

Second Solution: Use Equations (5-9) and (5-10) in MATLAB script to calculate the second solution. The procedure follows.

1. Enter Design Parameters
RS=15; XS=-20; RL=10; XL=-15; f=2e9

2. Use Equations (5-9) and (5-10) in MATLAB script to calculate the matching element values

B2=((RL*XS)-sqrt(RS*RL*(RS^2+XS^2-RS*RL)))/(RL*(RS^2+XS^2))
X2=(RL*XS-RS*XL)/RS+(RS-RL)/(B2*RS)
C1=-1/(2*pi*f*X2)
L2=-1/(2*pi*f*B2)

The calculation results show that B2= -0.075 and X2 = -2.795, therefore, the series element is a capacitor, C1 = 28.47 pF and the shunt element is an inductor, L2 = 1.065 nH.

To plot the response of the matching network, create a new schematic in and open a new schematic window. Add capacitor and inductor and connect the matching elements as shown in Figure 5-15.

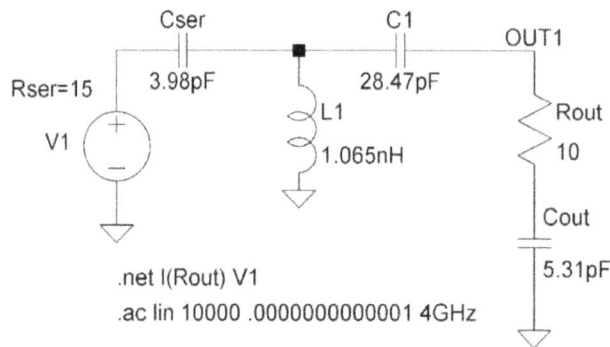

Figure 5-15 Schematic of the second matching L-network

Simulate the schematic and display the magnitude of the reflection coefficient, gamma, in a rectangular plot, as shown in Figure 5-16.

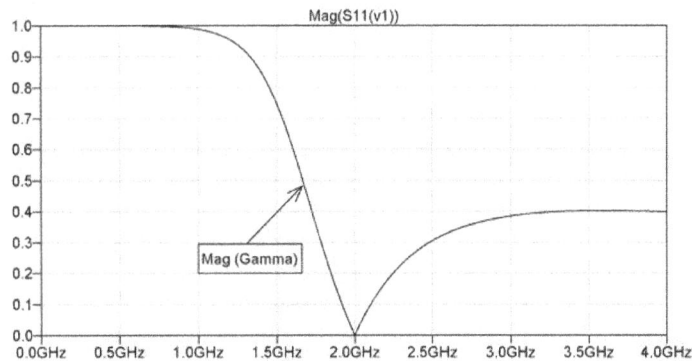

Figure 5-16 Magnitude of the input reflection coefficient

The simulated response in Figure 5-16 shows that the matching L-network perfectly matches the load impedance to the source impedance at 2 GHz.

Example 5.6: Design the third L-network that matches the complex load impedance, $Z_L = 10 - j15\ \Omega$, to a complex source impedance, $Z_S = 15 - j20\ \Omega$, at 2 GHz.

Third Solution: Use Equations (5-12) and (5-13) to calculate *B3* and *X3* for the third solution. The procedure follows.

1. Enter design parameters

RS=15; XS=-20; RL=10; XL=-15; f=2e9

2. Use Equations (5-12) and (5-13) in MATLAB script to calculate the matching element values

B3=((RS*XL)+sqrt(RS*RL*(RL^2+XL^2-RS*RL)))/(RS*(RL^2+XL^2))
X3=(RS*XL-RL*XS)/RL+(RL-RS)/(B3*RL)
L1=-1/(2*pi*f*B3)

L2=X3/(2*pi*f)

The calculation results show that B3 = -0.013 and X3 = 36.202, therefore, the shunt element is an inductor L1= 6.16 nH and the series element is another inductor L2=2.881 nH, as shown in Figure 5-17.

To plot the response of the matching network, create a new workspace in and open a new schematic window. Add capacitor and inductor and connect the matching elements as shown in Figure 5-17.

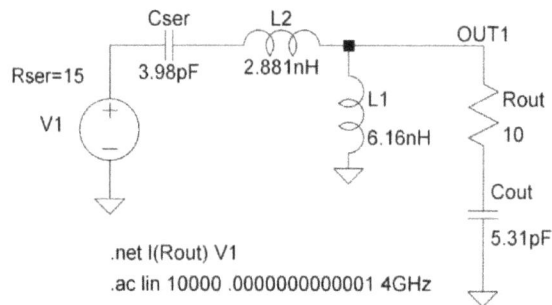

Figure 5-17: Schematic of the third matching L-network

Simulate the schematic and display S_{11} and S_{21} in a rectangular plot.

The simulated response in Figure 5-18 shows that the matching L-network perfectly matches the load impedance to the source impedance at 2 GHz.

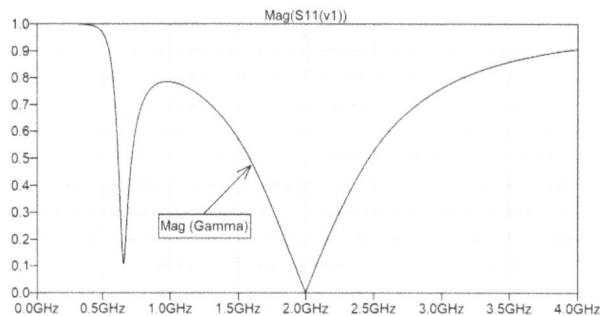

Figure 5-18 Magnitude of the input reflection coefficient

Example 5.7: Design the fourth L-network that matches the complex load impedance, $Z_L = 10 - j15 \ \Omega$, to a complex source impedance, $Z_S = 15 - j20 \ \Omega$, at 2 GHz.

Fourth Solution: Use Equations (5-14) and (5-15) to calculate *B4* and *X4*.

1. Enter design parameters

RS=15; XS=-20; RL=10; XL=-15; f=2e9

2. Use Equations (5-14) and (5-15) in MATLAB script to calculate the matching element values

B4=((RS*XL)-sqrt(RS*RL*(RL^2+XL^2-RS*RL)))/(RS*(RL^2+XL^2))
X4=(RS*XL-RL*XS)/RL+(RL-RS)/(B4*RL)
L1=-1/(2*pi*f*B4)
L2=X4/(2*pi*f)

The calculation results show that B4 = -0.079 and X4 = 3.798, therefore, the shunt element is an inductor L1=1.002 nH and the series element is another inductor L2=0.302 nH.

To plot the response of the matching network, create a new workspace in and open a new schematic window. Add capacitor and inductor and connect the matching elements as shown in Figure 5-19.

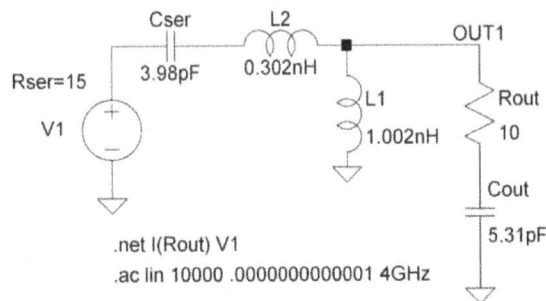

Figure 5-19 Schematic of the fourth matching L-network

Simulate the schematic and display S_{11} and S_{21} in a rectangular plot.

The simulated response in Figure 5-20 shows that the matching L-network perfectly matches the load impedance to the source impedance at 2 GHz.

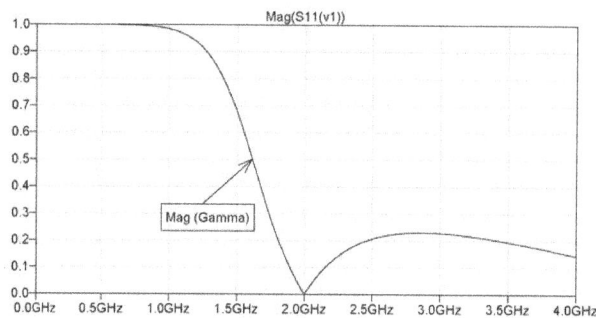

Figure 5-20 Magnitude of the input reflection coefficient

5.8 Matching a Complex Load to a Real Source Impedance

In amplifier design a common matching problem is the matching of a complex load impedance, $Z_L = R_L + jX_L$ to a real source impedance, $Z_S = R_S$. The complex impedance is usually the load and the real impedance is the characteristic impedance of transmission line connected to the source. In order to design the impedance matching networks we use both configurations of Figures 5-7 and 5-8. For the case of real source impedance the matching configuration of Figure 5-6 is redrawn in Figure 5-20.

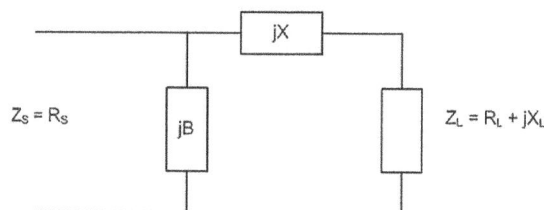

Figure 5-21 Matching complex load to resistive source-Case 1

To derive the analytical expressions for B and X we utilize the maximum power transfer condition and set the conjugate of the source impedance equal to input impedance of the matching network followed by the load impedance, as given in Equation (5-20).

$$R_S = \cfrac{1}{jB + \left(\cfrac{1}{jX + R_L + jX_L} \right)} \qquad (5\text{-}20)$$

Note that for the real source resistor its conjugate is equal to itself. The solutions for B and X in Equation (5-20) can be obtained by substituting $X_S = 0$ in Equations (5-7) through (5-10).

$$B_1 = \frac{+\sqrt{R_S - R_L}}{R_S\sqrt{R_L}} \qquad (5\text{-}21)$$

$$X_1 = +\sqrt{R_L(R_S - R_L)} - X_L \qquad (5\text{-}22)$$

And,

$$B_2 = \frac{-\sqrt{R_S - R_L}}{R_S\sqrt{R_L}} \qquad (5\text{-}23)$$

$$X_2 = -\sqrt{R_L(R_S - R_L)} - X_L \qquad (5\text{-}24)$$

Note that the solutions given in Equations (5-21) through (5-24) are only valid if $R_L < R_S$. To calculate B and X, when $R_L > R_S$, we use the second matching configuration shown in Figure 5-121,

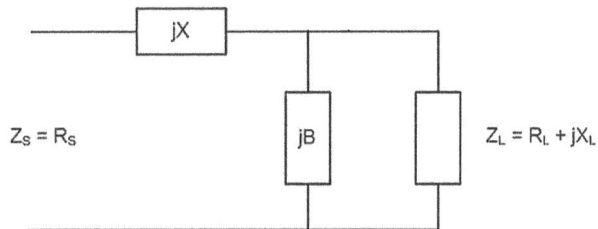

Figure 5-22 Matching complex load to resistive source –Case 2

Applying the maximum power condition to the matching network in Figure 5-18, we have:

$$R_S = jX + \cfrac{1}{jB + \left(\cfrac{1}{R_L + jX_L}\right)} \qquad (5\text{-}25)$$

The solutions for B and X in Equation (5-25) can be obtained by reusing Equations (5-12) and (5-14) and substituting $X_S = 0$ in Equations (5-13) and (5-15).

$$B_3 = \frac{R_S X_L + \sqrt{R_L R_S (R_L^2 + X_L^2 - R_L R_S)}}{R_S (R_L^2 + X_L^2)} \tag{5-26}$$

$$X_3 = \frac{R_S X_L}{R_L} + \frac{R_L - R_S}{B_3 R_L} \tag{5-27}$$

And

$$B_4 = \frac{R_S X_L - \sqrt{R_L R_S (R_L^2 + X_L^2 - R_L R_S)}}{R_S (R_L^2 + X_L^2)} \tag{5-28}$$

$$X_4 = \frac{R_S X_L}{R_L} + \frac{R_L - R_S}{B_4 R_L} \tag{5-29}$$

The conditions for the valid solutions are that the arguments of the square roots in Equations (5-26) through (5-29) be non-negative. Therefore, the two solutions obtained from Equations (5-26) through (5-29) are valid only if $R_L > R_S$. Combined conditions for valid solutions are summarized in Table 5-1.

Case #	First Condition	Second Condition	# of Solutions	Equations Used
1	$R_L < R_S$	$R_L^2 + X_L^2 - R_L R_S > 0$	4	(5-21) to (5-24) (5-26) to (5-29)
2	$R_L < R_S$	$R_L^2 + X_L^2 - R_L R_S < 0$	2	(5-21) to (5-24)
3	$R_L > R_S$	N/A	2	(5-26) to (5-29)

Table 5-1 Impedance matching conditions and the number of solutions

The solutions in Equations (5-26) through (5-29) can be simplified by normalizing the load impedance with respect to the source resistor. Therefore, if we let the source resistor be equal to Z_0, the normalized load resistance and reactance become,

$$r = \frac{R_L}{Z_0} \tag{5-30}$$

And

$$x = \frac{X_L}{Z_0} \tag{5-31}$$

The simplified equations are:

$$B_1 = \frac{\sqrt{\dfrac{(1-r)}{r}}}{Z_0} \tag{5-32}$$

$$X_1 = Z_0\left[\sqrt{r(1-r)} - x\right] \tag{5-33}$$

$$B_2 = -\frac{\sqrt{(1-r)}}{Z_0} \tag{5-34}$$

$$X_2 = -Z_0\left[\sqrt{r(1-r)} + x\right] \tag{5-35}$$

$$B_3 = \frac{x + \sqrt{r(r^2 + x^2 - r)}}{Z_0(r^2 + x^2)} \tag{5-36}$$

$$X_3 = Z_0\sqrt{\frac{(r^2 + x^2 - r)}{r}} \tag{5-37}$$

$$B_4 = \frac{x - \sqrt{r\left(r^2 + x^2 - r\right)}}{Z_0\left(r^2 + x^2\right)} \tag{5-38}$$

$$X_4 = -Z_0\sqrt{\frac{\left(r^2 + x^2 - r\right)}{r}} \tag{5-39}$$

Example 5-8: Design a single L-network that will match a real source impedance $Z_0 = 50 \ \Omega$ to a complex load impedance, $Z_L = 7 - j22 \ \Omega$, at a frequency of 1 GHz. Notice that for this example $R_L < Z_0$ and $R_L^2 + X_L^2 - R_L Z_0 = 183 > 0$, therefore, the matching network has four solutions.

Solution: We use Equations (5-32) and (5-33) in MATLAB script to calculate the element values of the matching network.

1. Enter design parameters and normalized load impedance

Z0=50; RL=7; XL=-22; f=1000e6
r=RL/Z0
x=XL/Z0

2. Use Equations (5-32) and (5-33) in MATLAB script to calculate the matching element values

B1=sqrt((1-r)/r)/Z0
X1=Z0*sqrt(r*(1-r))-x*Z0
LS1=X1/(2*pi*f)
CP2=B1/(2*pi*f)

The calculation results show that B1 = 0.05 and X1 = 39.349, therefore, the series element is an inductor L1= 6.263 nH and the shunt is a capacitor C2=7.889 pF, as shown in Figure 5-23. To plot the response of the matching network, create a new workspace in and open a new schematic window. Add capacitor and inductor and connect the matching elements as shown in Figure 5-23.

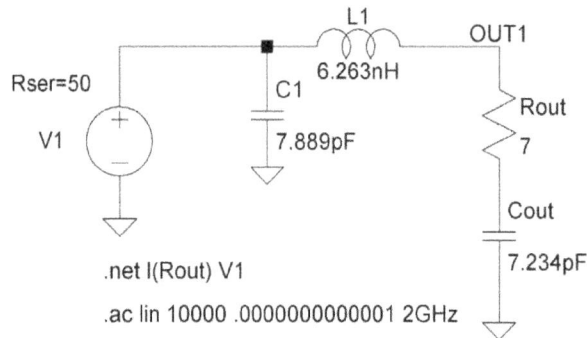

Figure 5-23 Schematic of the first matching L-network

Simulate the schematic and display S_{11} and S_{21} in a rectangular plot.
The simulated response in Figure 5-24 shows that the matching L-network perfectly matches the load impedance to the source impedance at 1 GHz.

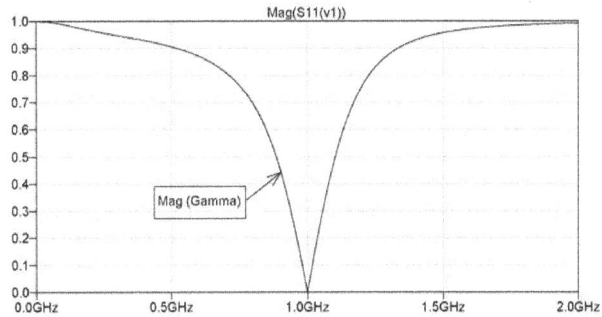

Figure 5-24 Magnitude of the input reflection coefficient

Example 5.9: Design the second L-network that matches a real source impedance, $Z_0 = 50\ \Omega$, to a complex load impedance, $Z_L = 7 - j22\ \Omega$, at a frequency of 1 GHz.

Solution: Use Equations (5-34) and (5-35) to calculate the element values of the second matching network. The procedure follows.

1. Enter design parameters and normalized load impedance

Z0−50; RL−7; XL−-22; f−1000e6
r=RL/Z0
x=XL/Z0

2. Use Equations (5-34) and (5-35) in MATLAB script to calculate the matching element values

B2=-sqrt((1-r)/r)/Z0
X2=-Z0*sqrt(r*(1-r))-x*Z0
L1=X2/(2*pi*f)
L2=-1/(2*pi*f*B2)

3. The calculation results show that B2 = -0.05 and X2 = 4.651, therefore, the series element is an inductor L1= 0.74 nH and the shunt element is another inductor L2= 3.211 nH, as shown in Figure 5-25.

To plot the response of the matching network, create a new workspace in and open a new schematic window. Add capacitor and inductor and connect the matching elements as shown in Figure 5-25.

Simulate the schematic and display S_{11} and S_{21} in a rectangular plot.

The simulated response in Figure 5-26 shows that the matching L-network perfectly matches the load impedance to the source impedance at 1 GHz.

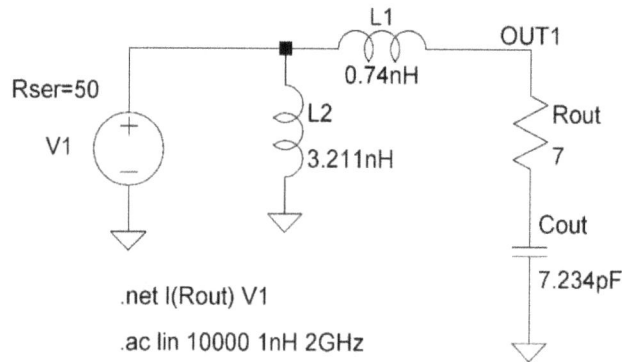

Figure 5-25 Schematic of the second matching L-network

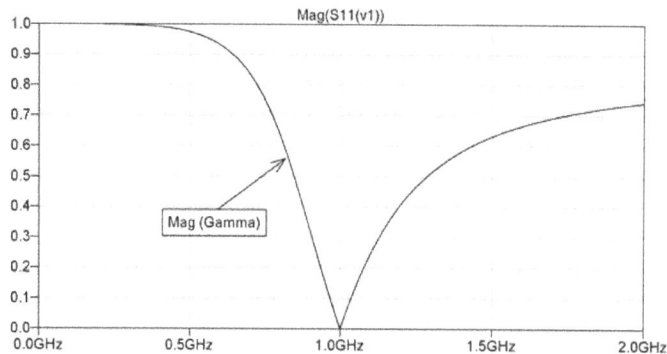

Figure 5-26 Magnitude of the input reflection coefficient

Example 5.10: Design the third L-network that matches a real source impedance, $Z_0 = 50 \ \Omega$, to a complex load impedance, $Z_L = 7 - j22 \ \Omega$, at a frequency of 1 GHz.

Solution: We use Equations (5-36) and (5-37) to calculate the element values of the matching network. The procedure follows.

1. Enter design parameters and normalize the load impedance

Z0=50; RL=7; XL=-22; f=1000e6
r=RL/Z0
x=XL/Z0

2. Use Equations (5-36) and (5-37) in MATLAB script to calculate the matching element values

B3=(x + sqrt(r*(r^2+x^2-r)))/(Z0*(r^2+x^2))
X3=Z0*sqrt((r^2+x^2-r)/r)
L2=X3/(2*pi*f)
L1=-1/(2*pi*f*B3)

The calculation results show that B3 = -0.032 and X3 = 36.154, therefore, the shunt element is an inductor L1= 5.008 nH and the series element is another inductor L2=5.754 nH, as shown in Figure 5-27.

To plot the response of the matching network, create a new workspace in and open a new schematic window. Add capacitor and inductor and connect the matching elements as shown in Figure 5-27.

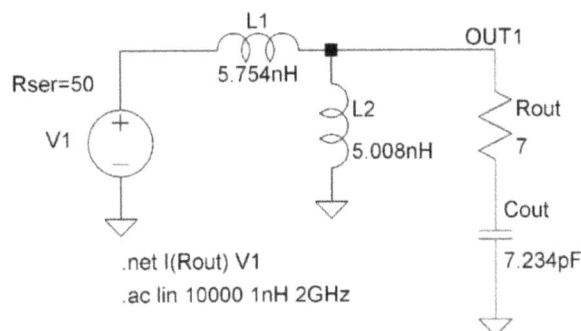

Figure 5-27 Schematic of the third matching L-network

Simulate the schematic from 500 to 1500 MHz and display the input return loss, S_{11}, and insertion loss, S_{21}, in dB.

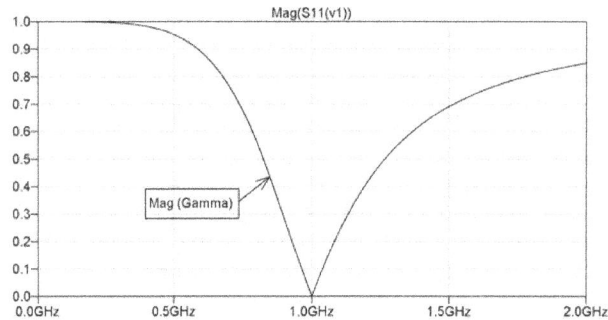

Figure 5-28 Magnitude of the input reflection coefficient

The simulated response in Figure 5-28 shows that the matching L-network perfectly matches the load impedance to the source impedance at 1 GHz.

Example 5.11: Design the fourth L-network that matches a real source impedance, $Z_0 = 50\ \Omega$, to a complex load impedance, $Z_L = 7 - j22\ \Omega$, at a frequency of 1 GHz.

Solution: For the fourth solution use Equations (5-38) and (5-39) to calculate the element values of the matching network. The procedure follows.

1. Enter design parameters and normalize the load impedance

Z0=50; RL=7; XL=-22; f=1000e6
r=RL/Z0
x=XL/Z0

2. Use Equations (5-38) and (5-39) to calculate the matching element values

B4=(x-sqrt(r*(r^2+x^2-r)))/(Z0*(r^2+x^2))
X4=-Z0*sqrt((r^2+x^2-r)/r)
C2=-1/(2*pi*f*X4)
L1=-1/(2*pi*f*B4)

The solution in the second configuration shows that B4 = -0.051 and X1 = -36.154, therefore, the shunt element is an inductor L1= 3.135 nH and the series element is a capacitor C2=4.402 pF.

To plot the response of the matching network, create a new workspace in and open a new schematic window. Add capacitor and inductor and connect the matching elements as shown in Figure 5-29.

Simulate the schematic and display S_{11} and S_{21} in a rectangular plot.

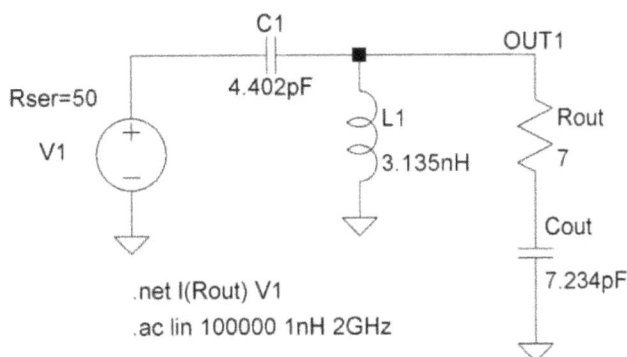

Figure 5-29 Schematic of the fourth matching L-network

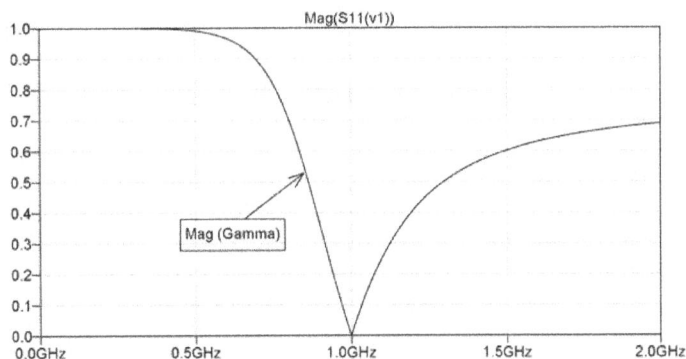

Figure 5-30 Magnitude of the input reflection coefficient

The simulated response in Figure 5-30 shows that the matching L-network perfectly matches the load impedance to the source impedance at 1 GHz.

5.9 Matching a Real Load to a Real Source Impedance

When source and load impedances are both real the first matching configuration of Figure 5-7 is redrawn in Figure 5-31.

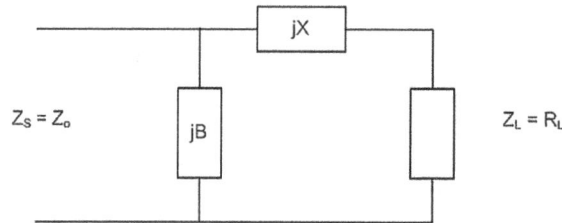

Figure 5-31 First matching configuration with $X_L = X_S = 0$

To design the matching network, apply the maximum power transfer condition and require that $Z_0^* = Z_{IN}$. Therefore,

$$Z_0 = \cfrac{1}{jB + \left(\cfrac{1}{jX + R_L} \right)} \qquad (5\text{-}40)$$

Substituting $r = \dfrac{R_L}{Z_0}$ in Equation (5-40), we get:

$$Z_0 = \cfrac{1}{jB + \left(\cfrac{1}{jX + rZ_0} \right)} \qquad (5\text{-}41)$$

The solutions for B and X in Equation (5-41) can be obtained by reusing Equations (5-32) and (5-34) and substituting $x = 0$ in Equations (5-33) and (5-35).

$$B_1 = \frac{\sqrt{(1-r)/r}}{Z_0} \qquad (5\text{-}42)$$

$$X_1 = Z_0\sqrt{r(1-r)} \qquad (5\text{-}43)$$

$$B_2 = -\frac{\sqrt{(1-r)/r}}{Z_0} \tag{5-44}$$

$$X_2 = -Z_0\sqrt{r(1-r)} \tag{5-45}$$

Note that the two solutions given by Equations (5-42) through (5-45) are only valid if r is less than 1 or $R_L < Z_0$. If r > 1, the second matching network configuration is used.

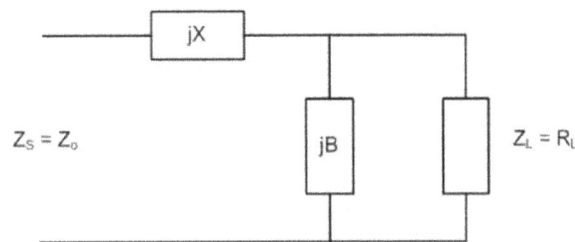

Figure 5-32 Second matching configuration with resistive load and source

To calculate the B and X values, we require that, $Z_0^* = Z_{IN}$, therefore,

$$Z_0 = jX + \frac{1}{jB + \left(\dfrac{1}{Z_0 r}\right)} \tag{5-46}$$

The solutions for B and X in Equation (5-46) can be obtained by substituting x = 0 in Equations (5-36) through (5-39).

$$B_3 = \frac{\sqrt{r-1}}{Z_0 r} \tag{5-47}$$

$$X_3 = Z_0\sqrt{r-1} \tag{5-48}$$

$$B_4 = \frac{-\sqrt{r-1}}{Z_0 r} \tag{5-49}$$

$$X_4 = -Z_0\sqrt{r-1} \qquad\qquad (5\text{-}50)$$

Note that the two solutions given by Equations (5-47) through (5-50) are only valid if r is greater than 1 or $R_L > Z_0$. The combined conditions for the validity of solutions are given below and summarized in Table 5-2.

Case 1 If r is less than 1 there are two L-networks that match the two impedances. The two solutions are given by Equations (5-42) through (5-45).

Case 2 If r is greater than 1 there are two L-networks that match the two impedances. The two solutions are given by Equations (5-47) through (5-50).

Case No	Condition	Solutions	Equations
1	$r < 1$	2	(5-42) to (5-45)
2	$r > 1$	2	(5-47) to (5-50)

Table 5-2 Impedance matching conditions and the number of solutions

Example 5.12: Design an L-network to match a 10 Ω load to a 50 Ω source resistor at 500 MHz.

Because the load resistor is smaller than the source resistor, the example has two solutions given in Equations (5-42) through (5-45).

First Solution: For the first solution use Equations (5-42) and (5-43) to calculate the element values of the matching L-network. The procedure follows.

1. Enter design parameters and normalized load impedance

Z0=50; RL=10; f=500e6
r=RL/Z0

2. Use Equations (5-42) and (5-43) in MATLAB script to calculate the matching element values
B1=sqrt((1-r)/r)/Z0

X1=Z0*sqrt(r*(1-r))
L1=X1/(2*pi*f)
C2=B1/(2*pi*f)

3. The calculation results show that B1 = 0.04 and X1 = 20, therefore, L1=6.366 nH and C2=12.73 pF.

To generate the schematic and plot the response of the matching network, create a new workspace in and open a new schematic window. Add capacitor and inductor and connect the matching elements as shown in Figure 5-33. Note that different reference designator in Figure 5-33 is used for the calculated capacitor above.

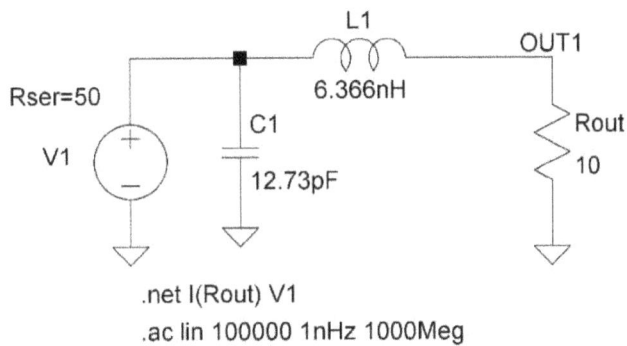

Figure 5-33 Schematic of the first matching L-network

Simulate the schematic and display the reflection coefficient in a rectangular plot.

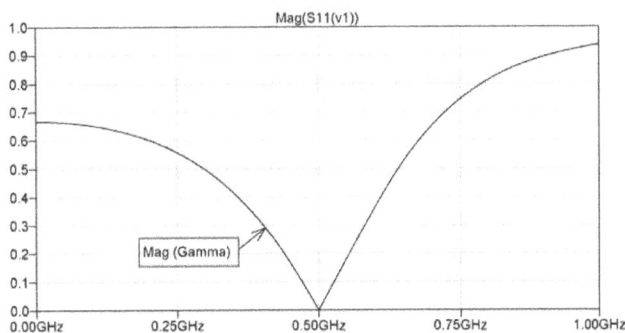

Figure 5-34 Magnitude of the input reflection coefficient

The simulated response in Figure 5-34 shows that the matching L-network perfectly matches the load impedance to the source impedance at 0.5 GHz.

Example 5.13: Design the second L-network to match a 10 Ω load to a 50 Ω source resistor at 500 MHz.

Second Solution: To design the matching L-network, use Equations (5-44) and (5-45) in Figure 5-31 and calculate the matching element values.

1. Enter design parameters and normalize the load impedance
Z0=50; RL=10; f=500e6
r=RL/Z0

2. Calculate matching element values
B2=-sqrt((1-r)/r)/Z0
X2=-Z0*sqrt(r*(1-r))
C1=-1/(2*pi*f*X2)
L2=-1/(2*pi*f*B2)

3. The calculation results show that B2 = -0.04 and X2 = -20, therefore, the series element is a capacitor, C=15.92 pF, and the shun element is an inductor, L= 7.958 nH.

To generate the schematic and plot the response of the matching network, create a new workspace in and open a new schematic window. Insert the S_Params Template and add capacitor and inductor from the Lumped-Components palette. Connect the matching elements as shown in Figure 5-35. Again, note that different reference designators have been used for the inductor and the capacitor in Figure 5-35.

Simulate the schematic and display the reflection coefficient in a rectangular plot.

The simulated response in Figure 5-36 shows that the matching L-network perfectly matches the load impedance to the source impedance at 0.5 GHz. That is the reflection coefficient at 0.5 GHz is equal to zero.

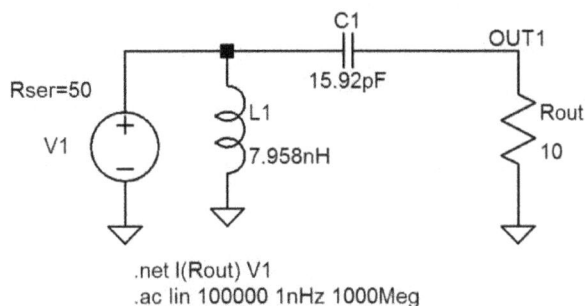

Figure 5-35 Schematic of the second matching L-network

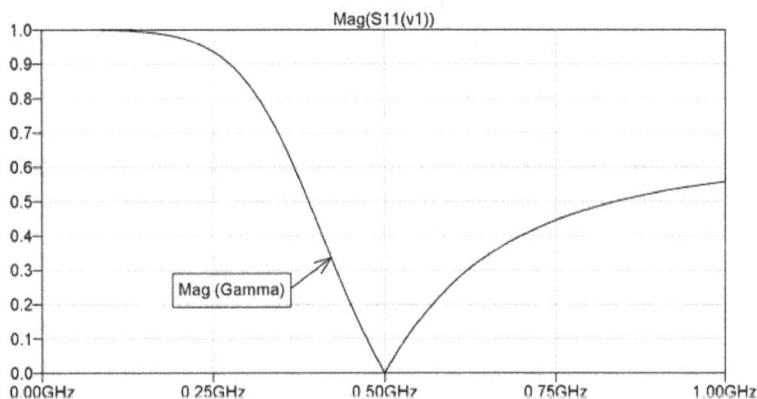

Figure 5-36 Magnitude of the input reflection coefficient

5.10 Introduction to Broadband Matching Networks

In the previous sections, the L-section matching networks achieved an impedance match at a fixed frequency capable of producing a 20 dB return loss over a narrow fractional bandwidth of less than 20%. Broadband networks are generally considered to have greater than 20% fractional bandwidths. In this section it is demonstrated that the bandwidth of a matching network can be increased by cascading L-networks. It is demonstrated that by cascading L-networks of equal Q factor, the bandwidth of a network can be increased. The design of equal-Q matching networks is based on the selection of intermediate, or virtual, resistors not necessarily 50 Ω, and then matching the load and source impedance to the virtual resistors. Successively adding additional L-networks of equal Q will continue to extend the bandwidth of the overall circuit.

5.11 Analytical Design of Broadband Matching Networks

This section demonstrates the importance of the selection of the proper intermediate network resistance that will result in the best broadband return loss. In example 5.14 the complex source and load impedance are matched to one specific intermediate resistor thus creating two, equal-Q, L-networks. The purpose is to show that this method provides a broader matching bandwidth at 20 dB return loss compared to the case when we chose a different resistor value. The Q of each L-network is the loaded Q factor defined by the source and load resistance ratio.

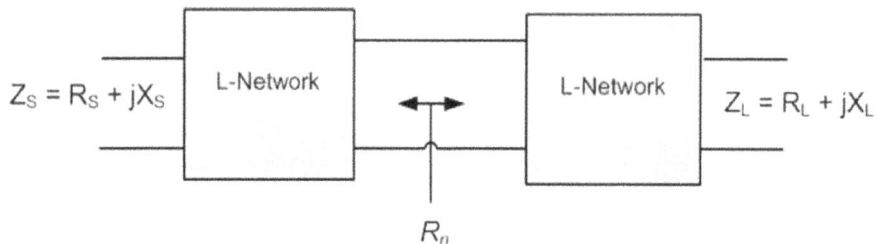

Figure 5-37 Cascaded L-networks with intermediate resistance, R_n

Example 5.14: A transmitter operates over a frequency range of 835 MHz to 1200 MHz. At its center frequency of 1 GHz the source impedance is $Z_S = 55 + j10 \ \Omega$ while the antenna input impedance is $Z_L = 20 + j15 \ \Omega$. Design a cascade of two L-networks that matches the transmitter output impedance to the antenna input impedance.

Solution: The first step in the solution is to calculate the intermediate resistor, R_n. The optimum intermediate resistor value is equal to square root of the product of the real part of the source and load impedance.

$$R_n = \sqrt{(55)(20)} = 33.166 \ \Omega$$

Use the Equation Editor to solve Equations (5-32) and (5-33) to calculate the element values of the load matching network, as shown in Figure 5-34. The calculations in Figure 5-34 show that the series element is an inductor, *LS1* = 0.195 nH and the shunt element is a capacitor, *CP2* = 3.893 pF.

1. Enter design parameters and normalized load impedance

Z0=33.166; RL=20; XL=15; f=1000e6
r=RL/Z0
x=XL/Z0

2. Calculate matching element values

B1=sqrt((1-r)/r)/Z0
X1=Z0*sqrt(r*(1-r))-x*Z0
LS1=X1/(2*pi*f)
CP2=B1/(2*pi*f)

The calculation results show that the series element is an inductor, L1 = 0.195 nH, and the shunt element is a capacitor, C1 = 3.893 pF.

To plot the response of the matching network, create a new schematic window in and add capacitor and inductor as shown in Figure 5-38.

Simulate the schematic and display the reflection coefficient in a rectangular plot.

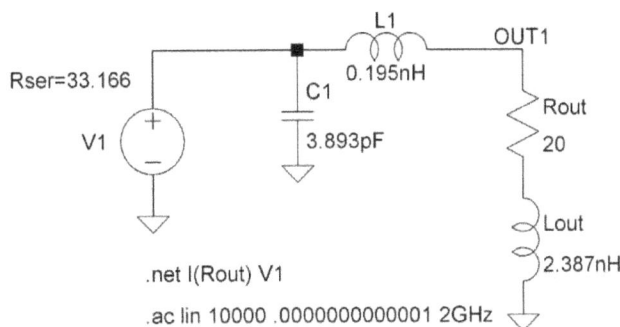

Figure 5-38 Schematic of the matching L-network

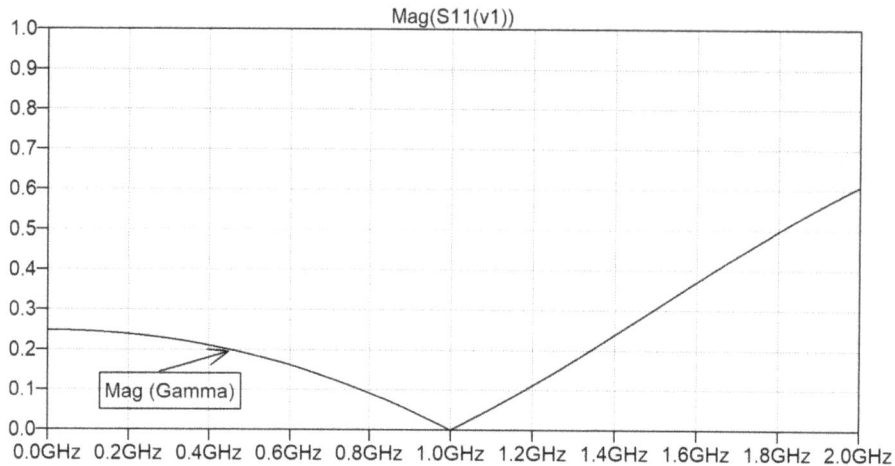

Figure 5-39 Magnitude of the input reflection coefficient

To match the source impedance to the intermediate resistor, we use Equations (5-36) and (5-37) and calculate the element values of the second matching network, as follows.

1. Enter design parameters and normalized load impedance

Z0=33.166; RL=55; XL=10; f=1000e6
r=RL/Z0
x=XL/Z0

2. Calculate matching element values

B3 = (x + sqrt(r*(r^2 + x^2-r)))/(Z0*(r^2+x^2))
X3 = Z0*sqrt((r^2 + x^2-r)/r)
LS1 = X3/(2*pi*f)
CP2 = B3/(2*pi*f)

The calculation results show that the shunt element is a capacitor, C1= 2.875 pF, and the series element is an inductor, L1=4.458 nH.

To plot the response of the matching network, create a new schematic window in and add capacitor and inductor as shown in Figure 5-40.

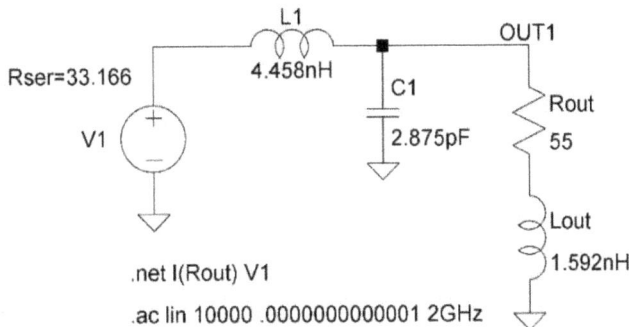

Figure 5-40 Schematic of the matching L-network

Simulate the schematic and display the reflection coefficient in a rectangular plot.

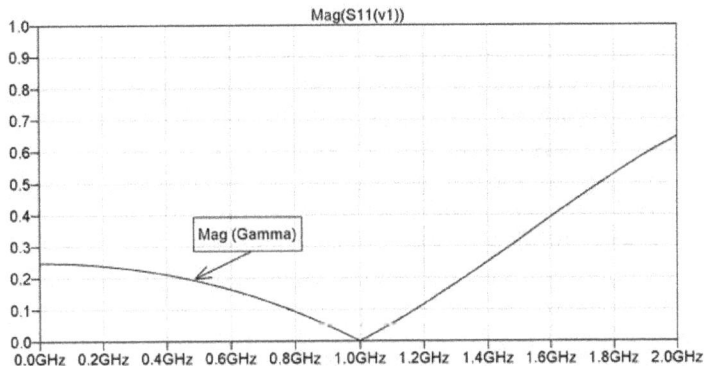

Figure 5-41 Magnitude of the input reflection coefficient

Finally, we cascade the two matching L-networks as shown in Figure 5-42.

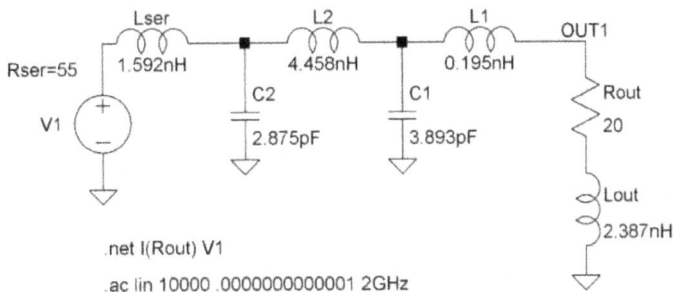

Figure 5-42 Schematic of the cascaded L-networks

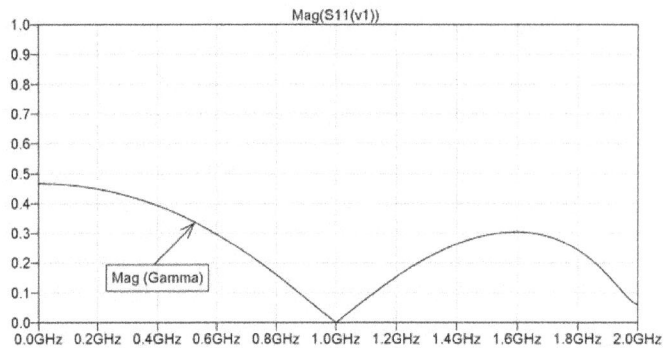

Figure 5-43 Magnitude of the input reflection coefficient

The simulated response in Figure 5-43 shows that the cascaded network perfectly matches the complex load impedance to the source impedance at 1 GHz.

Notice that the matching bandwidth at 20 dB return loss is:

$$BW_{20\,dB\ return\ loss} = 1210 - 830 = 380 \quad MHz$$

Therefore, the corresponding fractional bandwidths at 20 dB return loss is:

$$FBW_{20dB\,returnloss} = \frac{380}{\sqrt{(1210)(830)}} = 0.379 = 37.9\ \%$$

The Q factor of the each L-network is calculated by the following equation :

$$Q = \frac{1}{2}\sqrt{\frac{R_2}{R_1} - 1} \qquad\qquad R_2 > R_1$$

Where R_2 and R_1 are the input and output resistors of the L-network.

For Example 5.14 the Q factors of the source and load matching networks are:

$$Q_{source} = Q_{load} = \frac{1}{2}\sqrt{\frac{33.166}{20} - 1} = \frac{1}{2}\sqrt{\frac{55}{33.166} - 1} = 0.405$$

Example 5.15: Redesign the matching network of Example 5.14 by matching the source and load impedances to an intermediate resistance value of 50 Ω instead of 33.166 Ω. Compare the fractional bandwidth and Q factors with values for Example 5.14.

Solution: To redesign the matching network, change the intermediate resistor to 50 Ω and recalculate the matching element values.

Notice that the matching element values in both schematics have changed due to the change in the intermediate resistor. The cascaded schematic and simulated response are shown in Figures 5-44 and 5-45.

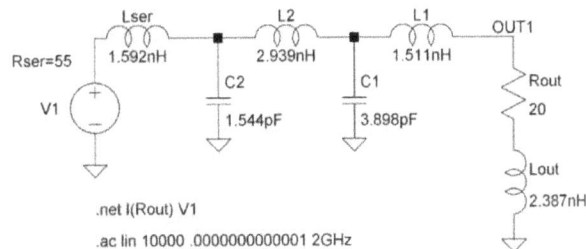

Figure 5-44 Schematic of the cascaded matching L-networks

The simulated response in Figure 5-45 shows that the matching L-network perfectly matches the load impedance to the source impedance at 1 GHz. The marker frequencies at 20 dB return loss shows that the bandwidth is:

$$BW_{20dB\ return loss} = 1126 - 874 = 252 \ MHz$$

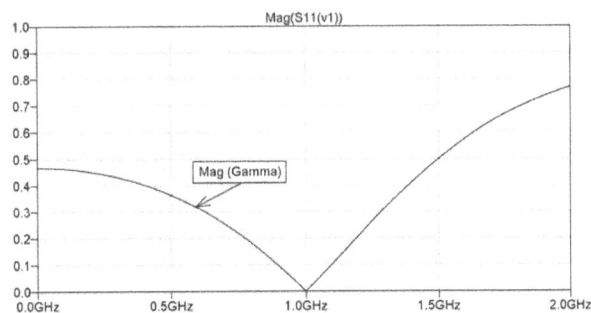

Figure 5-45 Magnitude of the input reflection coefficient

Therefore, the corresponding fractional bandwidths at 20 dB return loss is:

$$FBW_{20dBreturnloss} = \frac{252}{\sqrt{(1126)(874)}} = 0.253 = 25.4\ \%$$

A comparison of the fractional bandwidths for the two examples show that the matching network in Figure 5-41 provides over 50 % more matching bandwidth, at 20 dB return loss, than the matching network in Figure 5-44.

We can show that the lesser bandwidth is due to the fact that the source and load matching networks in the second example do not have the same Q factors, as shown by the following calculations:

$$Q_{source} = \frac{1}{2}\sqrt{\frac{55}{50} - 1} = 0.158$$

$$Q_{load} = \frac{1}{2}\sqrt{\frac{50}{20} - 1} = 0.612$$

The equal Q factors in Example 5.14 cause the cascaded matching network to transfer more power to the load while the unequal Q factors in example 5.15 cause the matching network to have less power transferred.

5.12 Broadband Impedance Matching Using N-Cascaded L-Networks

In Example 5.14 we showed that cascading two equal-Q L-networks convert a narrowband matching L-network into a broadband matching network. Because the Q factor is related to the ratio of the source and load resistance on each L-network, the Q can be further reduced by applying multiple L-networks in cascade. In practice as the number of L-networks becomes greater than five, the element values become difficult to physically realize. In this section we show that by using a network of four equal-Q L-networks in cascade, we can lower the individual Q factor and provide an even greater matching bandwidth for the Example of 5.14.

Example 5.16: Redesign the matching network of Example 5.14 with four equal-Q L-networks. Compare the Q and bandwidth of this example with Example 5.14.

Solution: This example uses the same steps developed for Example 5.4-1. The general equation for the calculation of any number of intermediate resistors is given by Equation (5-51).

$$R_n = R_S (r)^{n/N} \qquad n = 1, 2, 3, \, to \, N - 1 \qquad (5\text{-}51)$$

where, R_n is the intermediate resistor values
R_S is the source resistor value
r is the normalized load resistor, r = R_L/R_S
N is the number of cascading networks

Now we use Equation (5-51) to calculate intermediate resistors for N = 4.
1. Enter design parameters and normalize the load impedance
RS=55; RL=20;
r=RL/RS; N=4
2. Use MATLAB script to calculate intermediate resistor values
R1=RS*r^(1/N)
R2=RS*r^(2/N)
R3=RS*r^(3/N)

The calculation results show that R1 = 42.71 Ω, R2 = 33.166 Ω, and R3=25.755 Ω.

Design procedure for the matching L-networks are shown here.

1. Enter design parameters and normalized load impedance

R3=25.755; RL=20; XL=15; f=1000e6: r=RL/R3: x=XL/R3

2. Using MATLAB script to calculate the matching element values

B1=sqrt((1-r)/r)/R3
X1=R3*sqrt(r*(1-r))-x*R3

CS1=-1/(2*pi*f*X1)
CP2=B1/(2*pi*f)

The calculation results show that C1= 37.26 pF and C2= 3.315 pF, as shown in Figure 5-46.

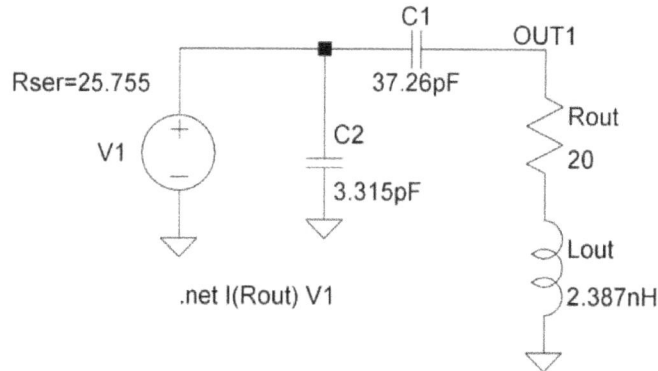

Figure 5-46 Schematic of the first matching L-network

Similarly the calculation of the matching L-network, between R3 and R2, and the associated schematic are shown as follows.

1. Enter design parameters and normalized load impedance
R2=33.166; RL=25.755; f=1000e6
r=RL/R2

2. Calculate matching element values

B1 = sqrt((1 - r)/r)/R2
X1 = R2*sqrt(r*(1 - r))
LS1 = X1/(2*pi*f)
CP2 = B1/(2*pi*f)

The calculation results show that L= 2.199 nH and C1= 2.574 pF, as shown in Figure 5-47.

Next, the calculation of the matching L-networks, between R2 and R1, is shown here

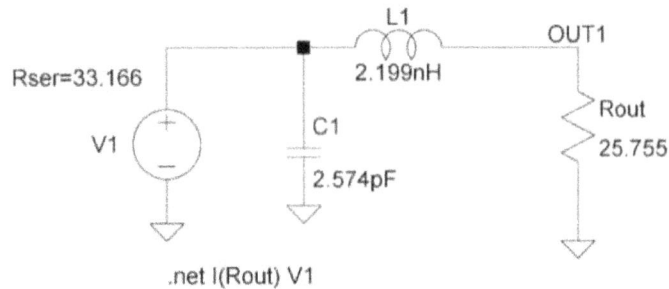

Figure 5-47 Schematic of the second matching L-network

1. Enter design parameters and normalized load impedance

R1 = 42.71; RL = 33.166; f = 1000e6

r = RL/R1

2. Calculate matching element values

B1 = sqrt((1 - r)/r)/R1

X1 = R1*sqrt(r*(1 - r))

LS1 = X1/(2*pi*f)

CP2 = B1/(2*pi*f)

The calculation results show that L1= 832 nH and C1=1.999 pF, as shown in Figure 5-48.

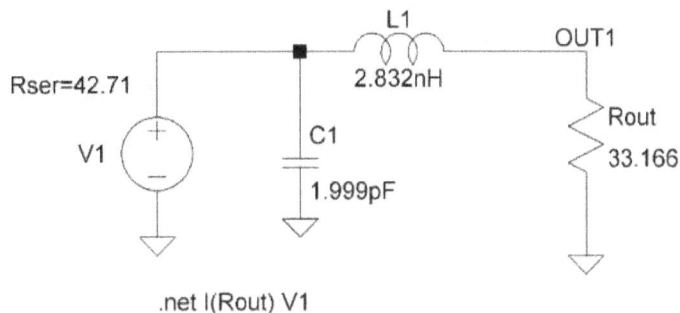

Figure 5-48 Schematic of the third matching L-network

Finally, the calculation of the matching L-network between R1 and the source impedance is shown in here.

1. Enter design parameters and normalized load impedance

R1 = 42.71; RL = 55; XL = 10; f = 1000e6
r = RL/Z0
x = XL/Z0

2. Calculate matching element values

B3 = (x + sqrt(r*(r^2 + x^2 - r)))/(R1*(r^2 + x^2))
X3 = R1*sqrt((r^2 + x^2 - r)/r)
LS1 = X3/(2*pi*f)
CP2 = B3/(2*pi*f)

The calculation results show that L1=3.907 nH and C1= 2.119 pF, as shown in Figure 5-49.

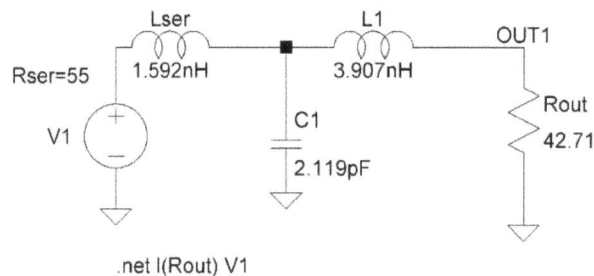

Figure 5-49 Schematic of the fourth matching L-network

Cascade the four L-networks and identify the nodes, from right to left, as shown in Figure 5-50.

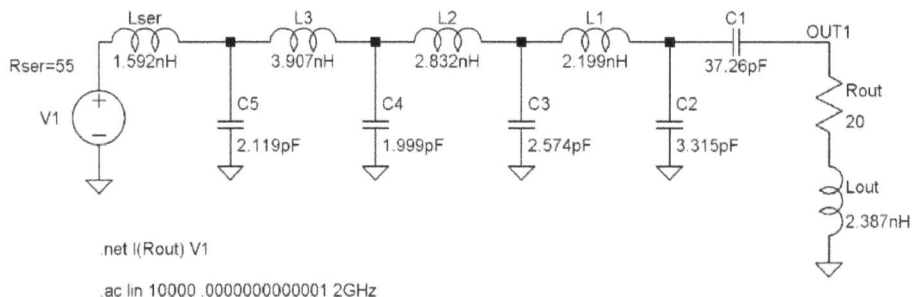

Figure 5-50 Schematic of the matching network using 4 L-networks

The simulated response (Figure 5-51) shows that the matching bandwidth at 20 dB return loss (Mag(S11(v1)) = Mag (Gamma) = 0.1) is:

$$BW_{20dB\ return\ loss} = 1334 - 821 = 513\ MHz$$

Therefore, the fractional bandwidth at 20 dB return loss is:

$$FBW_{20dB\ returnloss} = \frac{513}{\sqrt{(1334)(821)}} = 0.490 = 49.0\%$$

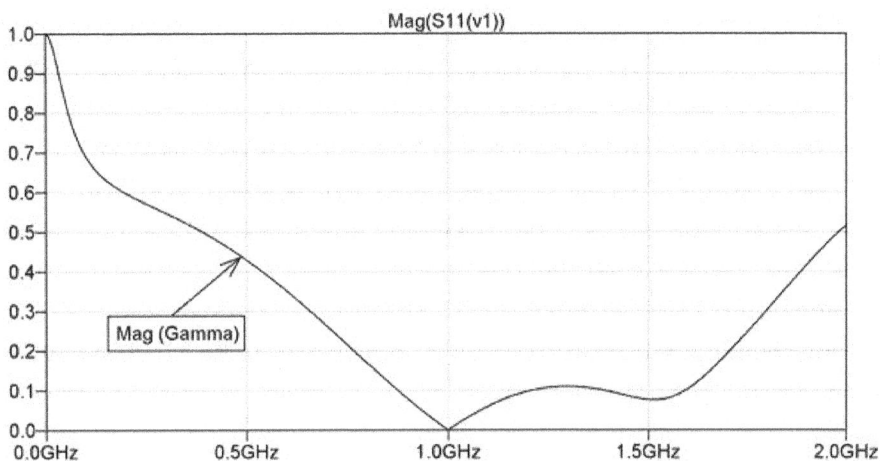

Figure 5-51 Magnitude of the input reflection coefficient

The simulated response in Figure 5-51 shows that the matching L-network perfectly matches the load impedance to the source impedance at 1 GHz.

The comparison between the fractional bandwidth of this example with the fractional bandwidth of example 5.15 shows that cascading four equal-Q matching networks increases the fractional bandwidth by more than 93% over the network with two equal-Q matching networks (i.e., [(0.49-0.254)/0.254]=0.93)).

As we stated earlier, the reason for wider matching bandwidth is that all four individual matching networks in Figure 5-50 have the same Q factor as shown in the following calculations.

$$Q_{4L} = \frac{1}{2}\sqrt{\frac{55}{42.71}-1} = \frac{1}{2}\sqrt{\frac{42.71}{33.166}-1} = \frac{1}{2}\sqrt{\frac{33.166}{25.755}-1} = \frac{1}{2}\sqrt{\frac{25.755}{20}-1} = 0.268$$

5.13 Effect of Finite Q on the Matching Networks

Figure 5-52 shows the revised schematic of a fifth-order matching network for a given Q factor for each inductor and a capacitor. To complete the matching circuit design, the components need to be converted to their physical equivalents by accounting for the finite Q factor. This can be accomplished by incorporating physical properties of inductors and capacitors in their corresponding LTspice models. Since the loss of a real inductor can be modeled by a series resistance, Rs, knowing the Q factor of the inductor at a specified frequency, the resistor value can be calculated from the expression $Q_L = \omega L/R_s$. Similarly, the loss of a capacitor can be modeled by a parallel resistor. Furthermore, since the Q of a capacitor can be expressed as $Q_C = \omega R_p C$, knowing the Q factor at a specified frequency, the parallel resistor can be easily determined from the expression for Q_c. Once R_s and R_p for each inductor and a capacitor is determined, each value can be entered into LTspice model of the specific component. Figure 5-52 shows the revised schematic where Q factor of 150 and 100 at 50 MHz are assigned to capacitors and inductors, respectively.

The network response with the initial Qcap and Qind values is shown in Figure 5-53. We can see that the insertion loss < 1 dB and return loss > 20 dB can be achieved between 60 MHz and 164 MHz. It can be shown that if a capacitor Q of 50 and an inductor Q of 40 are used in Figure 5-52, above specifications cannot be met.

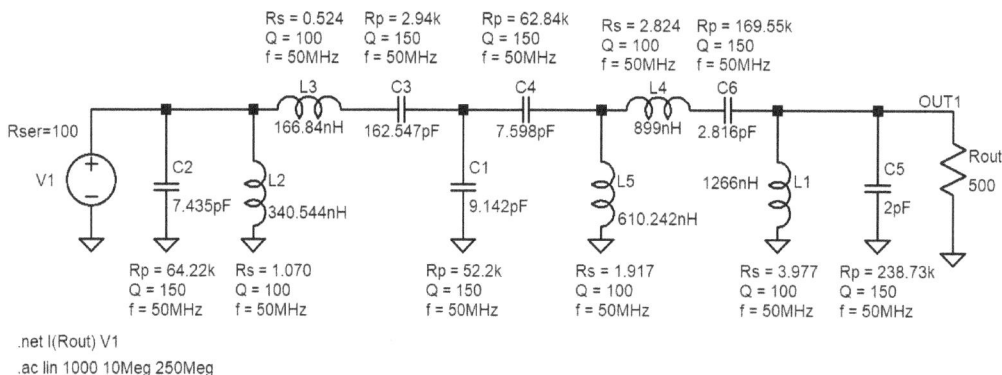

Figure 5-52 Fifth order matching schematic with Q factor variables

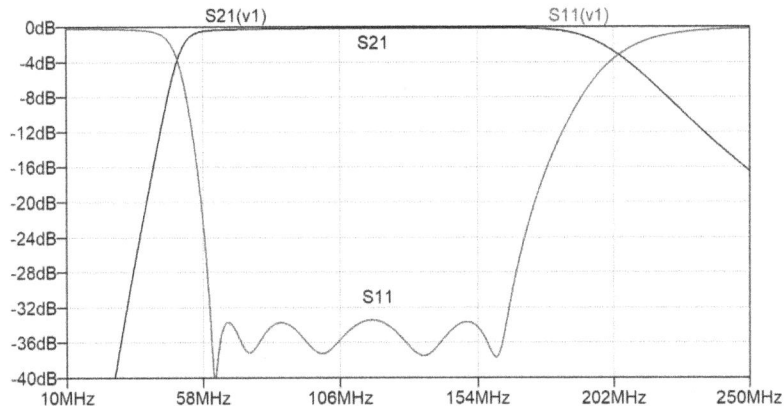

Figure 5-53 Response of Figure 5-52 with Qcap=150 and Qind=100

Problems

5-1.   For a load of 55 Ω with a 10 pF series capacitance we want to have maximum power transfer at 1 GHz. Find a source inductance that cancels the reactance of the load at this frequency.

5-2.   Design all the L-networks that will match a complex load impedance, $Z_L = 10 - j10$ Ω, to a complex source impedance, $Z_S = 25 + j25$ Ω, at a frequency of 1 GHz. Verify all the solutions by plotting the response of the matching networks.

5-3.   Design all the L-networks that will match a source impedance, $Z_S = 30$ Ω, to a complex load impedance, $Z_L = 25 + j25$ Ω, at a frequency of 1 GHz. Verify all the solutions by plotting the response of the matching networks.

5-4.   Design all the L-networks that will match a load impedance, $Z_L = 15$ Ω, to a source impedance, $Z_S = 75$ Ω, at a frequency of 0.65 GHz. Verify all the solutions by plotting the response of the matching networks.

5-5.   The output impedance of a transmitter operating at a frequency of 2.4 GHz is: $Z_S = 30 + j10$ Ω. Design all the matching L-networks such that the maximum power is delivered to the antenna whose input impedance is $Z_L = 10 + j10$ Ω.

5-6.   Redesign the matching network in Problem 5.5 with four equal-Q L-networks. Compare the Q and bandwidth of the two networks.

5-7   For a load and source resistor ratio of 4, find the minimum number of cascaded L sections needed to achieve a loaded Q of 0.5.

References and Further Readings

[1]   Ali A. Behagi, *RF and Microwave Circuit Design*, A Design Approach Using (**ADS**), Techno Search, Ladera Ranch, CA 2015

[2]   Ali Behagi and Manou Ghanevati, *Fundamentals of RF and Microwave Circuit Design,* Ladera Ranch, CA 2017

[3]   R. K. Feeney and D. R. Hertling, *RF/Wireless Principles & Practice*, Georgia Institute of Technology, 1998.

[4]   Guillermo Gonzales, *Microwave Transistor Amplifiers – Analysis and Design*, Second Edition, Prentice Hall Inc., Upper Saddle River, NJ.

[5]   Randy Rhea, *The Yin-Yang of Matching: Part 2 – Practical Matching Techniques*, High Frequency Electronics, March 2006

[6]   Steve C. Cripps, *RF Power Amplifiers for Wireless Communications*, Artech House Publishers, Norwood, MA. 1999.

[7]   David M. Pozar, *Microwave Engineering*, Third Edition, John Wiley & Sons, New York, 2005

[8]   R. Ludwig, P. Bretchko, *RF Circuit Design*, Theory and applications, Prentice Hall, Upper Saddle River, NJ, 2000

[9]   Jerry Sevick, *Transmission Line Transformers*, The American Radio Relay League, Newington, CT., 1990

Chapter 6

Distributed Matching Network Design

6.1 Introduction

Distributed networks are comprised of transmission line elements rather than discrete resistors, inductors, and capacitors. These transmission lines can take various forms as covered in chapter 2. At microwave frequencies even lumped elements behave like distributed components. At microwave frequencies distributed networks are more realizable than lumped element networks. As a general rule, any electrical part larger than one tenth of the signal wavelength should be considered as a distributed element. At RF and microwave frequencies even a short length of wire can create effects that are not predictable by the lumped element analysis.

In this chapter we discuss both narrowband and broadband distributed matching networks. Several examples are given to show how distributed matching networks are designed. Distributed matching networks discussed in this chapter include quarter-wave and single-stub matching networks.

6.2 Quarter-Wave Matching Networks

At RF and microwave frequencies the impedance matching between a resistive source and a resistive load is easily achieved by a 90 degree transmission line, known as a quarter-wave network. In this section the properties of the quarter-wave networks are defined and their application in matching two resistors are investigated.

6.3 Analysis of Quarter-Wave Matching Networks

In chapter 2 we showed that the input impedance of a quarter-wave network with characteristic impedance Z_0 terminated in a resistive load R_L is given by:

$$Z_{IN} = \frac{Z_0^{\ 2}}{R_L}$$

Therefore, the characteristic impedance Z_0 can be written as:

$$Z_O = \sqrt{R_L Z_{IN}} \qquad\qquad (6\text{-}1)$$

In chapter 5, we also showed that the maximum power transfer from a source with resistance R_S to a network terminated in a resistive load R_L is achieved only if the input impedance of the network is equal to the source resistance:

$$Z_{IN} = R_S$$

Therefore, the characteristic impedance of quarter-wave matching network, must satisfy the following equation.

$$Z_O = \sqrt{R_S R_L} \qquad\qquad (6\text{-}2)$$

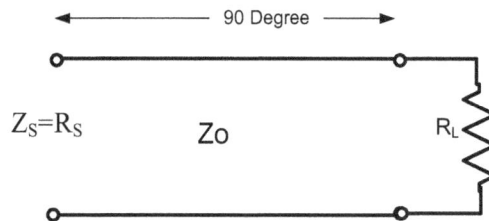

Figure 6-1 Quarter-wave network terminated in resistor R_L

Equation (6-2) states that the characteristic impedance of the quarter-wave network, matching R_L to R_S, must be equal to the square root of the product of source and load resistors.

If we normalize the load resistor R_L with respect to R_S,

$$r = \frac{R_L}{R_S} \qquad\qquad (6\text{-}3)$$

The characteristic impedance of the quarter-wave network becomes a function of R_S and r.

$$Z_O = R_S \sqrt{r} \qquad (6\text{-}4)$$

Notice that for $R_L > R_S$, the characteristic impedance Z_0 is greater than R_S while for $R_L < R_S$, the characteristic impedance Z_0 is less than R_S. The fractional bandwidth of a network, FBW, is defined in Equation (6-5) where f_H and f_L are the upper and lower frequencies of the bandwidth and the center frequency f_0 is equal to $\sqrt{f_H f_L}$, respectively.

$$FBW = \frac{f_H - f_L}{\sqrt{f_H f_L}} \qquad (6\text{-}5)$$

The fractional bandwidth of a quarter-wave matching network is given in Equation (6-6)

$$FBW = 2 - \frac{4}{\pi} \cdot \cos^{-1}\left(\frac{2\Gamma_m \sqrt{r}}{\sqrt{1 - \Gamma_m^{\,2}} \,|1 - r|} \right) \qquad (6\text{-}6)$$

Where Γ_m is the magnitude of the reflection coefficient.

Equation (6-6) shows that the fractional bandwidth of a quarter-wave matching network depends upon the magnitude of the input reflection coefficient, Γ_m, and the mismatch ratio, r. The solutions to Equation (6-6) are only valid if,

$$\frac{2\Gamma_m \sqrt{r}}{\sqrt{1 - \Gamma_m^{\,2}} \,|1 - r|} \leq 1$$

At 3 dB return loss the reflection coefficient is $\Gamma_m = 0.707$, therefore, Equation (6-6) reduces to:

$$FBW_{3dB} = 2 - \frac{4}{\pi} \cdot \cos^{-1}\left(\frac{2\sqrt{r}}{|1-r|}\right) \tag{6-7}$$

At 3 dB return loss Equation (6-7) has valid solutions only if,

$$\frac{2\sqrt{r}}{|1-r|} \leq 1 \quad or \quad r^2 - 6r + 1 \geq 0 \tag{6-8}$$

Similarly, if we define the bandwidth at $\Gamma_m = 0.1$, corresponding to 20 dB return loss, as a good matching bandwidth it is insightful to evaluate the fractional bandwidth associated with this 20 dB return loss. From Equation (6-6) the fractional bandwidth at $\Gamma_{in} = 0.1$, is:

$$FBW_{20dB} = 2 - \frac{4}{\pi} \cdot \cos^{-1}\left(\frac{0.2\sqrt{r}}{\sqrt{0.99}\,|1-r|}\right) \tag{6-9}$$

At 20 dB return loss Equation (6-9) has valid solutions only if,

$$\frac{0.2\sqrt{r}}{\sqrt{0.99}\,|1-r|} \leq 1 \quad or \quad 99r^2 - 202r + 99 \geq 0 \tag{6-10}$$

The loaded quality factor, Q_L, of the quarter-wave matching network is defined as the inverse of the fractional bandwidth at 3 dB return loss; therefore, the loaded Q factor can be calculated from Equation (6-11).

$$Q_L = \frac{1}{FBW_{3dB}} = \frac{1}{2 - \frac{4}{\pi} \cdot \cos^{-1}\left(\frac{2\sqrt{r}}{|1-r|}\right)} \tag{6-11}$$

Equation (6-11) shows that the validity condition in Equation (6-8) for the 3 dB fractional bandwidth is the same for the Q factor except that whenever the 3 dB fractional bandwidth tends towards infinity the Q factor tends towards zero. The Q factor given in Equation (6-11) is not to be confused with the unloaded transmission line Q factor as covered in chapter 4. This is actually an external Q factor as it relates to the loaded Q of the overall network. If f is the center (design) frequency, the bandwidth of the circuit is then calculated from Equation (6-12).

$$BW = (f) \cdot (FBW) \tag{6-12}$$

6.4 Analytical Design of Quarter-Wave Matching Networks

In this subsection, based on the equations developed in section 6.2.1, two quarter-wave matching networks for $R_L = 2\ \Omega$ and $R_L = 150\ \Omega$ are designed. For each case the loaded Q factor and bandwidth is calculated at 3 and 20 dB return loss.

Example 6.1 Design a quarter-wave network intended to match a 50 Ω source to a 2 Ω load at 100 MHz. Calculate the Q factor and the fractional bandwidths at 3 and 20 dB return loss. Compare the calculations with the simulation results.

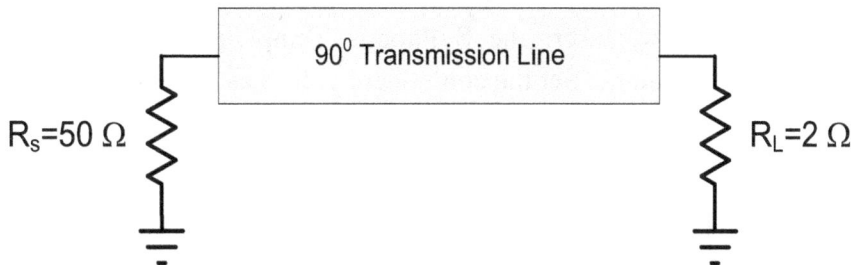

Figure 6-2 Matching quarter-wave transformer ($R_L < R_s$)

Solution: First we solve the problem analytically. Using the equations developed in section 6.3, the characteristic impedance, loaded Q factor, and the bandwidths of the matching network are calculated as follows.

1. Enter design parameters and normalize the load impedance
RS=50; RL=2; f=100e6: r=RL/RS

2. Calculate Z1, BW, FBW, and Q

Z1=RS*sqrt(r)

FBW20dB = 2 - (4/PI)*acos((0.2*sqrt(r))/(sqrt(0.99)*(abs(1 - r))))

BW20dB = (f*FBW20dB)

FBW3dB = 2 - (4/PI)*acos((2*sqrt(r))/(abs(1 - r)))

BW3dB = (f*FBW3dB)

Q = 1/FBW3dB

The calculation results show that the Z1 = 10 Ω, Q factor is 1.827, and the bandwidths at 3 and 20dB return loss are 54.72 MHz and 5.333 MHz, respectively.

The fractional bandwidth at 3 dB is 54.7% and at 20 dB return loss is only 5.333% indicating a narrowband matching network. This is characteristic of a narrowband matching network with a load resistor that is much smaller than the source resistor.

To measure the same parameters in LTspice, create a new workspace and open a new schematic window. Insert the S_Params Template and place the TLIN component on the schematic. Set the component values as shown in Figure 6-3.

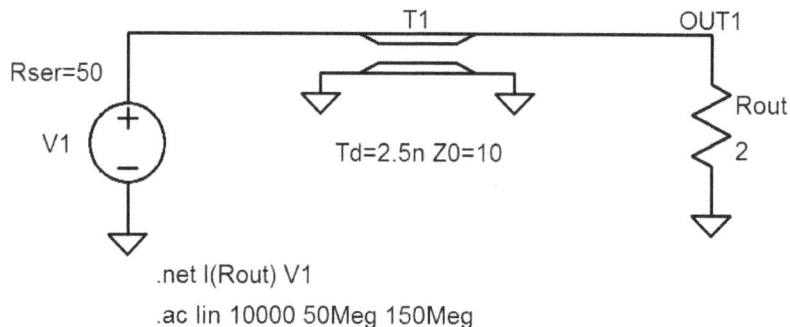

Figure 6-3 Schematic of the quarter-wave matching network

Simulate the schematic and display the input reflection coefficient, S_{11}, and forward transmission, S_{21} in a rectangular plot. To measure the bandwidths manually, place markers at 3 and 20 dB return loss as shown in Figure 6-4.

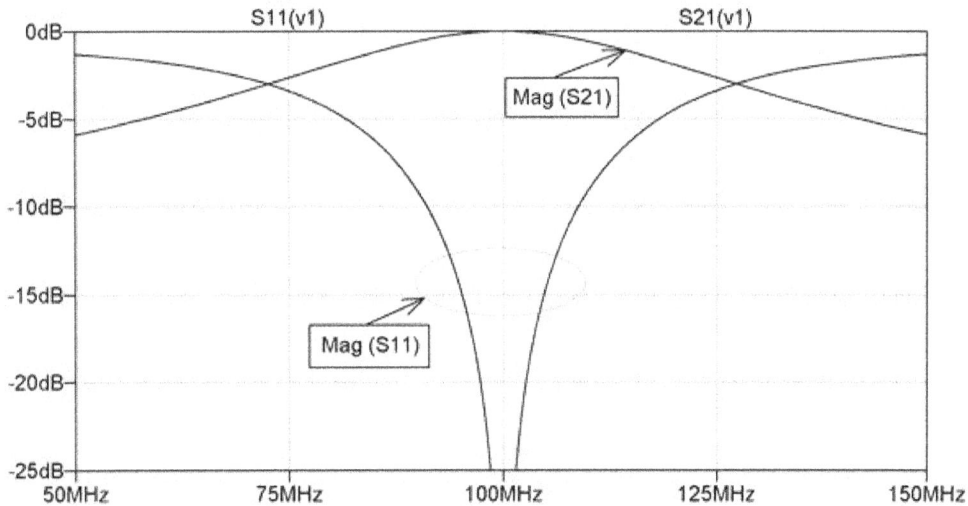

Figure 6-4 Magnitude of the transmission and input reflection loss

Measuring the bandwidth at 3 and 20 dB return loss, the corresponding values are:

$$BW_{-3dB} = 127.3 - 72.7 = 54.6 \qquad MHz$$

$$BW_{-20dB} = 102.7 - 97.3 = 5.4 \qquad MHz$$

The Q factor of the matching network is calculated by using Equation (6-9),

$$Q = \frac{1}{FBW_{-3dB}} = \frac{\sqrt{(127.3)(72.7)}}{127.3 - 72.7} = 1.76$$

Note that the measured Q factor and bandwidths shows a close agreement with the calculated values.

Example 6.2 Design a quarter-wave network intended to match a 50 Ω source to a 150 Ω load. The design frequency is 100 MHz. Compare the calculated Q factor and the fractional bandwidths, at 3 and 20 dB return loss, with the measurements.

Solution: The schematic of the quarter-wave matching network is shown in Figure 6-5.

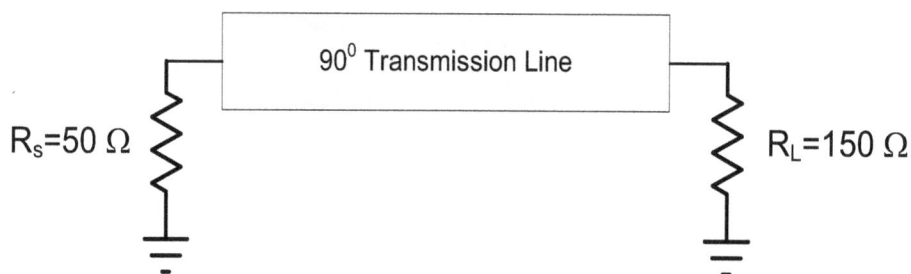

Figure 6-5 Quarter-wave matching network (R_L > R_S)

Using Equation (6-2) the characteristic impedance of the quarter-wave matching network is:

$$Z_1 = \sqrt{(50)(150)} = 86.6 \ \Omega$$

The schematic of the matching network is shown in Figure 6-6.

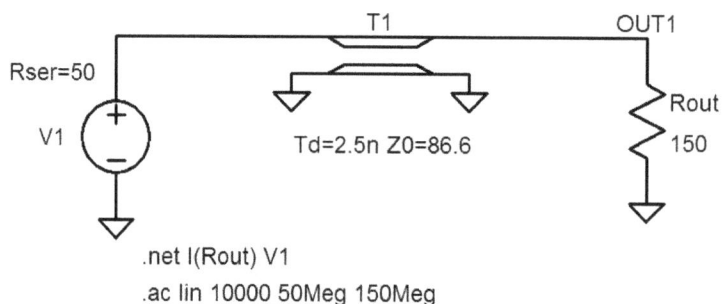

.net I(Rout) V1
.ac lin 10000 50Meg 150Meg

Figure 6-6 Schematic of the matching quarter-wave network

Using Equations developed in section 6.3, calculate the characteristic impedance, the Q factor, and the bandwidths of the matching network at 3 and 20 dB return loss. The procedure follows.

1. Enter design parameters and normalized load impedance

RS = 50; RL = 150; f = 100e6; r = RL/RS

2. Calculate Z1, FBW, BW and Q factor

Z1 = RS*sqrt(r)

FBW20dB = 2-(4/PI)*acos((0.2*sqrt(r))/(sqrt(0.99)*(abs(1 - r))))

BW20dB = (f*FBW20dB)

FBW3dB = 2-(4/PI)*acos((2*sqrt(r))/(abs(1 - r)))

BW3dB = (f*FBW3dB)

Qe = 1/FBW3dB

The calculation results show that the bandwidth at 20 dB return loss is 22.28 MHz. This is over four times the bandwidth that was achieved with the 2 Ω load impedance. Notice that the overall Q of the network has significantly increased due to higher load impedance.

Next simulate the matching network in Figure 6-6 from 50 MHz to 150 MHz and display the input reflection coefficient, S_{11}, and forward transmission, S_{21} in a rectangular plot. Place markers to measure the bandwidths at 3 and 20 dB return loss as shown in Figure 6-7.

As Figure 6-7 shows the measured bandwidth at 20 dB return loss is:

$$111.1 - 88.86 = 22.24 \text{ MHz}$$

This agrees well with the calculated bandwidth.

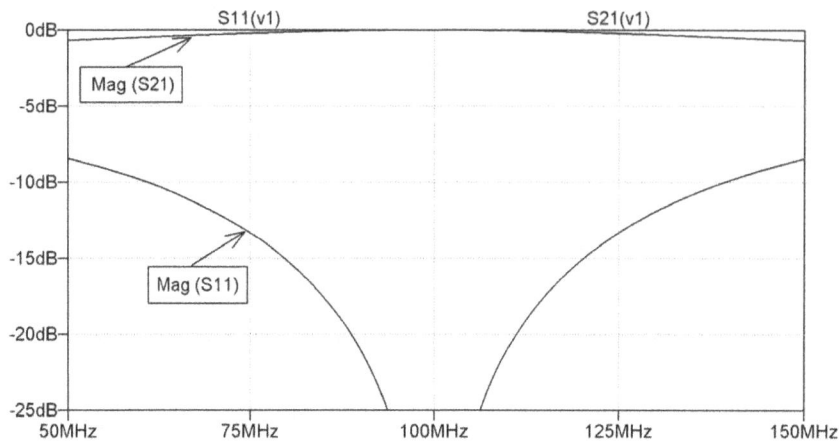

Figure 6-7 Magnitude of the transmission and input reflection loss

6.5 Quarter-Wave Network Matching Bandwidth

Equation (6-6) shows that the achievable bandwidth in a quarter-wave matching network is related to the ratio of the load to the source impedance (mismatch ratio) as well as the value of the input reflection coefficient. It is insightful to examine the relationship between these quantities when one of the parameters is swept in value.

6.6 Quarter-Wave Network Matching Bandwidth and Power Loss

The fractional bandwidth and power loss of a quarter-wave matching network can be calculated in an Equation Editor as a function of the input reflection coefficient, Γ_{IN}, and the normalized load resistor, r. The procedure is listed here.
1. Enter equations for input reflection coefficient, power loss, and the conversion of reflection coefficient to return loss in dB.
2. Enter the desired values for the source and load resistors
3. Normalize the load resistor with respect to source resistor

4. Use Equation (6-6) to calculate the fractional bandwidth.

Example 6.3: For a 50 Ω source and 2 Ω load resistors, calculate the fractional bandwidth and powerloss from $\Gamma = 0.1$ to $\Gamma = 0.707$.

Solution: Follow the above procedure to calculate the fractional bandwidth and power loss when $R_S = 50\ \Omega$ and $R_L = 2\Omega$.

1. Enter reflection coefficient, power loss, and return loss Equations

Vswr = Sweep1_Data.VSWR1

ReflCoef=(vswr-1)/(vswr+1)

Powerloss=(1-(1-(abs(ReflCoef)^2)))*100

RLdB=-20*log(abs(ReflCoef))

2. Enter the source and load resistor values

RS=50; RL=2; r=RL/RS

3. Calculate the Fractional Bandwidth

FBW=(2-(4/PI)*acos((2*(ReflCoef)*sqrt(r))/(sqrt((1-(ReflCoef)^2))*abs(1-r))))

Sweep the parameters to display the fractional bandwidth and power loss for reflection coefficients from 0.1 to 0.707, as shown in Table 6-1.

Table 6-1 shows that the fractional bandwidths at 0.1 reflection coefficient, corresponding to 20 dB return loss, is 5.3% with 1% power loss while at 0.707 reflection coefficient, corresponding to 3 dB return loss, the fractional bandwidth is 54.7% with 49.985% power loss. Notice that the fractional bandwidths, at 3 and 20 dB return loss, are the same as calculated in Example 6.2-1.

	ReflCoef	FBW ...	RLdB	Powerloss
1	0.1	0.053	20	1
2	0.11	0.059	19.172	1.21
3	0.12	0.064	18.417	1.44
4	0.13	0.07	17.721	1.69
5	0.14	0.075	17.077	1.96
6	0.15	0.081	16.478	2.25
7	0.16	0.086	15.917	2.56
8	0.17	0.092	15.391	2.89
9	0.18	0.097	14.895	3.24
10	0.19	0.103	14.425	3.61
11	0.2	0.108	13.979	4
12	0.3	0.167	10.458	9
13	0.4	0.233	7.959	16
14	0.5	0.309	6.021	25
15	0.6	0.405	4.437	36
16	0.707	0.547	3.012	49.985

Table 6-1 Fractional bandwidth (in Radian, col. 3) and power loss for $R_L = 2\ \Omega$

Example 6.4: For a 50 Ω source and 150 Ω load resistors, calculate the fractional bandwidth and power loss from $\Gamma = 0.1$ to $\Gamma = 0.707$.

Solution: Following the procedure, the calculations are shown in Table 6-2. Sweep the parameters and display the result, as shown in Table 6-2.

1. Enter reflection coefficient, powerloss, and return loss Equations

vswr=Sweep1_Data.VSWR1

ReflCoef=(vswr-1)/(vswr+1)

Powerloss=(1-(1-(abs(ReflCoef)^2)))*100

RLdB=-20*log(abs(ReflCoef))

2. Enter the source and load resistor values and normalized load

RS=50; RL=150; r=RL/RS

3. Calculate the Fractional Bandwidth

FBW=(2-(4/PI)*acos((2*(ReflCoef)*sqrt(r))/(sqrt((1-(ReflCoef)^2))*abs(1-r))))

The reflection coefficient, power loss, and return loss are given in the following Table.

Table 6-2 shows that the fractional bandwidth at $\Gamma = 0.1$, corresponding to 20 dB return loss, is 22.3% with a 1% power loss. The fractional bandwidth is the same as calculated in the previous example. The higher fractional bandwidth in this example, compared to its value in the previous example, is due to the lower ratio of the load to source resistor.

	ReflCoef	FBW (rad)	RLdB	Powerloss
1	0.1	0.223	20	1
2	0.11	0.246	19.172	1.21
3	0.12	0.269	18.417	1.44
4	0.13	0.292	17.721	1.69
5	0.14	0.315	17.077	1.96
6	0.15	0.339	16.478	2.25
7	0.16	0.362	15.917	2.56
8	0.17	0.386	15.391	2.89
9	0.18	0.411	14.895	3.24
10	0.19	0.435	14.425	3.61
11	0.2	0.46	13.979	4
12	0.3	0.733	10.458	9
13	0.4	1.091	7.959	16
14	0.5	2	6.021	25
15	0.6	1.#QO	4.437	36
16	0.707	1.#QO	3.012	49.985

Table 6-2 Fractional bandwidth and power loss measurements for R_s=50 Ω and R_L=150 Ω

6.7 Single-Stub Impedance Matching Networks

The single-stub, also known as line and stub, matching network is a popular narrowband transmission line technique used to match real or complex load impedance to real source impedance. This technique is frequently used in distributed matching circuit designs using microstrip or stripline. The single-stub matching network, shown in Figure 6-8, consists of a series transmission line connected directly to the load impedance and a shunt stub (short-circuited or open-circuited transmission line) attached to the source impedance. The characteristic impedance of both matching elements has the same value as the source impedance, R_s. This section demonstrates both analytical and graphical techniques to design fixed frequency single-stub matching networks.

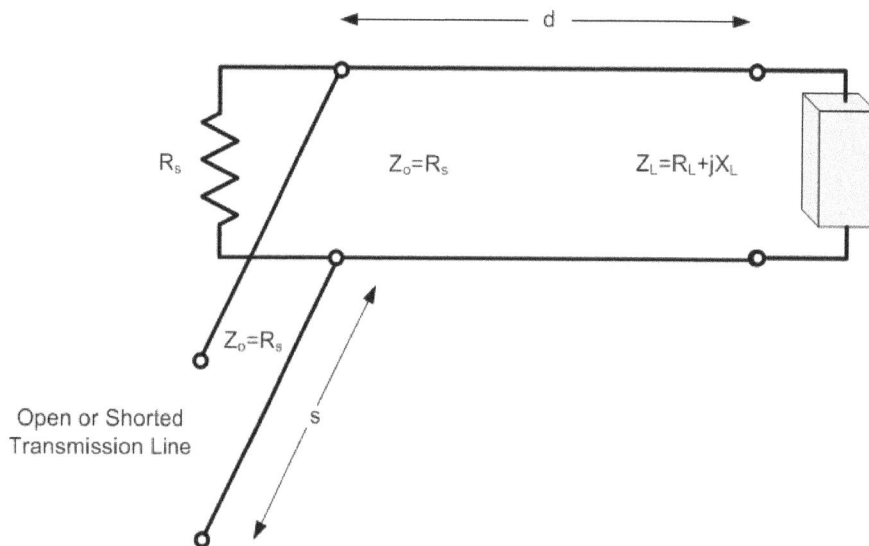

Figure 6-8 Single-stub matching network

To start the design of the single-stub matching network, attach a transmission line section, with characteristic impedance equal to R_S, to the load and determine its electrical length in such a way that the normalized admittance at a distance d from the load falls on the unit conductance circle of the Smith Chart. Calculation of the electrical lengths for the series line and shunt stub are given separately.

6.8 Analytical Design of Series Transmission Line

The input impedance of a lossless transmission line of length d and characteristic impedance Z_0 terminated in an arbitrary load Z_L, was given in chapter 2 in Equation (2-46) and repeated here for convenience.

$$Z_{IN} = Z_o \frac{Z_L + jZ_o \tan \beta d}{Z_o + jZ_L \tan \beta d}$$

Setting $Z_0 = R_S$ and tan $\beta d = t$, the input admittance of the network can be written as:

$$Y_{IN} = \frac{1}{Z_{IN}} = \frac{R_S + jZ_L t}{R_S(Z_L + jR_S t)} = G_{IN} + jB_{IN} \tag{6-13}$$

Substituting the normalized load impedance, $Z_L/R_s = r + jx$ into Equation (6-13) and separating its real and imaginary parts we get:

$$G_{IN} = \frac{r(1+t^2)}{R_S(r^2 + x^2 + t^2 + 2xt)} \tag{6-14}$$

And,

$$B_{IN} = \frac{xt^2 + (r^2 + x^2 - 1)t + x}{R_S(r^2 + x^2 + t^2 + 2xt)} \tag{6-15}$$

The value of d, which implies t, can be obtained by setting the input conductance, G_{IN}, equal to source conductance:

$$\frac{r(1+t^2)}{R_S(r^2 + x^2 + t^2 + 2xt)} = \frac{1}{R_S} \tag{6-16}$$

Equation (6-16) can be rearranged as:

$$(r-1)\cdot t^2 - 2xt - \left(r^2 + x^2 - r\right) = 0 \tag{6-17}$$

Notice that the quadratic Equation (6-17) has two solutions for t. For a wider bandwidth and lower loss usually the smaller value of t is selected. The two solutions for t are:

$$t_1 = \frac{x + \sqrt{r(r^2 + x^2 - 2r + 1)}}{r - 1} \tag{6-18}$$

And,

$$t_2 = \frac{x - \sqrt{r(r^2 + x^2 - 2r + 1)}}{r - 1} \tag{6-19}$$

With $\tan \beta d = t$, and $\beta \lambda = 2\pi$, we have $d = \frac{\lambda}{2\pi} \tan^{-1} t$ and the two solutions for d are:

$$d_1 = \frac{\lambda}{2\pi} \tan^{-1} t_1 \qquad t_1 \geq 0 \tag{6-20}$$

$$d_2 = \frac{\lambda}{2\pi} \tan^{-1} t_2 \qquad t_2 \geq 0 \tag{6-21}$$

To specify the lengths of d_1 and d_2 in electrical degrees, we get:

$$d_1 = \frac{360}{2\pi} \tan^{-1} t_1 \qquad t_1 \geq 0 \tag{6-22}$$

$$d_2 = \frac{360}{2\pi} \tan^{-1} t_2 \qquad t_2 \geq 0 \tag{6-23}$$

Because at every half wavelength the input impedance of a transmission line repeats, there are an infinite number of transmission line lengths that matches the load to source impedance. Usually the shorter length is selected to improve the matching bandwidth. If t_1 or t_2 is negative, we add half a wavelength to each line to get positive d_1 and d_2.

$$d_1 = \frac{360(\pi + \tan^{-1} t_1)}{2\pi} \qquad t_1 < 0 \tag{6-24}$$

$$d_2 = \frac{360(\pi + \tan^{-1} t_2)}{2\pi} \qquad t_2 < 0 \qquad (6\text{-}25)$$

6.9 Design of the Open-Circuited Shunt Stub

To calculate the electrical length of the shunt stub, substitute t_1 and t_2 in Equation (6-15) to determine B_1 and B_2.

$$B_1 = \frac{xt_1^2 + (r^2 + x^2 - 1)t_1 + x}{R_S(r^2 + x^2 + t_1^2 + 2xt_1)} \qquad (6\text{-}26)$$

$$B_2 = \frac{xt_2^2 + (r^2 + x^2 - 1)t_2 + x}{R_S(r^2 + x^2 + t_2^2 + 2xt_2)} \qquad (6\text{-}27)$$

The electrical lengths of the open circuited stubs are found by setting the susceptance of the stubs equal to the negative of the input susceptance.

$$so_1 = \frac{-\lambda\left(\tan^{-1}(R_S B_1)\right)}{2\pi} \qquad (6\text{-}28)$$

$$so_2 = \frac{-\lambda\left(\tan^{-1}(R_S B_2)\right)}{2\pi} \qquad (6\text{-}29)$$

If either stub length in Equation (6-28) or (6-29) is negative, add one half wavelength to obtain a positive stub length. For short-circuited stubs, the two solutions are:

$$ss_1 = \frac{\lambda\left(\tan^{-1}\left(\frac{1}{R_S B_1}\right)\right)}{2\pi} \qquad (6\text{-}30)$$

$$SS_2 = \frac{\lambda \left(tan^{-1} \left(\dfrac{1}{R_S B_2} \right) \right)}{2\pi} \tag{6-31}$$

6.10 Single-Stub Impedance Matching Network Design

Example 6.5: Design an open-circuited single-stub network to match a load resistance $Z_L = 2 - j5$ Ω to a resistive source $R_S = 50$ Ω at 100 MHz. Calculate the fractional bandwidths of the matching network at 3 and 20 dB return loss. Display the response and compare the calculations with measurements.

Solution: The design example is shown in Figure 6-9.

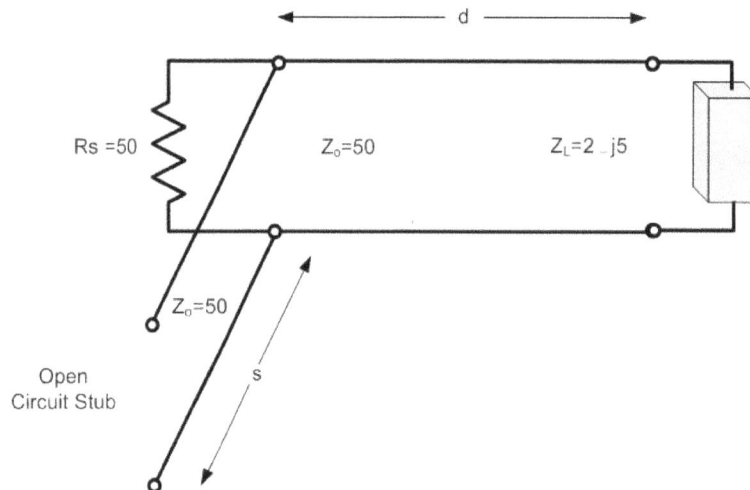

Figure 6-9 Complex load to resistive source matching network

Using the equations developed in section 6.9, calculate the electrical lengths of the line and stub matching transmission lines. The calculation in MATLAB script follows.

1. Enter design parameters and normalized load impedance
RS=50; RL=2; XL=-5; f=100e6

r = RL/RS

x = XL/RS

2. Calculate t1, t2, d1 and d2, B1, B2, so1 and so2.

t1 = (x + sqrt (r*(r^2+x^2-2*r+1)))/(r-1)

t2 = (x – sqrt (r*(r^2+x^2-2*r+1)))/(r-1)

d1 = 360*(atan (t1))/(2*pi)

d2 = 360*(atan (t2))/(2*pi)

B1 = (x*t1^2 + (r^2 + x^2-1)*t1 - x)/(RS*(r^2 + x^2 + t1^2 + 2*x*t1))

B2 = (x*t2^2 + (r^2 + x^2-1)*t2 - x)/(RS*(r^2 + x^2 + t2^2 + 2*x*t2))

so1 = 360*(pi – atan (RS*B1))/(2*pi)

so2 = -360*atan (RS*B2)/(2*pi)

The calculation results show that the problem has two solutions. Either solution can be used in the design of the single-stub matching network. However, for both line and stub, usually the shorter line lengths are chosen. For this example we select the shorter electrical lengths in the second solution; d2 for the line and so2 for the open-circuited stub. The schematic of the single-stub matching network is shown in Figure 6-10.

Simulate the schematic and display S_{11} and S_{21} in a rectangular plot.

Notice that the matching network is perfectly matched and the 3 dB bandwidth is 11.1 MHz.

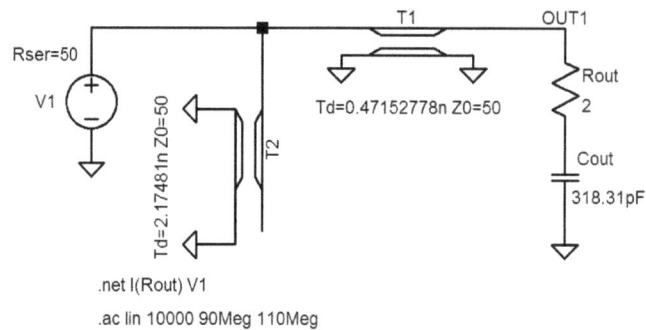

Figure 6-10 Schematic of the single-stub matching network

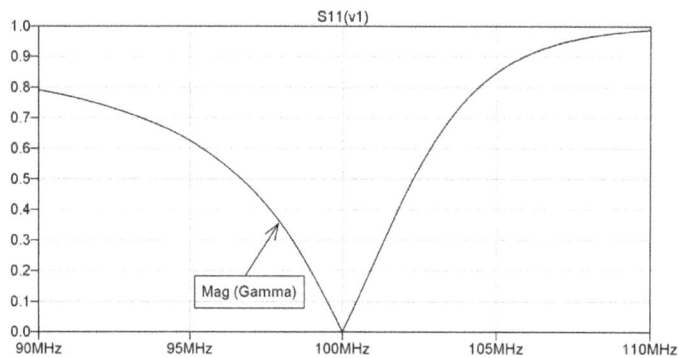

Figure 6-11 Response of the single-stub matching network

6.11 Graphical Design of the Open-Circuited Stub

In this section a 50 Ω source will be matched to a 5 - 25j Ω load at a frequency of 1000 MHz. Both open-circuited and short-circuited shunt stubs will be considered.

Example 6.6 Consider matching the 5 - j25 Ω load impedance to 50 Ω source resistor.

Solution: Create a new schematic in LTspice. Wire up the components and set the component values as shown in Figure 6-12.

Because we intend to place a shunt element after the series transmission line, enable the admittance circles on the Smith Chart. Make the length of the line tunable and tune it to move the load impedance (point A) to intersect the unit conductance circle on the top half of the Smith chart (point B. As the schematic shows, an electrical length of $42.5°$ places the impedance on the unit conductance circle at point B.

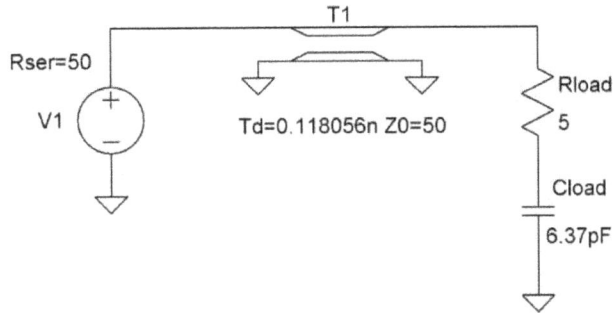

Figure 6-12 Adding series transmission line (electrical length=42.5°)

The movement of the impedance from point A to point B is displayed in Figure 6-13.

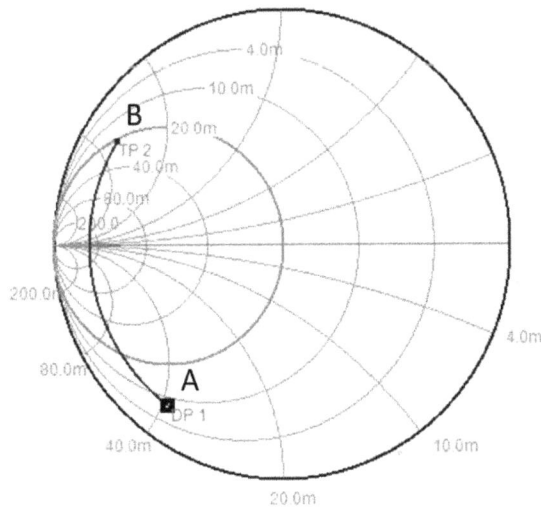

Figure 6-13 Moving from point A to B on the unit conductance circle

At point B the normalized impedance is $z = (4.4 + j14)/50 = 0.088 + j0.28 \ \Omega$. To move point B to the center of Smith chart, add an open-circuited shunt transmission line and tune the length of the line until the impedance moves to the center of the chart (50Ω). As the schematic of Figure 6-14 shows, a 73° length of transmission line would be required.

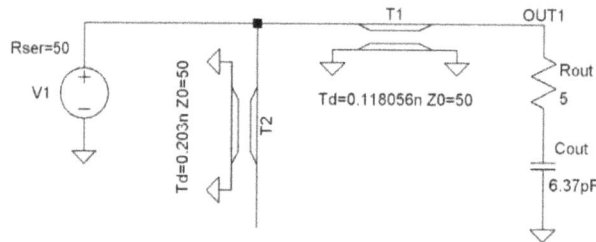

Figure 6-14 Adding open-circuited shunt transmission line (73°)

The movement of point B to the center of the Smith chart is shown in Figure 6-15.

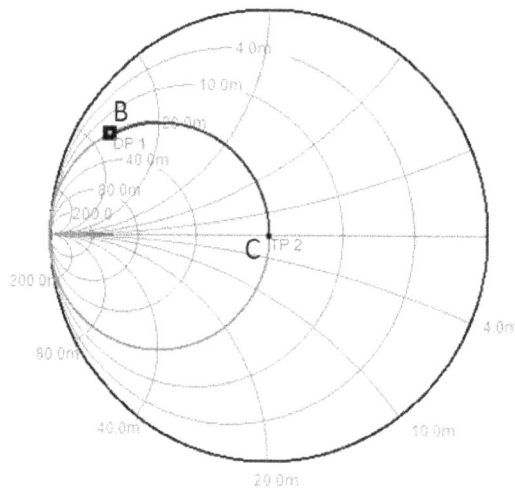

Figure 6-15 Moving point B to the center of Smith chart (73°)

6.12 Graphical Design Using a Short-Circuited Stub

In this Section a short-circuited shunt stub will be used.

Example 6.7 Consider matching the 5 - j25 Ω load impedance to 50 Ω source resistor using short-circuited shunt stubs.

Solution: A short-circuited shunt transmission line can be used to perform the function of the shunt stub. Using a short-circuited shunt transmission line, the series transmission line should intersect the unit conductance circle on the bottom half of the Smith Chart. Add a series transmission line of $11°$ electrical length to move the impedance at point A to intersect the unit conductance circle at point B as shown in Figure 6-17. Then add the short-circuited shunt transmission line as shown in Figure 6-18 or 6-19. Tune the length to $17°$ to move the impedance to the center of the Smith Chart.

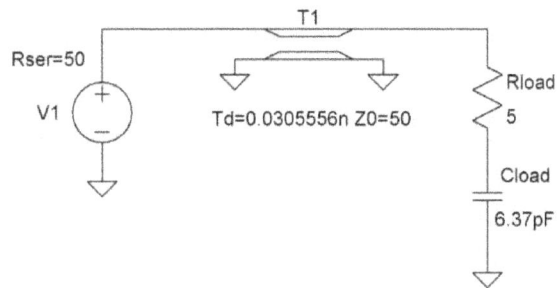

Figure 6-16 Adding a series transmission line ($11°$) to 5-j25 Ω

Simulate the schematic and notice the movement of the impedance from point A to point B.

To move point B to the center of Smith chart, add a short-circuited shunt transmission line and tune the length of the line until the impedance moves to the center of the chart (50 Ω). As the schematic of Figure 6-18 shows, a $17°$ length of transmission line would be required.

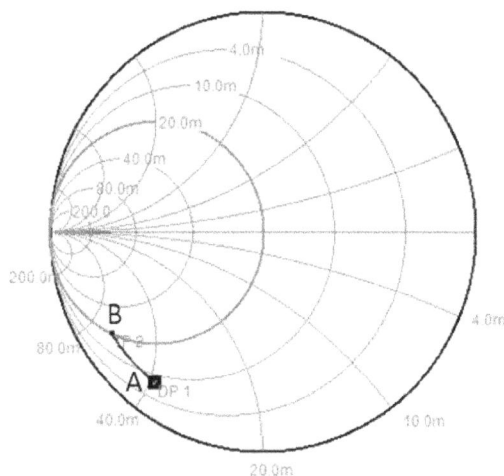

Figure 6-17 Adding shunt transmission line ($11°$) to 5-j25 Ω

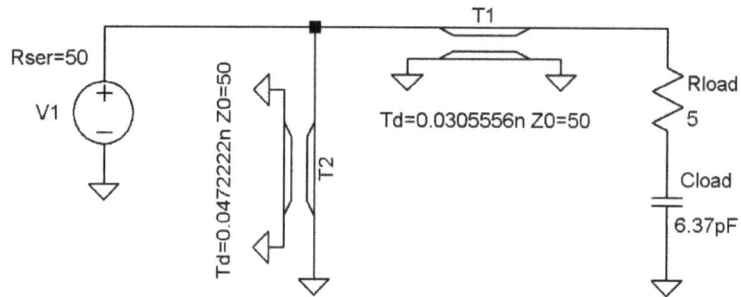

Figure 6-18 Adding short-circuited shunt transmission line (17°)

Simulate the schematic and notice the movement of the impedance from point B to the center of Smith chart.

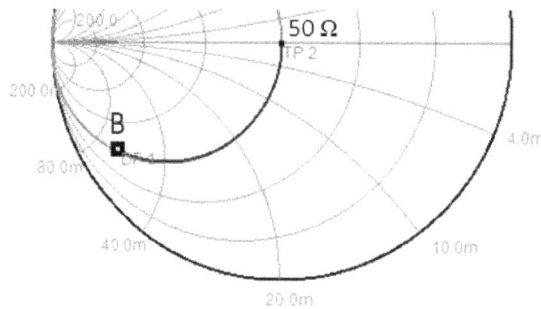

Figure 6-19 Adding short-circuited shunt transmission line (17°)

6.13 Design of Cascaded Single-Stub Matching Networks

When the impedance of the load and the source are both complex functions, we can define a virtual resistor and design single-stub networks to match the complex impedances to the virtual resistor. The final matching network is obtained by cascading the single-stub matching networks. The procedure is demonstrated in the following example.

Example 6.8: Design single-stub networks to match a complex load $Z_L = 10 - j5$ Ω to a complex source $Z_S = 50 - j15$ Ω at 100 MHz. Calculate the electrical lengths of the lines and the fractional bandwidths of the matching network at 3 dB and 20 dB return loss. Display the simulated response and verify the calculations with simulation.

Solution: The following procedure is used to match the two complex impedances.

1. Find the intermediate resistor, $R_1 = \sqrt{R_L R_S} = \sqrt{(10)\,(50)} = 22.36\ \Omega$
2. Design a single-stub matching network between $R1$ and the source impedance
3. Design a second single-stub matching network between R_1 and the load impedance

Cascade the two matching networks
1. Enter design parameters and normalized load impedance

RS = 22.36; RL = 50; XL = -15; f = 100e6; r = RL/RS; x =XL/RS

2. Use MATLAB script to calculate t1, t2, d1, d2, B1, B2, so1, and so2

t1 = (x + sqrt (r*(r^2 + x^2 - 2*r + 1)))/(- 1)

t2 = (x − sqrt (r*(r^2 + x^2-2*r + 1)))/(r - 1)

d1 = 360*(atan (t1))/(2*pi)

d2 = 360*(pi + atan (t2))/(2*pi)

B1 = (x*t1^2 + r^2+x^2 - 1)*t1 - x)/(RS*(r^2 + x^2 + t1^2 + 2*x*t1))

B2 = (x*t2^2 + (r^2 + x^2 - 1)*t2 - x)/(RS*(r^2 + x^2 + t2^2 + 2*x*t2))

so1 = 360*(pi − atan (RS*B1))/(2*pi)

so2 = -360*atan (RS*B2)/(2*pi)

First Solution: For the first solution we select d2 and so2. The schematic is shown in Figure 6-20.

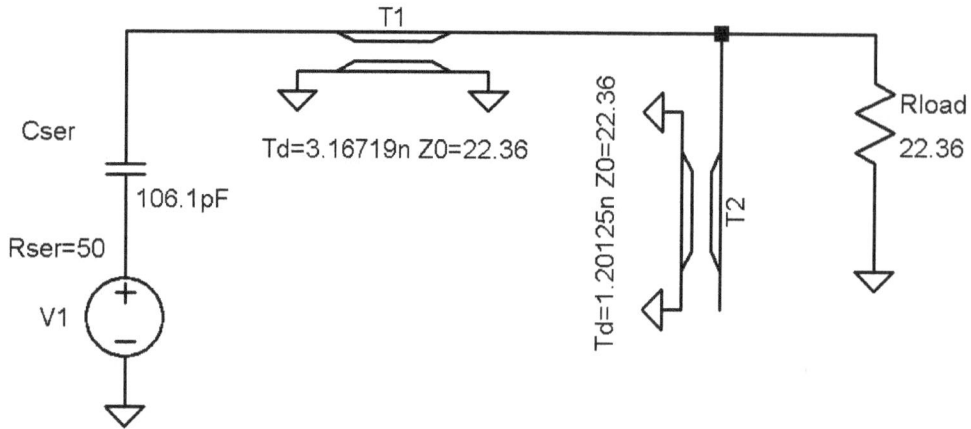

Figure 6-20 Schematic of the first matching network

Second Solution: To design the second matching network use the following procedure.

1. Enter design parameters and normalize the load impedance
RS = 22.36; RL = 10; XL = -5; f = 100e6; r = RL/RS; x = XL/RS

2. Calculate t1 and t2, d1 and d2, B1 and B2, so1 and so2,
t 1 = (x + sqrt (r*(r^2 + x^2 - 2*r + 1)))/(r - 1)

t2 = (x-sqrt (r*(r^2 + x^2 - 2*r + 1)))/(r - 1)

d1 = 360*(pi + atan (t1))/(2*pi)

d2 = 360*(atan (t2))/(2*pi)

B1 = (x*t1^2 + (r^2+x^2-1)*t1-x)/(RS*(r^2+x^2+t1^2+2*x*t1))

B2 = (x*t2^2+(r^2+x^2-1)*t2-x)/(RS*(r^2+x^2+t2^2+2*x*t2))

so1 = 360*(pi-atan (RS*B1))/(2*pi)

so2 = -360*atan (RS*B2)/(2*pi)

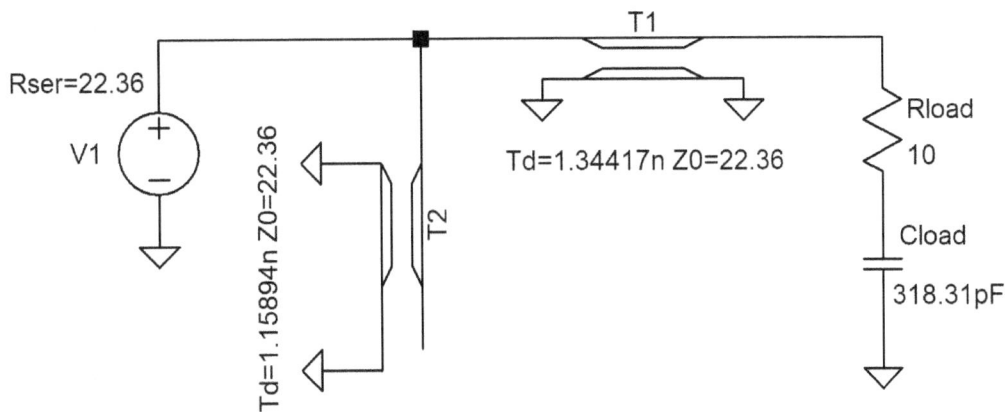

Figure 6-21 Schematic of the second matching network

Now connect both line and stub matching networks in cascade to obtain the complete matching network, as shown in Figure 6-22.

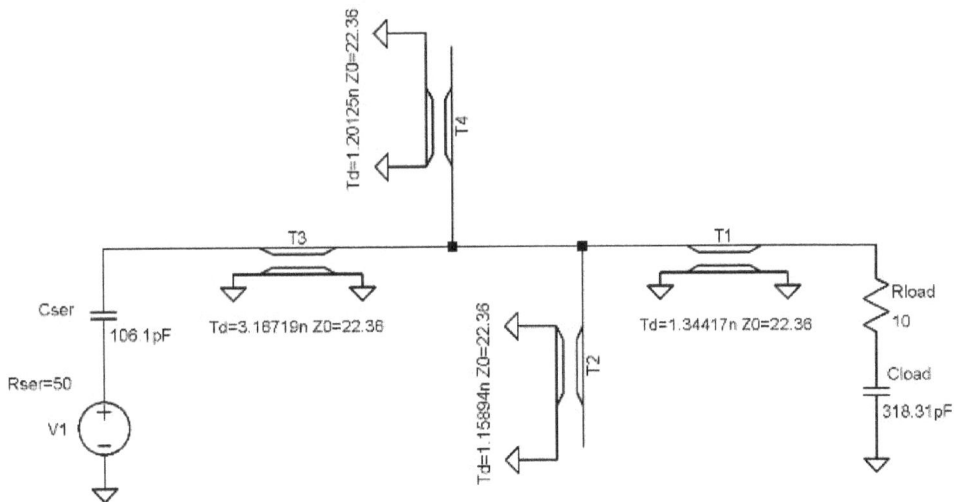

Figure 6-22 Cascading single-stub matching networks

The simulated response of the complete matching network is shown in Figure 6-23. Notice that the cascaded single-stub matching network has about 100% fractional bandwidth at the 3 dB return loss and 47% fractional bandwidth at the 12 dB return loss.

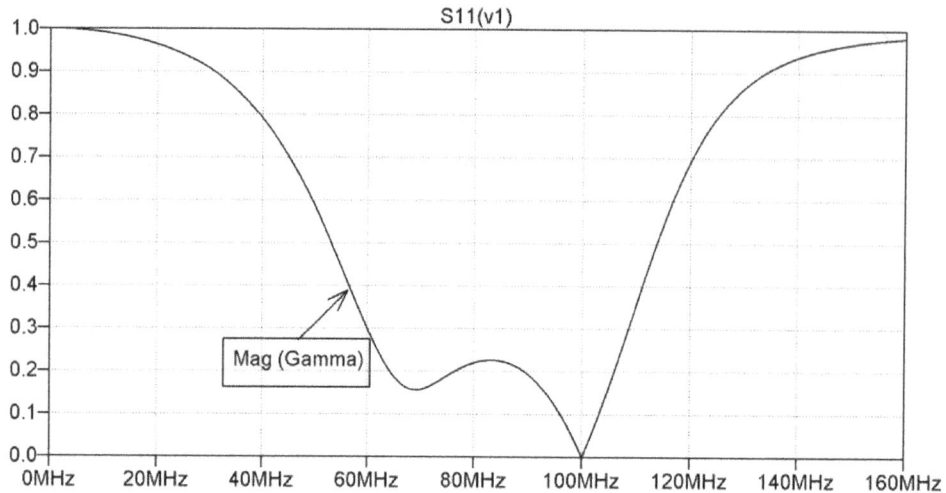

Figure 6-23 Response of the cascaded matching network

6.14 Broadband Quarter-Wave Matching Network Design

In Examples 6.1 and 6.2, the bandwidth of a single quarter-wave transformer matching network is less than 10% which is considered to be a narrowband matching network. We can increase the bandwidth by cascading two or more quarter-wave transformers to achieve a broadband matching network. To analytically design a broadband matching network with N quarter-wave transformers first use Equation (6-32) in an Equation Editor to calculate the characteristic impedance of each quarter-wave transformer then cascade all the sections into one matching network. Let Rs and R_L be the source and load impedances to be matched by the N quarter-wave transformation network. The characteristic impedance of each section can be calculated from Equation (6-32).

$$Z_n = R_S(r)^{(2n-1)/2N} \quad n = 1, 2, \ldots, N \qquad (6\text{-}32)$$

Where $r = R_L/R_S$ is the normalized load resistance and N is the number of quarter-wave transformers.

Example, 6.9 Design a three-section quarter-wave transformer network to match a load resistance $R_L = 2\ \Omega$ to a resistive source $R_S = 50\ \Omega$ at 100 MHz. Calculate the characteristic impedance of the quarter-wave transformers, the Q factor and the fractional bandwidths of the matching network at 3 and 20 dB return loss. Display the simulated response and compare the measurements with the measurements of single quarter-wave transformer matching network.

Solution: Using Equation (6-32) the characteristic impedances of the 3-section quarter-wave transformers are:

$$Z1 = R_S(r)^{1/2N} = 50(0.04)^{1/6} = 29.24$$

$$Z2 = R_S(r)^{3/2N} = 50(0.04)^{3/6} = 10.00$$

$$Z3 = R_S(r)^{5/2N} = 50(0.04)^{5/6} = 3.42$$

The intermediate resistors, R_n, can be obtained from Equation (6-33).

$$R_n = R_S(r)^{\frac{n}{N}} \quad n = 1, 2, 3,\ldots, N\text{-}1 \qquad (6\text{-}33)$$

Therefore, the two intermediate resistors are:

$$R_1 = 50(0.04)^{\frac{1}{3}} = 17.1 \ \Omega$$
$$R_2 = 50(0.04)^{\frac{2}{3}} = 5.848 \ \Omega$$

The schematic of the three-section quarter-wave matching network is shown in Figure 6-24.

1. Enter design parameters and normalize the load impedance

RS = 50; RL = 2; f = 100e6; r = RL/RS

N = 3

3. Calculate characteristic impedances and intermediate impedances

Z1=RS*r^(1/(2*N))

$$Z2=RS*r^{(3/(2*N))}$$

$$Z3=RS*r^{(5/(2*N))}$$

$$R1=RS*r^{(1/N)}$$

$$R2=RS*r^{(2/N)}$$

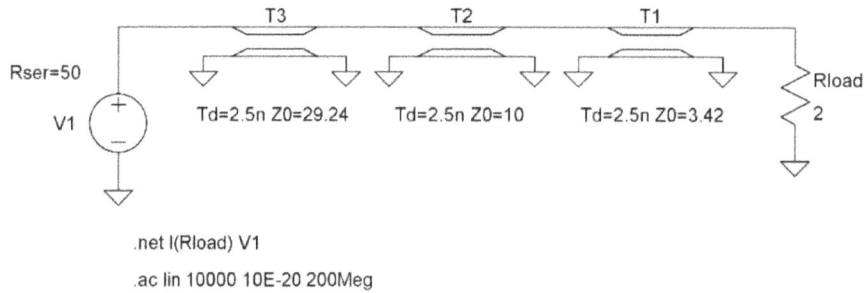

Figure 6-24 Three-section quarter-wave matching network

The simulated response of the matching network is shown in Figure 6-25.

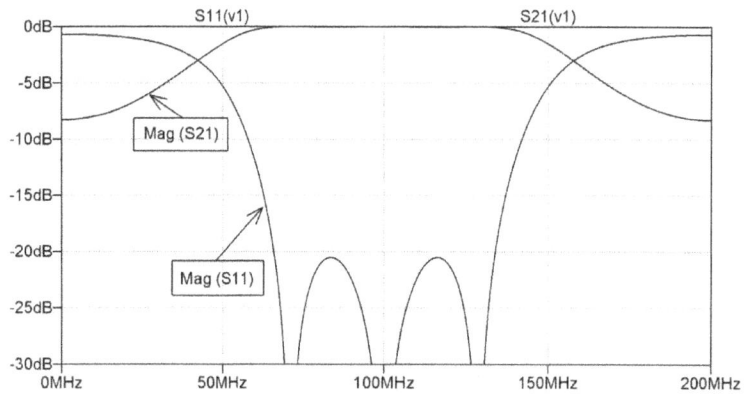

Figure 6-25 Response of the cascaded matching network R

The measured Q factor for the matching network is,

$$Q = \frac{1}{1.408} = 0.71$$

The wide bandwidth and lower Q factor is an indication that by adding two more quarter-wave sections the Q factor has reduced to less than one half and the 3-dB bandwidth has more than doubled compared to a single quarter-wave matching network.

Example 6.10: Design a broadband quarter-wave network to match a complex load, $Z_L = 10 - j5$ Ω to 50 Ω source impedance at 100 MHz. Calculate the bandwidth at 20 dB return loss and compare with the bandwidth of a single-stub matching network.

Solution: To design a broadband matching network between a resistive source and complex load impedance; first design a 5 section matching network between the real source and the real part of the load impedance. Then replace the quarter-wave transformer adjacent to the load with a single-stub that matches the complex load to the real resistor. Calculation of the broadband matching network between R_L and R_S is shown in the MATLAB script as follows.

1. Enter design parameters and normalize the load impedance

RS = 50; RL = 10; f = 100e6

r = RL/RS

N=5

2. Calculate characteristic impedances and intermediate resistors

Z1=RS*r^(1/(2*N))

Z2=RS*r^(3/(2*N))

Z3=RS*r^(5/(2*N))

Z4=RS*r^(7/(2*N))

Z5=RS*r^(9/(2*N))

R1=RS*r^(1/N)

R2=RS*r^(2/N)

R3=RS*r^(3/N)

R4=RS*r^(4/N)

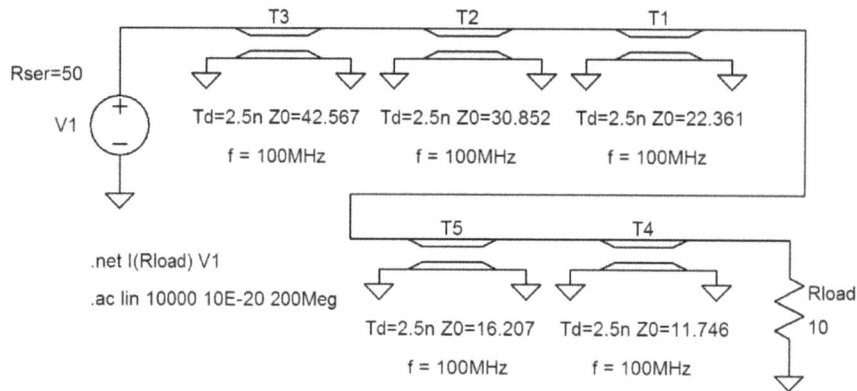

Figure 6-26 Five-section quarter-wave transformer matching network

Next replace the quarter-wave section adjacent to the load with a single-stub matching network. The design of the single-stub matching network is shown in the following Equation Editor. Note that the source resistor, R4 = 13.797 Ω, was calculated in the Equations. The single-stub matching network is shown in Figure 6-27.

1. Enter design parameters and normalize the load impedance

RS = 13.797; RL = 10; XL = -5; f = 100e6; r = RL/RS; x = XL/RS

2. Calculate t1 and t2, d1 and d2, B1 and B2, so1 and so2,

t1 = (x + sqrt(r*(r^2 + x^2 - 2* r+ 1)))/(r - 1)

t2 = (x - sqrt(r*(r^2 + x^2-2*r + 1)))/(r - 1)

d1 = 360*(pi + atan (t1))/(2*pi)

d2 = 360*(atan (t2))/(2*pi)

B1 = (x*t1^2 + (r^2 + x^2 - 1)*t1 - x)/(RS*(r^2 + x^2 + t1^2 + 2*x*t1))

B2 = (x*t2^2 + (r^2 + x^2 - 1)*t2 - x)/(RS*(r^2 + x^2 + t2^2 +2 *x*t2))

so1 = 360*(pi - atan(RS*B1))/(2*pi)

so2 = -360*atan(RS*B2)/(2*pi)

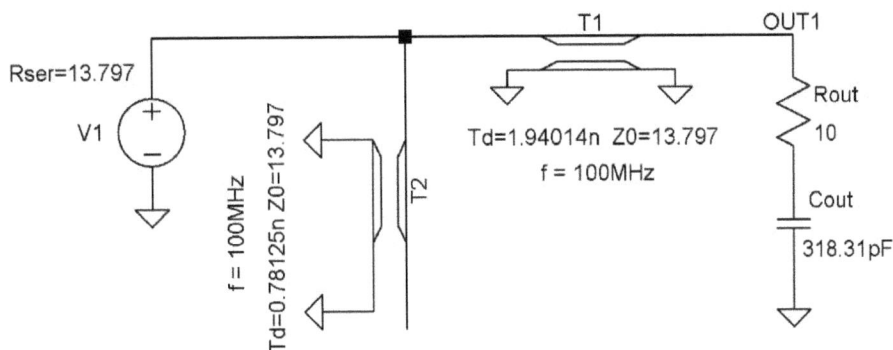

Figure 6-27 Schematic of the single-stub matching network

2. Now cascade the line and stub with four quarter-wave networks to form the final design of the matching network in Figure 6-28.

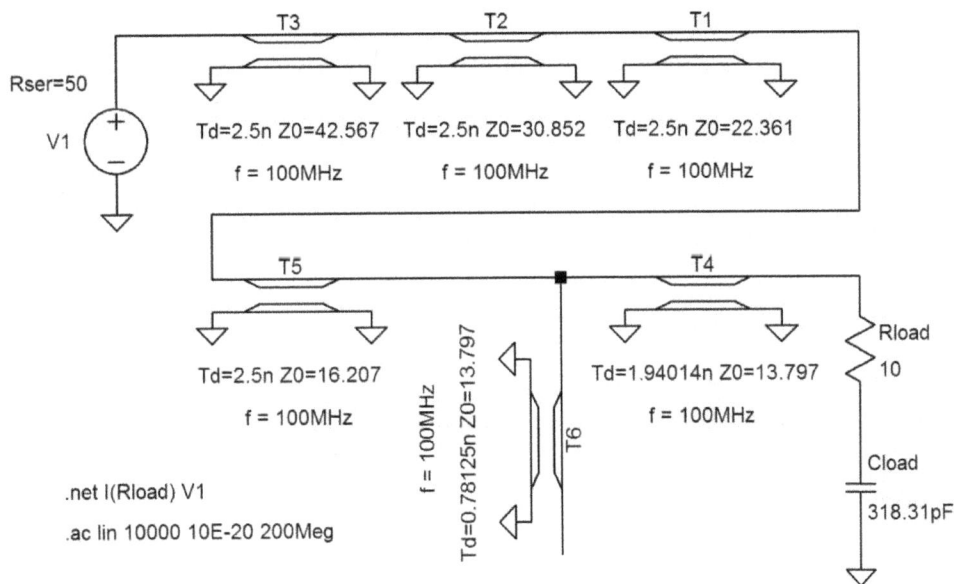

Figure 6-28 Schematic of the broadband matching network

The simulated response of the matching network is shown in Figure 6-29.

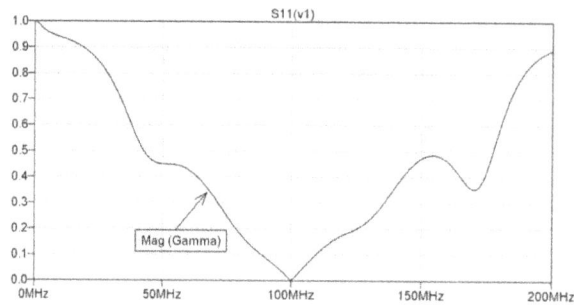

Figure 6-29 Response of the broadband matching network

Example 6.11: Design a broadband network to match a complex load, $Z_L = 150 - j30$ Ohm to 50 Ω source impedance at 100 MHz. Display the simulated response and measure the bandwidth at 20 dB return loss. Compare the results with the singe-stub matching network.

Solution: Use the same method as in Example 6.10 to design the broadband matching network. The design of the matching network with five quarter-wave sections is shown in Figure 6-30.

1. Enter design parameters and normalize the load impedance

RS=50; RL=150; f=100e6; r=RL/RS; N=5

2. Calculate characteristic impedances and intermediate resistors

$Z1 = RS*r\hat{}(1/(2*N))$

$Z2 = RS*r\hat{}(3/(2*N))$

$Z3 = RS*r\hat{}(5/(2*N))$

$Z4 = RS*r\hat{}(7/(2*N))$

$Z5 = RS*r\hat{}(9/(2*N))$

$R1 = RS*r\hat{}(1/N)$

$R2 = RS*r\hat{}(2/N)$

$R3 = RS*r\hat{}(3/N)$

R4 = RS*r^(4/N)

The schematic of the matching network is shown in Figure 6-30. The simulated response of the matching network is shown in Figure 6-31.

Next replace the quarter-wave transformer adjacent to the load with a single-stub matching network. The design of the single-stub matching network is shown in the following Equation Editor. Note that the source resistor, R4 = 13.797 Ω, was calculated.

1. Enter design parameters and normalize the load impedance

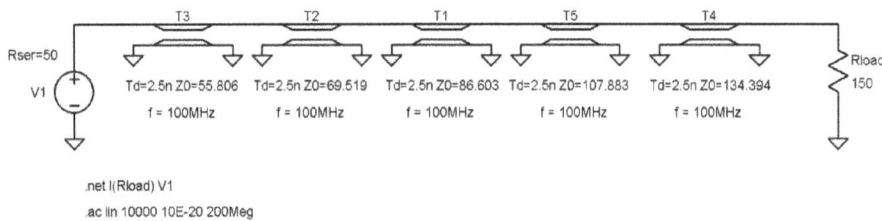

Figure 6-30 Five-section quarter-wave matching network

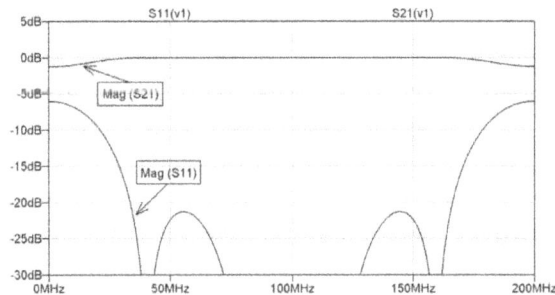

Figure 6-31 Response of the matching

RS=120.411; RL=150; XL=-30; f=100e6; r=RL/RS; x=XL/RS

2. Calculate t1 and t2, d1 and d2, B1 and B2, so1 and so2,

$t1 = (x + \mathrm{sqrt}\ (r*(r^2 + x^2 - 2*r + 1)))/(r - 1)$

$t2 = (x - \mathrm{sqrt}\ (r*(r^2 + x^2 - 2*r + 1)))/(r - 1)$

$d1 = 360*(\mathrm{atan}\ (t1))/(2*pi)$

$d2 = 360*(pi + atan\ (t2))/(2*pi)$

$B1 = (x*t1^2 + (r^2 + x^2 - 1)*t1 - x)/(RS*(r^2 + x^2 + t1^2 + 2*x*t1))$

$B2 = (x*t2^2 + (r^2 + x^2 - 1)*t2 - x)/(RS*(r^2 + x^2 + t2^2 + 2*x*t2))$

$so1d = 360*(pi-atan\ (RS*B1))/(2*pi)$

$so2d = -360*atan\ (RS*B2)/(2*pi)$

The single-stub matching network is shown in Figure 6-32.

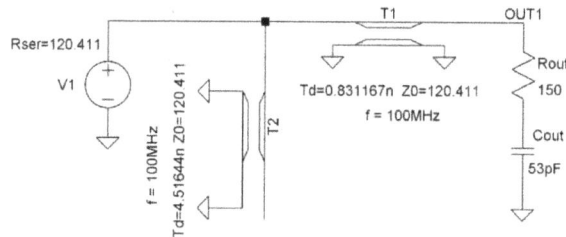

Figure 6-32 Schematic of the single-stub matching network

Now cascade the two matching networks to form the final design of broadband matching network, as shown in Figure 6-33.

The simulated response of the quarter-wave matching network is shown in Figure 6-34.

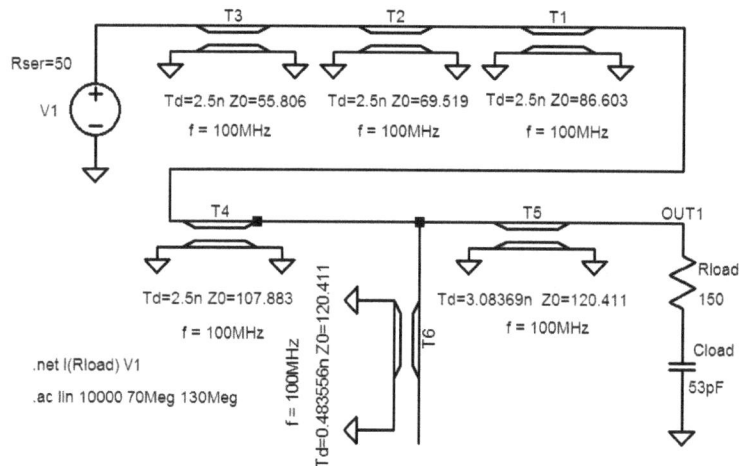

Figure 6-33 Schematic of the broadband matching network

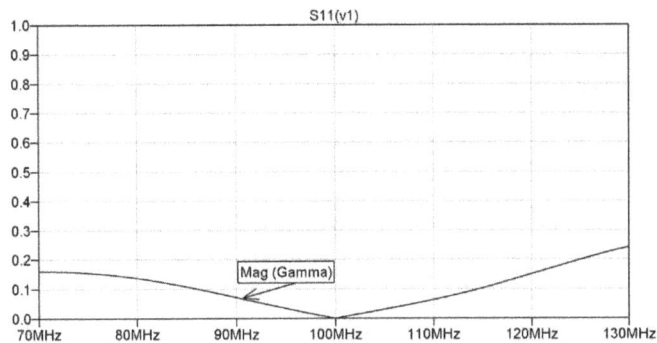

Figure 6-34 Response of the broadband matching network

Figure 6-34 shows that the bandwidths of the broadband matching network at 20 dB return loss are:

$$BW_{20dB} = 113.8 - 86.0 = 27.8 \text{ MHz}$$
$$FBW_{20dB} = \frac{27.8 x 100}{\sqrt{(113.8).(86.0)}} = 28.1 \text{ \%}$$

The same numbers for the narrowband single-stub matching network are:

$$BW_{20dB} = 102.8 - 97.1 = 5.7 \text{ MHz}$$
$$FBW_{20dB} = \frac{5.7 x 100}{\sqrt{(102.8).(97.1)}} = 5.7 \text{ \%}$$

Notice that the fractional bandwidth at 20 dB return loss is about 28.1% as opposed to only 5.7% for the narrowband matching network. This is an indication that, by adding four quarter-wave sections to the single-stub matching network, we have increased the matching bandwidth at 20 dB return loss nearly five times over a single-stub matching network.

6.15 Distributed Power Dividers

Power dividers are widely used in distributing power and exciting radiating elements in an array antenna, or as power dividers and power combiners in balanced power amplifiers. [7] They are considered as three-port networks. Unlike resistive power dividers, where extremely wide bandwidth can be achieved at the price of power loss in the resistive divider network, lossless power dividers can me made using distributing networks consisting transmission lines. For an equal-split resistive power divider, the scattering matrix are given by [5]

$$[S] = \begin{bmatrix} S_{11} & S_{12} & S_{13} \\ S_{21} & S_{22} & S_{23} \\ S_{31} & S_{32} & S_{33} \end{bmatrix} = \begin{bmatrix} 0 & 0.5 & 0.5 \\ 0.5 & 0 & 0.5 \\ 0.5 & 0.5 & 0 \end{bmatrix} \tag{6-34}$$

Examination of the scattering matrix above reveals that all ports are matched as $S_{11} = S_{22} = S_{33} = 0$. However, $S_{21} = S_{31} = 0.5$ or $20*\log(0.5) = -6$ dB. Therefore, the power level at ports 2 and 3 (although equal) are 6 dB lower than the input power at port 1. This is attributed to the loss in the resistive network. The isolation between ports 2 and 3 can be determined from S_{23} or S_{32}, indicating poor isolation of only 6 dB ($20*\log(0.5) = -6$ dB).

Figure 6-35 shows a lossless three-port junction used as a power divider [7] We can divide the input power P_1 such that power at ports 2 and 3 are fractions of P_1, namely $P_2 = \alpha P_1$ and $P_3 = (1 - \alpha) P_1$. To have a matched condition at port 1, we require that $Z_1 = Z_2 \| Z_3 = Z_2 Z_3 /(Z_2 + Z_3)$ or $Y_1 = Y_2 + Y_3$. Under matched condition, input power is equal to the incident power at port 1, which is related to the incident voltage and line impedance at port 1

$$P_1 = \frac{1}{2}\frac{|V_1^+|^2}{Z_1} = \frac{1}{2} Y_1 |V_1^+|^2 = \frac{1}{2} Y_2 |V_1^+|^2 + \frac{1}{2} Y_3 |V_1^+|^2 \tag{6-35}$$

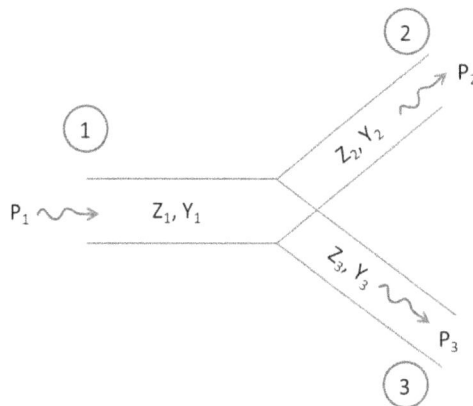

Figure 6-35 A lossless three-port junction used as a power divider

To obtain the desired power division at port 2 and 3, we then require that

$$\frac{Y_2}{Y_3} = \frac{P_2}{P_3} = \frac{\alpha\, P_1}{(1-\alpha)P_1} = \frac{\alpha}{(1-\alpha)} \tag{6-36}$$

Suppose that we want to split the input power such that $P_2 = P_1/4$ and $P_3 = P_1$ (3/4), that is $\alpha = \frac{1}{4}$. Therefore, $Y_2/Y_3 = 1/3$ or $Y_3 = 3Y_2$. As a result, $Z_3 = (1/3)\, Z_2$. Furthermore, $Y_1 = Y_2 + Y_3 = Y_2 + 3Y_2 = 4Y_2$. So, $Z_1 = (1/4)\, Z_2$, or $Z_2 = 4Z_1$. Finally, $Z_3 = (1/3)\, Z_2 = (4/3)\, Z_1$.

Example 6.12: Design a power divider with the following characteristics: $Z_1=50$ ohms and $\alpha = 0.25$, such $P_2 = 0.25P_1$ and $P_3=0.75P_1$. Simulate the circuit from 0.5 GHz to 1.5 GHz.

(a)   Show that the power at port 3 is $0.75P_1$ or -1.249 dB (=10*log(0.75)) relative to the input power.

(b)   Show that the power divider is matched at port 1.

Solution (a): The schematic diagram of the power divider is shown in Figure 6-36. The line impedance $Z_2 = 4Z_1 = 200\ \Omega$ and line impedance $Z_3 = (3/4)\, Z_1 = 66.67\ \Omega$.

Figure 3-37(a) depicts the power at port 3 relative to port 1. As expected, power at port 3 is about 1.25 dB below the power at port 1. **Note that S_{21} here refers to S_{31}.**

Solution (b): Figure 3-37(b) depicts the value of S_{11}, showing the matching condition, as expected.

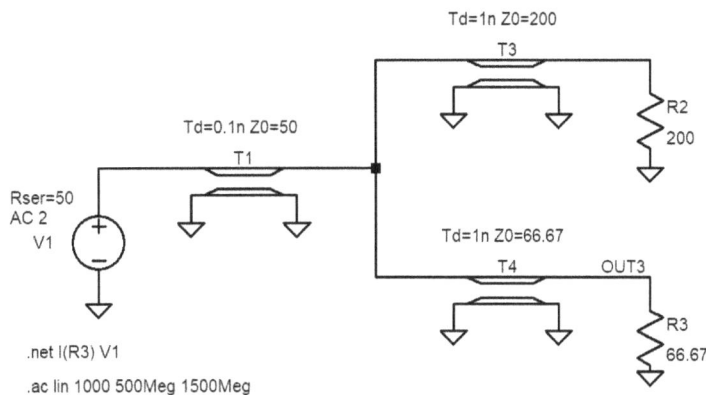

Figure 6-36 A lossless power divider with $Z_1 = 50\ \Omega$

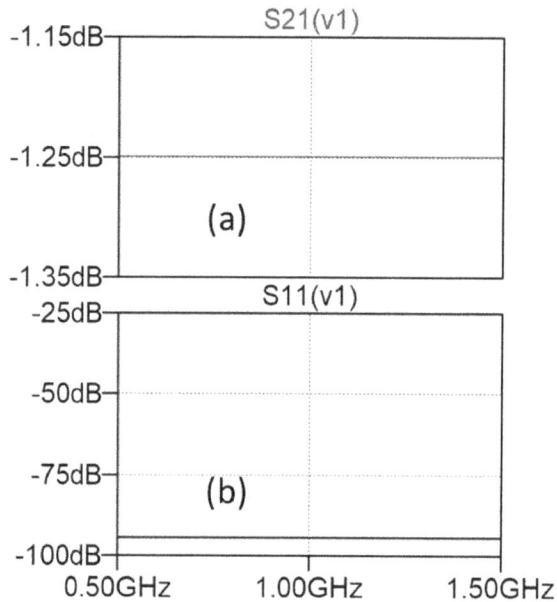

Figure 6-37 Simulation of Figure 3-36: (a) S_{31} and (b) S_{11}

Although the above circuit is lossless and provides good input matching, it does not provide good output matching and good isolation between output ports, as investigated in Problem 6-7. The addition of a shunt susceptance at the junction in Figure 6-35 can remedy the above issues while still providing good match at the input.

The development of an N-way power divider that splits the input power into N ports while providing good isolation between outputs is attributed to Wilkinson. Wilkinson used resistors between the output ports to achieve this isolation. The schematic diagram of a three-port Wilkinson power divider is shown in Figure 6-38. The goal is to split the input power between port 2 and 3 such that $P_3 = K^2 P_2$ while maintain input / output matching and achieving isolation between output ports. When ports 2 and 3 are terminated in matched loads, there is no current flow in resistor R. In that case, the output voltage $V2^-$ at port 2 must equal to the output voltage $V3^-$ at port 3. Therefore, to achieve desired power ratio, we require that that $K^2 |V2^-|^2/Z_{L2} = |V3^-|^2/Z_{L3}$, and as a result $Z_{L2} = K^2 Z_{L3}$. Following the same analysis as in [7], we can derive below equations

$$Y_{in} = Y_o = \frac{Z_{L2}}{Z_2^2} + \frac{Z_{L3}}{Z_3^2} \tag{6-37}$$

$$(K^2 Z_3^2 + Z_2^2) Z_{L3} = \frac{Z_2^2 Z_3^2}{Z_o} \tag{6-38}$$

$$Z_2 = K^2 Z_3 \tag{6-39}$$

$$Y_2 = \frac{Y_o}{1 + K^2} \tag{6-40}$$

To determine the relation between R and various line impedances, port 1 can be terminated in a matched load resulting in a two-port network. To decouple ports 2 and 3, G (=1/R) must then be equal to $-Y_{23}$ resulting in

$$R = \frac{Z_2 Z_3}{Z_o} \tag{6-41}$$

And finally, Z_2, Z_3, Z_{L2}, and Z_{L3} can be expressed in terms of K and R.

$$Z_2 = K \sqrt{R Z_o} \tag{6-42}$$

$$Z_3 = \frac{1}{K} \sqrt{R Z_o} \tag{6-43}$$

$$Z_{L2} = \frac{K^2 R}{K^2 + 1} \tag{6-44}$$

$$Z_{L3} = \frac{R}{K^2 + 1} \tag{6-45}$$

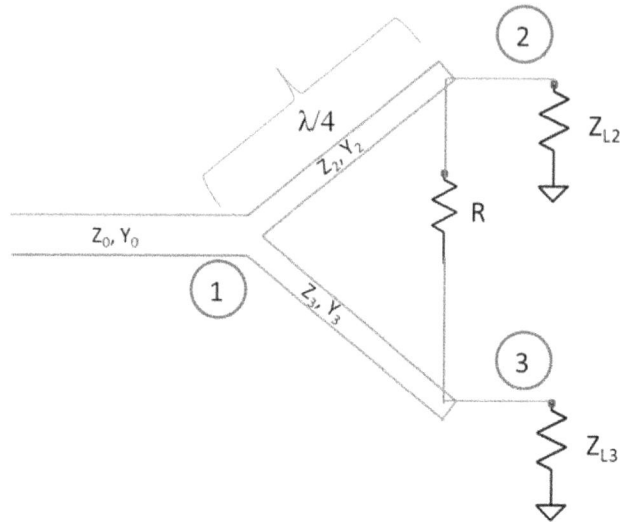

Figure 6-38 General schematic diagram of a three-port Wilkinson power divider

Example 6.13: Design a 3-dB Wilkinson power divider at 1 GHz such that Zo = 50 Ω. Show that the phase of S_{21} or S_{31} is equal to -90°. What is the isolation between ports 2 and 3.

Solution: To achieve 3-dB power split, $K^2 = 1$. Also, for $Z_{L2} = Z_{L3} = Z_o$, from Eq. 6-44 we obtain R = 2 Z_o = 100 Ω. The resulting schematic diagram is shown in Figure 6-39.

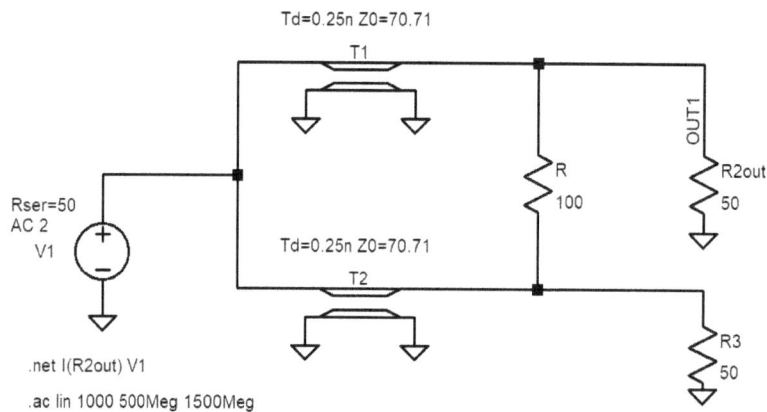

Figure 6-39 Schematic diagram of the 3-dB Wilkinson power divider at 1 GHz, configured to simulate S_{11} and S_{21}.

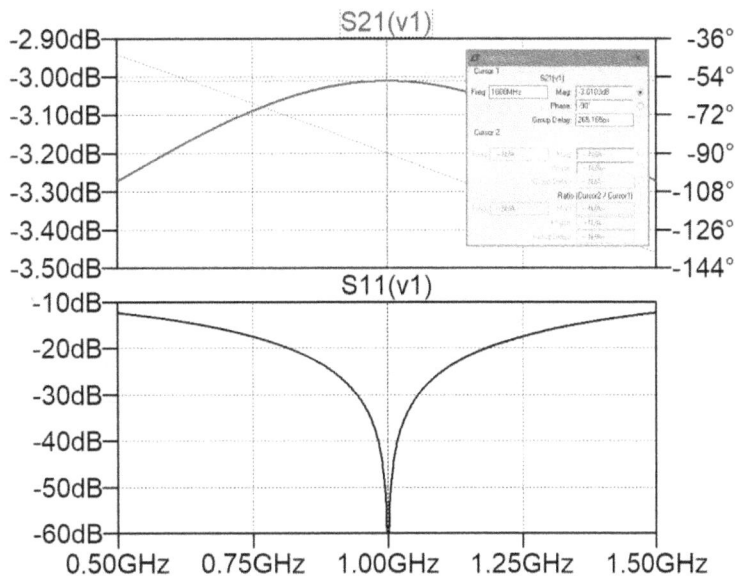

Figure 6-40 Simulated S_{21} and S_{11} for Figure 6-39.

Example 6.14: Determine the output return loss at port 3 and isolation between ports 2 and 3. What is the general form of the scattering matrix for this 3-dB Wilkinson power divider?

Solution: The schematic diagram for simulation of S33 and S32 is depicted in Figure 6-41. The simulated plots for S33 and S23 are shown in Figure 6-42. Note that since the excitation voltage is at port 3 and the output is at port 2, S11 represents S33, the output return loss at port 3. Similarly, S21 represents S23, the isolation between port 3 and port 2. We observe that S33 = 0 (a matched condition) and S23 = 0 (excellent isolation between ports 3 and 2).

Due to the symmetry in the circuit, we expect that $S_{22} = 0$ and $S_{32} = 0$. The scattering matrix for the 3-dB Wilkinson power divider using the results of Examples 6-13 and 6-14 is then given by

$$[S] = \begin{bmatrix} S_{11} & S_{12} & S_{13} \\ S_{21} & S_{22} & S_{23} \\ S_{31} & S_{32} & S_{33} \end{bmatrix} = \begin{bmatrix} 0 & -j/\sqrt{2} & -j/\sqrt{2} \\ -j/\sqrt{2} & 0 & 0 \\ -j/\sqrt{2} & 0 & 0 \end{bmatrix}$$

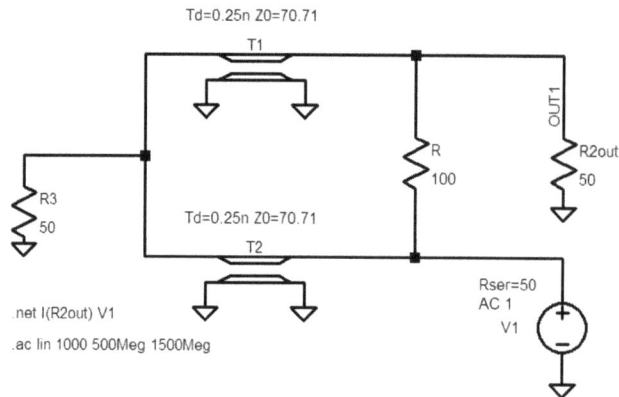

Figure 6-40 Schematic diagram of the 3-dB Wilkinson power divider at 1 GHz, configured to simulate S_{33} and S_{23}.

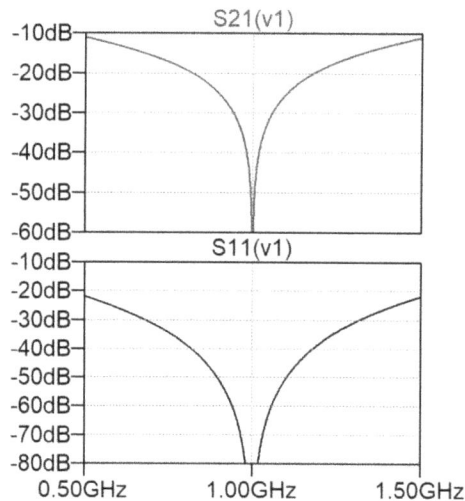

Figure 6-41 Simulated S_{33} and S23 from for Figure 6-40

Problems

6-1.   Design a quarter-wave transmission line to match a load resistance $R_L = 5$ Ω to a resistive source $R_S = 50$ Ω at 500 MHz. Calculate the characteristic impedance of the quarter-wave line, the Q factor and the fractional bandwidths of the matching network at 3 and 20 dB return loss. Display the simulated response and compare the calculations with measurements. For the quarter-wave matching network, calculate the fractional bandwidth and power loss from $\Gamma = 0.1$ to $\Gamma = 0.707$.

6-2.   Design a single-stub network to match a load resistance $Z_L = 5 - j5 \, \Omega$ to a resistive source $R_S = 75 \, \Omega$ at 850 MHz. Calculate the electrical lengths of the matching line and stub and the fractional bandwidths of the matching network at 3 and 20 dB return loss. Display the simulated response and compare the calculations with measurements.

6-3.   Design a single-stub network to match a load resistance $Z_L = 100 + j20 \, \Omega$ to a resistive source $R_S = 40 \, \Omega$ at 800 MHz. Calculate the electrical lengths of the matching line and stub and the fractional bandwidths of the matching network at 3 dB and 20 dB return loss. Display the simulated response and compare the calculations with measurements.

6-4.   Design a single-stub network to match a complex load $Z_L = 20 + j5 \, \Omega$ to a complex source $Z_S = 50 + j20 \, \Omega$ at 1000 MHz. Calculate the electrical lengths of the matching line and stub and the fractional bandwidths of the matching network at 3 dB and 20 dB return loss. Display the simulated response and verify the calculations with measurements.

6-5.   Design a three-section quarter-wave network to match a load resistance $R_L = 10 \, \Omega$ to a resistive source $R_S = 50 \, \Omega$ at 900 MHz. Calculate the characteristic impedance of the quarter-wave line, the Q factor and the fractional bandwidths of the matching network at 3 and 20 dB return loss. Display the simulated response and compare the measurements with the singe quarter-wave matching network.

6-6.   Design a broadband network to match a complex load, $Z_L = 25 + j15 \, \Omega$ to 75 Ω source impedance at 500 MHz. Measure the bandwidth at 20 dB return loss and compare it with the results of singe-stub matching network.

6-7.   For the schematic diagram of Example 6.12, determine S22, S33, and S32. Comment on poor output return loss and isolation between ports 2 and 3.

6-8.   Design a 3-dB Wilkinson power divider at 1 GHz given Zo = 75 Ω. Simulate and plot S_{11}, S_{22}, S_{33}, and S_{32}.

6-9.   Design a 3-dB Wilkinson power divider at 4 GHz given Zo = 50 Ω. Simulate and plot S_{11}, S_{22}, S_{33}, and S_{32}. Comment on the bandwidth of this power divider.

References and Further Readings

[1]   Ali A. Behagi, *RF and Microwave Circuit Design*, A Design Approach Using (**ADS**), Techno Search, Ladera Ranch, CA 2015

[2]   Guillermo Gonzales, *Microwave Transistor Amplifiers – Analysis and Design*, Second Edition, Prentice Hall Inc., Upper Saddle River, NJ.

[3]   Randy Rhea, *The Yin-Yang of Matching: Part 1 – Basic Matching Concepts*, High Frequency Electronics, March 2006

[4]   Steve C. Cripps, *RF Power Amplifiers for Wireless Communications*, Artech House Publishers, Norwood, MA. 1999.

[5]   David M. Pozar, *Microwave Engineering*, Third Edition, John Wiley & Sons, New York, 2005

[6]   R. Ludwig, P. Bretchko, *RF Circuit Design*, Theory and Applications, Prentice Hall, Upper Saddle River, NJ, 2000

[7]   Robert E. Collin, *Foundations for Microwave*, Second Edition, McGraw-Hill, Inc., New York, 1996.

Chapter 7

Single Stage Amplifier Design

7.1 Introduction

Modern amplifier design involves the use of high frequency transistors to use DC power to control and amplify RF energy. These transistors are fundamentally of the bipolar and field effect transistor (FET) variety. There are many subtypes of bipolar and FET devices that the reader is encouraged to explore. For the many types of transistors the impedance matching techniques covered in chapters 5 and 6 are very useful for the design and simulation of linear transistor amplifiers. There are many different ways in which transistor amplifiers are impedance matched depending on the function or purpose of the amplifier circuit. In this chapter we design two single stage amplifiers that require two different impedance matching techniques.

Maximum Gain Amplifier Design
Low Noise Figure Amplifier Design

Most amplifier designs actually involve selective mismatching of the transistor to its source and load impedance to accomplish its intended purpose. Only the maximum gain amplifier requires a conjugate matching network design. The specific matching techniques for each category of amplifiers are listed below:

1. Maximum Gain Amplifier

In a maximum gain amplifier the input and the output are simultaneously conjugate matched to achieve maximum gain.

2. Low Noise Figure Amplifier

In a low noise amplifier the transistor's input is mismatched from 50 Ω to achieve a specific Noise Figure.

7.2 Maximum Gain Amplifier Design

This section covers the design of the maximum gain amplifier at 1 GHz using the Avago's AT-41486 transistor. AT-41486 is a general purpose NPN bipolar transistor that offers excellent high frequency performance. The AT-41486 is housed in a low cost surface mount .085" diameter plastic package. The AT-41486 bipolar transistor is fabricated using Avago's 10 GHz Self-Aligned-Transistor (SAT) process. The pin connection is shown in Figure 7-1.

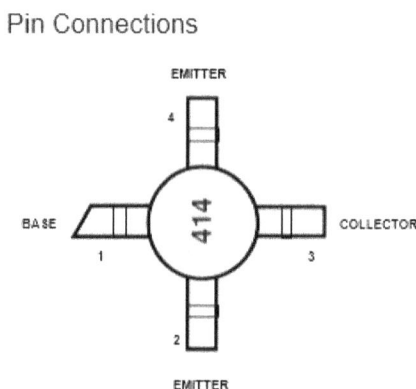

Figure 7-1 AT-41486 pin connections. (Courtesy of Data Sheet Lib)

The maximum gain amplifier is conjugately matched at the input and output resulting in very good return loss over a narrow bandwidth. A device can only be conjugately matched if it is unconditionally stable. A device is unconditionally stable at a given frequency if the real parts of its input and output impedance is greater than zero for all passive source and load impedance values. Otherwise, the device is potentially unstable. If the device is potentially unstable in the band of interest, a conjugate impedance match cannot be realized unless we add additional circuitry to stabilize the device.

Example 7-1: Set up the AT-41486 transistor for maximum gain amplifier with the following specifications.

(a) Create a symbol for the AT41486 transistor Spice model.

(b) Use the transistor spice model and bias the transistor at Vce = 8 V and Ic = 25 mA.

(c) Plot the DC-IV curves and determine the transistor operating point

Solution (a): We first create a symbol for the device in LTspice and save it with .asy extension. The following steps are used to generate the transistor symbol.

1) Start LTspice IV and select File > New Symbol.
2) Select Edit > Add Pin/Port > OK to add ports. Repeat the step for a total of three ports.
3) Select Draw > Line to draw lines to connect the Ports.
4) Select Draw > Text to open the Edit Text on the Symbol window. Type NPN and press OK.
5) Repeat the above step to add Unnn, as shown in Figure 7-2.

Figure 7-2 Transistor symbol

6) Press Ctrl-A to open the Symbol Attribute Editor window, as show in Figure 7-3.
7) In the Symbol Type box select Cell. Type X for Prefix and NPN for Value. Type Bipolar NPN transistor for Description, as shown in Figure 7-4. Press OK.
8) Select File > Save as AT-41486.asy in your LTspice directory.

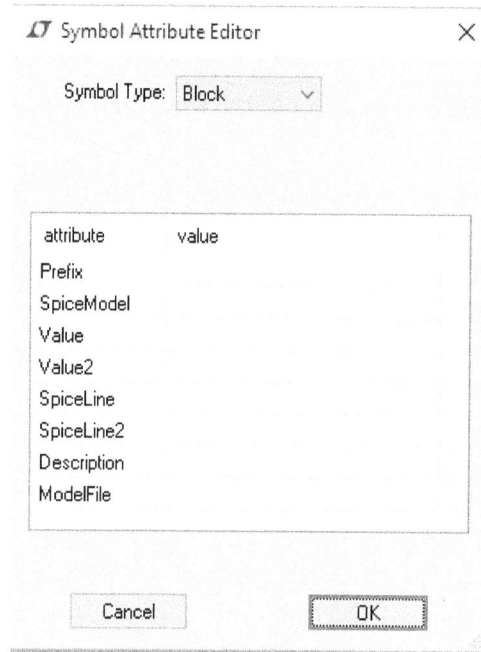

Figure 7-3 Symbol Attribute Editor

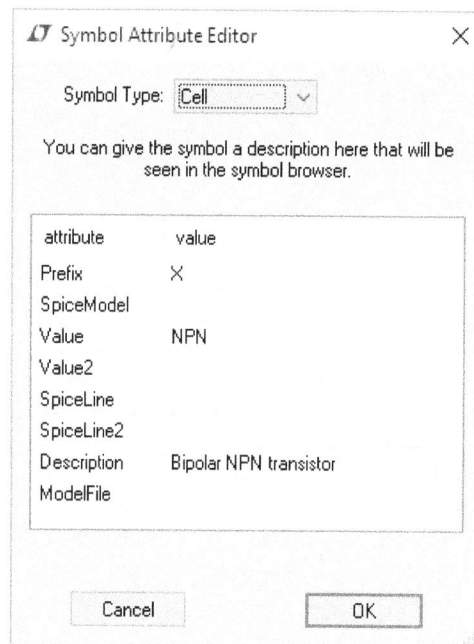

Figure 7-4 Transistor Attributes

Solution (b): Now that the AT-41486 symbol has been created we use the following steps to simulate the schematic and plot the DC-IV curves

1) Open a new schematic in LTspice by selecting File > New Schematic.
2) Select Edit > Component to open the Select Component Symbol window.
3) Select npn > OK to insert the Bipolar NPN transistor, as shown in Figure 7-5.

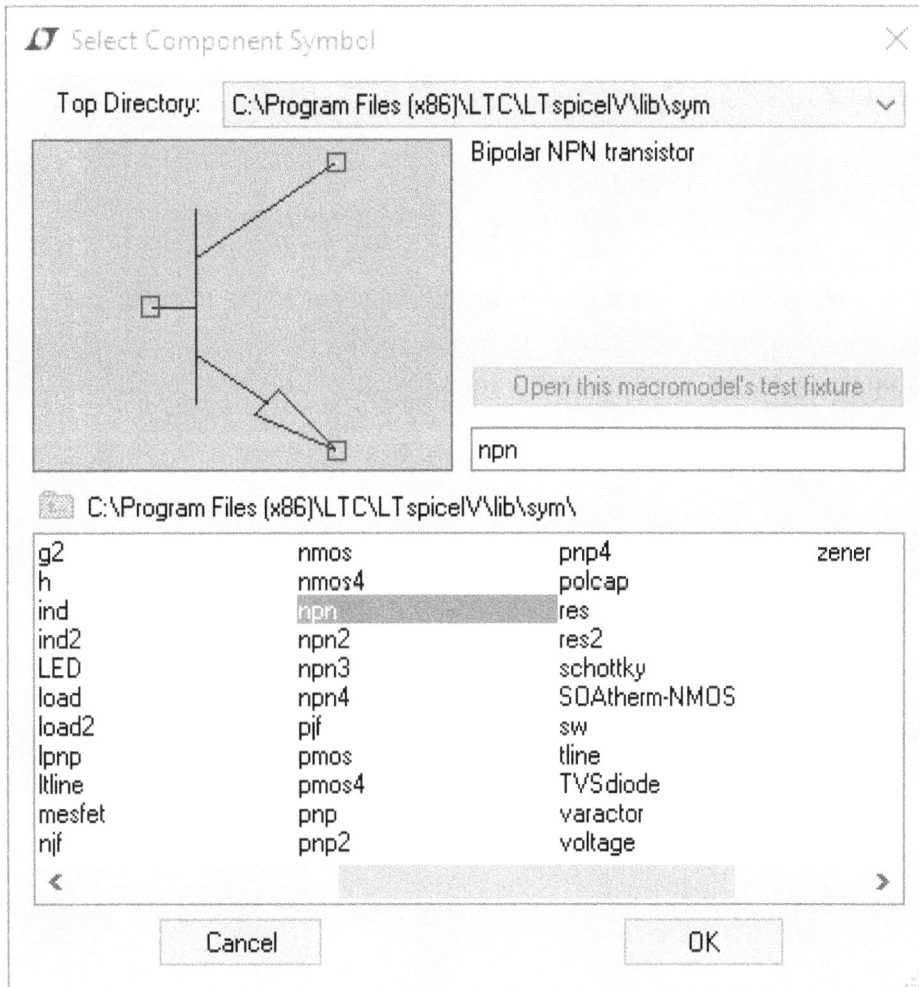

Figure 7-5 Selecting Bipolar NPN Transistor

4) Right click on Q1 and change to U1, then click OK.
5) Right click on NPN and change to AT41486, then click OK.

6) Select Edit > SPICE Directive to open the following Edit Text on the Schematic window.

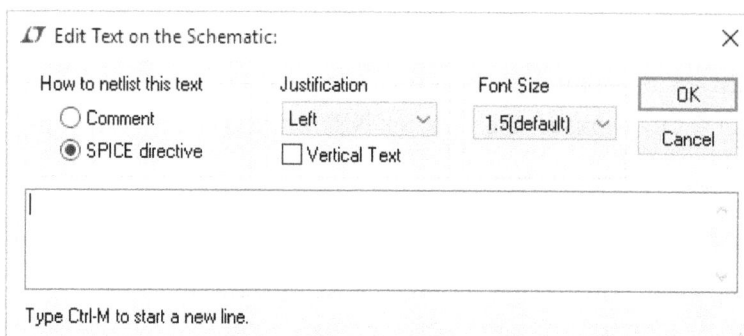

Figure 7-6 Edit Text on the Schematic window

7) Type in the following spice model for the transistor in the window provided. If you already have the transistor Spice model saved somewhere in your computer, you may copy and paste the spice model inside the window.

```
*SPICE model for AT-41486
*
.SUBCKT AT41486 60    20    40
LL1    20    25    .55NH
* PI NETWORK TO SIMULATE TRANSMISSION LINE T1
C1T1   25    0     .06PF
LT1    25    30    .48NH
C2T1   30    0     .06PF
*
LLB    30    35    .55NH
CCEB   30    50    .02PF
LL2    40    45    .06NH
* PI NETWORK TO SIMULATE TRANSMISSION LINE T2
C1T2   45    0     .04PF
LT2    45    50    .08NH
C2T2   50    0     .04PF
*
LLE    50    55    .1NH
```

CCEC   50   70   .03PF
LL3   60   65   .25NH
* PI NETWORK TO SIMULATE TRANSMISSION LINE T3
C1T3   65   0   .04PF
LT3   65   70   .30NH
C2T3   70   0   .04PF
*
CCBC   30   70   .03PF
* CALL DIE MODEL
XDIE   70   35   55   AT414
.ENDS

* DIE MODEL (excludes bond wires)
.SUBCKT AT414   75   20   85
CCB   20   60   .032PF
DCD1   20   60   DMOD   572
RRB1   20   25   1.07
DCD2   25   60   DMOD   680
RRB2   25   30   3.2
DCD3   30   60   DMOD   340
RRB3   30   35   2.7
RRC   60   75   5
RRE   80   85   .24
CCE   60   85   .032PF
QINT   60   35   80   QDIS   420
.ENDS
*
.MODEL DMOD   D(IS=1E-25, CJO=2.45E-16, VJ=.76, M=.53, BV=45,
IBV=1E-9)
.MODEL QDIS   NPN (BF=100, BR=2.5, IS=1.65E-18, VA=20, TF=12PS,
+   CJE=2.4E-15, VJE=1.01, MJE=0.6, PTF=25, XTB=1.818,
+   VTF=6, ITF=3E-4, IKF=1.3E-4, XTF=4, NF=1.03, ISE=5E-15,
+   NE=2.5)

To attach DC voltage sources V1 and V2 to the base and collector of the transistor do the following:

1) Select Edit > Component > voltage > OK. The first voltage source is V1.
2) Repeat to insert the second voltage source V2.
3) Wire up the schematic as shown in Figure 7-7.
4) To simulate the schematic select Simulate > Edit Simulation Cmd to open the Edit Simulation Command.
5) Select DC sweep and then select V1 for the 1st Source and Linear for Type of Sweep, if already is not selected.
6) Type in 0.78 for the Start Value, 0.85 for the Stop Value, and 0.012 for the Increment, as shown in Figure 7-8.
7) Select V2 for the 2nd Source and Type in 0 for the Start Vale, 10 for the Stop Value, and 0.5 for the Increment, as shown in Figure 7-9.

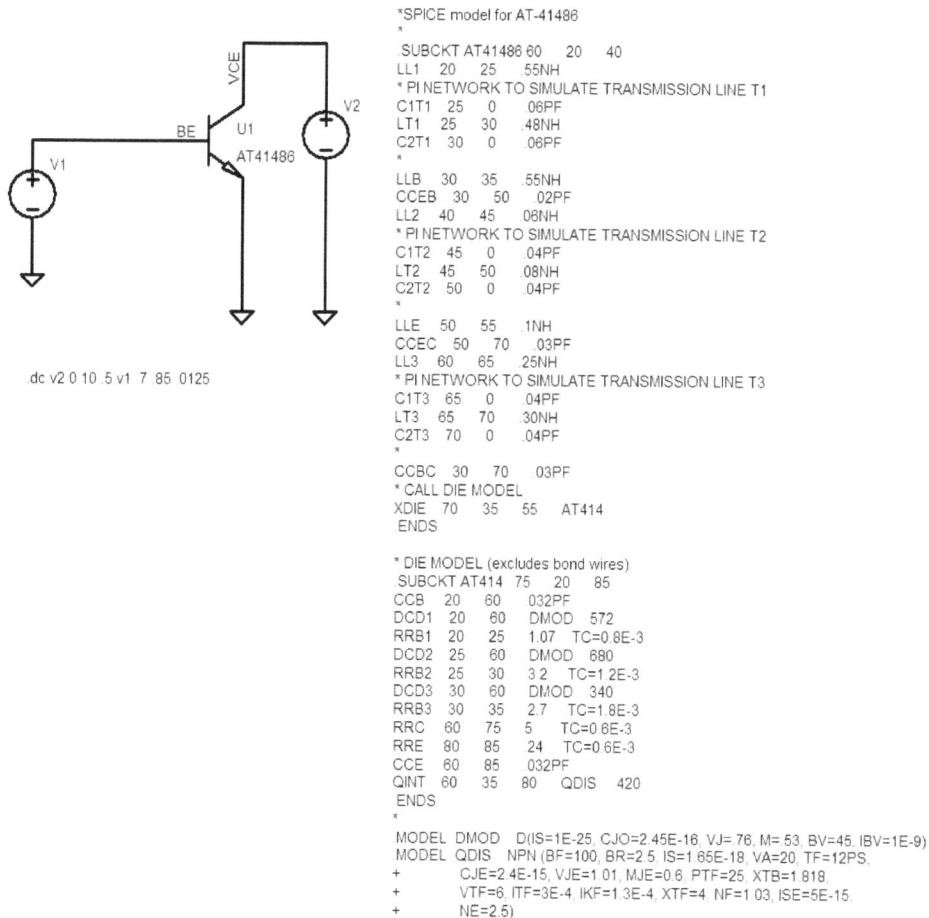

```
*SPICE model for AT-41486
*
.SUBCKT AT41486 60   20    40
LL1    20   25    .55NH
* PI NETWORK TO SIMULATE TRANSMISSION LINE T1
C1T1   25   0     .06PF
LT1    25   30    .48NH
C2T1   30   0     .06PF
*
LLB    30   35    .55NH
CCEB   30   50    .02PF
LL2    40   45    .06NH
* PI NETWORK TO SIMULATE TRANSMISSION LINE T2
C1T2   45   0     .04PF
LT2    45   50    .08NH
C2T2   50   0     .04PF
*
LLE    50   55    .1NH
CCEC   50   70    .03PF
LL3    60   65    .25NH
* PI NETWORK TO SIMULATE TRANSMISSION LINE T3
C1T3   65   0     .04PF
LT3    65   70    .30NH
C2T3   70   0     .04PF
*
CCBC   30   70    .03PF
* CALL DIE MODEL
XDIE   70   35    55    AT414
.ENDS

* DIE MODEL (excludes bond wires)
.SUBCKT AT414  75   20    85
CCB    20   60    .032PF
DCD1   20   60    DMOD   572
RRB1   20   25    1.07  TC=0.8E-3
DCD2   25   60    DMOD   680
RRB2   25   30    3.2   TC=1.2E-3
DCD3   30   60    DMOD   340
RRB3   30   35    2.7   TC=1.8E-3
RRC    60   75    5     TC=0.6E-3
RRE    80   85    24    TC=0.6E-3
CCE    60   85    .032PF
QINT   60   35    80    QDIS   420
.ENDS
*
.MODEL DMOD   D(IS=1E-25, CJO=2.45E-16, VJ=.76, M=.53, BV=45, IBV=1E-9)
.MODEL QDIS   NPN (BF=100, BR=2.5, IS=1.65E-18, VA=20, TF=12PS,
+       CJE=2.4E-15, VJE=1.01, MJE=0.6, PTF=25, XTB=1.818,
+       VTF=6, ITF=3E-4, IKF=1.3E-4, XTF=4, NF=1.03, ISE=5E-15,
+       NE=2.5)
```

.dc v2 0 10 .5 v1 .7 .85 .0125

Figure 7-7: AT-41486 transistor Spice model with the schematic

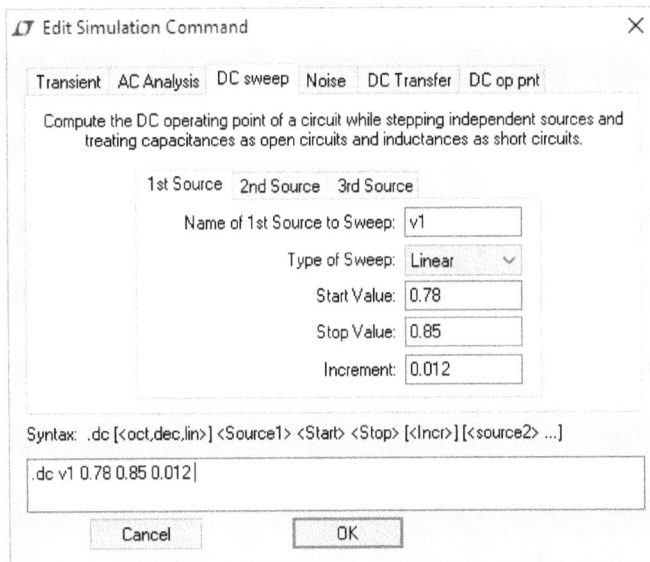

Figure 7-8: Selecting V1 values

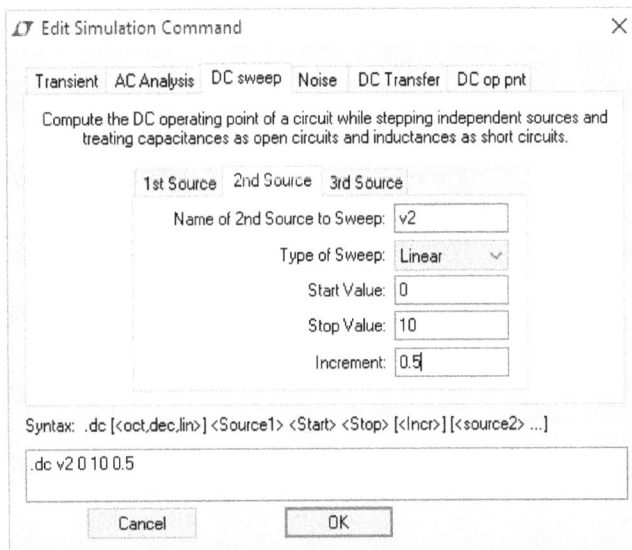

Figure 7-9: Selecting V2 values

8) To simulate the schematic select Simulate > Run to open the Plot window.
9) Right click on the Plot window and select Visible Traces to open the Select Visible Waveforms window.
10) Select Vbe and ix[U1:C] as shown in Figure 7-10.

11) Press OK to display V(be) and the collector current in Figure 7-11.

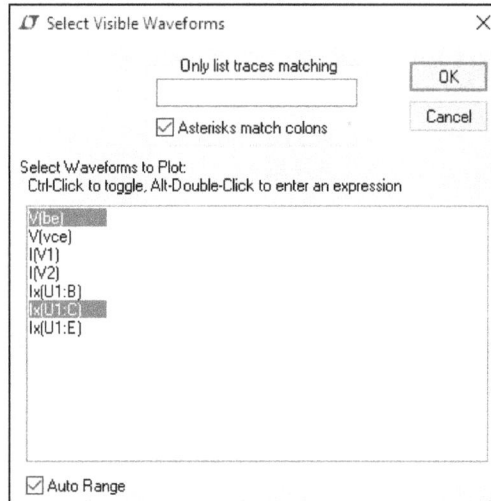

Figure 7-10: Selection of V(be) and collector current

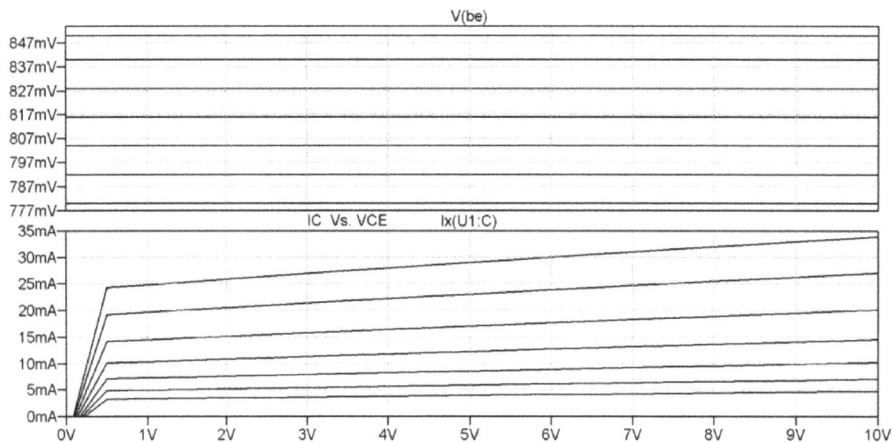

Figure 7-11: DC-IV plots of AT-41486 transistor

From Figure 7.11 it can be seen that a collector current of 25 mA at Vce of 8 VDC is approximately achieved with a Vbe of about 0.839 V.

Solution (c): We can obtain more accurate operating bias conditions by adjusting the Vbe increments until the desired Ic value of 25 mA is reached. To achieve this goal we use the following steps.

1) From the schematic diagram select Simulate > Edit Simulation Cmd to open the Edit Simulation Command. Select DC op pnt and then OK to place .op syntax on the schematic. Right click on V1 and enter 0.8393 as DC value for V1 (Vbe).

2) Right click on V2 and enter 8 as DC value for V2 (Vce).

3) Select Simulate > Run to simulate the schematic.

4) A table of all Operating Point currents and voltages will appear. The value of Ic (i.e., I(V2)) is about 24.87 mA.

5) Continue to incrementally increase Vbe value until Ic is 25 mA. It can be seen that collector current of 25 mA is achieved with Vbe of 0.8393 V, as shown in Table 7-1.

```
            --- Operating Point ---

V(be):          0.8393          voltage
V(vce):         8               voltage
I(V2):          -0.0250326      device_current
I(V1):          -0.000248777    device_current
Ix(u1:60):      0.0250326       subckt_current
Ix(u1:20):      0.000248777     subckt_current
Ix(u1:40):      -0.0252814      subckt_current
```

Table 7-1 DC operating point for AT41486

Table 7-1 shows that the Vbe value of 0.8393 Volt generates a collector current value of 25 mA at Vce of 8 Volt. Ix denotes intrinsic currents at various nodes.

7.3 Transistor Stability Considerations

Most Microwave transistors are potentially unstable at some frequency. This does not necessarily make them undesirable for use as an amplifier. A potentially unstable transistor does not mean that it will definitely oscillate in a circuit. Referring to the device in Figure 7-12 there may exist some combination of input or output reflection coefficient, Γ_S or Γ_L, that, when presented to the transistor, may indeed make the device oscillate. An oscillation condition is indicated by $|\Gamma_{IN}| > 1$ or $|\Gamma_{OUT}| > 1$. This can also be viewed as a positive value for the Return Loss, S_{11} or S_{22}, in dB. This is referred to as negative resistance which is actually a design goal in the design of microwave oscillators. The conditions for unconditional stability for an RF transistor are given by equations (7-1) through (7-4) [2].

$$|\Gamma_S| < 1 \tag{7-1}$$

$$|\Gamma_L| < 1 \tag{7-2}$$

$$|\Gamma_{IN}| = \left| S_{11} + \frac{S_{12}S_{21}\Gamma_L}{1 - S_{22}\Gamma_L} \right| < 1 \tag{7-3}$$

$$|\Gamma_{OUT}| = \left| S_{22} + \frac{S_{12}S_{21}\Gamma_S}{1 - S_{11}\Gamma_S} \right| < 1 \tag{7-4}$$

Typically there exists a select set of reflection coefficient values for Γ_S and Γ_L that will give the device this negative resistance characteristic. One method of dealing with this problem is to select values for Γ_S and Γ_L that avoid these unstable regions of impedance. This technique is demonstrated in Section 7.4. Another technique is to introduce a network that forces the device to be unconditionally stable for any values of Γ_S and Γ_L. This technique is often preferable as it will be shown that a transistor can be made unconditionally stable over a very wide frequency range. This is important because even though a device may be stable in its desired frequency band, it may become unstable at some frequency outside of the band of operation.

A simultaneous numerical solution of Equations (7-1) through (7-4) for all values of Γ_S and Γ_L can prove that there are two conditions for unconditional stability. The stability factor, K, and stability measure, B1, are given by the following Equations.

$$K = \frac{(1 - |S_{11}|^2 - |S_{22}|^2 + |(S_{11}) \cdot (S_{22}) - (S_{12}) \cdot (S_{21})|^2)}{(2 \cdot |(S_{12}) \cdot (S_{21})|)} > 1 \tag{7-5}$$

$$B_1 = 1 + |S_{11}|^2 - |S_{22}|^2 - |(S_{11}) \cdot (S_{22}) - (S_{12}) \cdot (S_{21})|^2 > 0 \tag{7-6}$$

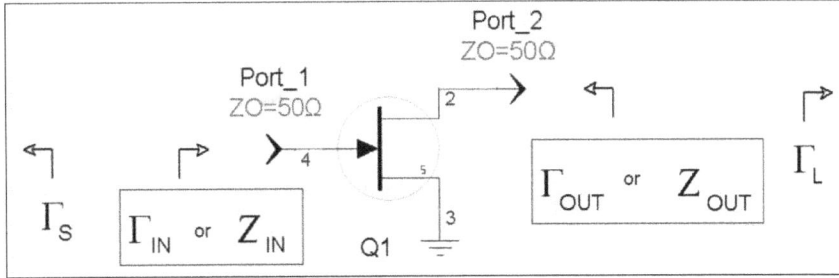

Figure 7-12 Transistor input and output reflection coefficients

When the transistor is unconditionally stable a simultaneous conjugate match can be defined. The conjugate match exists when $\Gamma_{IN} = \Gamma_S{}^*$ and $\Gamma_{OUT} = \Gamma_L{}^*$.

The simultaneous conjugate match reflection coefficients are often referred to as: Γ_{MS} and Γ_{ML}. The maximum gain that can be obtained with a simultaneous conjugate matched amplifier is determined from the transducer gain equation, G_{Tmax}.

$$G_{T\max} = \frac{\left(1 - |\Gamma_{MS}|^2\right)|S_{21}|^2\left(1 - |\Gamma_{ML}|^2\right)}{\left|\left(1 - S_{11}\Gamma_{MS}\right)\left(1 - S_{22}\Gamma_{ML}\right) - S_{12}S_{21}\Gamma_{ML}\Gamma_{MS}\right|^2} = \frac{|S_{21}|}{|S_{12}|}\left(K - \sqrt{K^2 - 1}\right) \quad (7\text{-}7)$$

When K is exactly equal to one, G_{Tmax} becomes the Maximum Stable Gain, G_{MSG}. G_{MSG} is the maximum value that G_{Tmax} can achieve.

$$G_{MSG} = \frac{|S_{21}|}{|S_{12}|} \quad\quad\quad (7\text{-}8)$$

7.4 Stabilizing the Device

As Equations (7-5) through (7-6) show, the stability factor K and the stability measure B_1 are functions of the transistor S parameters. The LTspice software can be used to plot the stability parameters K and B_1 as functions of frequency.

Example 7.2: (a) Measure and display the stability factor, K, and the stability measure, B_1, for the AT41486 transistor biased at 25 mA collector current and 8

V collector to base voltage. **(b)** Add stabilizing circuits to ensure that the transistor is unconditionally stable in the frequency range of interest.

Solution (a): Now that the transistor operating point has been established in Table 7-1, we need the transistor S parameters to check the stability of the transistor in the frequency range of interest. Once the S-parameters are determined, we can use Equations (7-5) through (7-7) in MATLAB script to plot the stability parameters K and B_1.

The linear S parameters of the AT-41486 in PDF format is given on the Avago website as follows.

AT-41486 Typical Scattering Parameters,
Common Emitter, Z_O = 50 Ω, T_A = 25°C, V_{CE} = 8 V, I_C = 25 mA

Freq. GHz	S_{11} Mag.	S_{11} Ang.	S_{21} dB	S_{21} Mag.	S_{21} Ang.	S_{12} dB	S_{12} Mag.	S_{12} Ang.	S_{22} Mag.	S_{22} Ang.
0.1	.50	-75	32.0	40.01	142	-41.3	.009	54	.85	-17
0.5	.55	-158	23.2	14.38	97	-34.1	.020	48	.51	-24
1.0	.57	177	17.5	7.50	78	-29.9	.032	61	.46	-24
1.5	.57	161	14.1	5.07	65	-27.3	.043	62	.44	-28
2.0	.59	148	11.5	3.75	53	-24.8	.058	59	.43	-35
2.5	.61	139	9.6	3.02	45	-22.9	.072	58	.40	-41
3.0	.65	128	8.0	2.52	34	-21.6	.083	57	.38	-49
3.5	.70	121	6.7	2.17	24	-20.1	.099	56	.36	-59
4.0	.74	113	5.7	1.92	14	-18.8	.115	52	.34	-72
4.5	.78	107	4.7	1.72	3	-17.6	.132	47	.32	-87
5.0	.78	102	3.7	1.53	-8	-16.6	.149	42	.31	-106
5.5	.78	96	2.7	1.36	-19	-15.4	.169	36	.31	-125
6.0	.76	91	1.6	1.21	-29	-14.5	.188	31	.33	-144

A model for this device is available in the DEVICE MODELS section.

Table 7-2 AT-41486 linear S parameters

The maximum gain of the device under simultaneous conjugate match at 1 GHz is GTmax = 21.61 which is equal 13.34 dB.

$$10(\log (21.61)) = 13.34 \text{ dB}$$

Since we are using the nonlinear spice model of the transistor in this example we need to generate the S-parameters and plot the stability parameters K and B_1.

1) Create a new schematic in LTspice and place the AT 41486 transistor on the schematic by selecting Edit > Component to open the Select Component Symbol window.

2) Select Edit > Component to open the Select Component Symbol window.

3) Select npn > OK to insert the Bipolar NPN transistor.

4) Attach an 8 VDC voltage source to the collector and 0.8393 VDC voltage source to the base.

5) Insert a new voltage source V3 from LTspice library in the schematic. Right click on the voltage source and select Advanced to open the Independent Voltage Source window.

6) Type in 1 for the AC Amplitude and 50 for the Series Resistance.

7) Attach a 50-Ohms load resistor. Right click on the load resistor designator and change R1 to Rout.

8) Insert 1 µF DC blocking capacitors and 1 µH AC blocking inductors as shown in the schematic.

9) Wire up the schematic as shown in Figure 7-13.

10) Select Edit > SPICE Directive to open the Edit Text on the Schematic window shown in Figure 7-6.

11) Copy and paste the transistor spice parameters.

12) Wire up the schematic as shown in Figure 5-13.

13) Select Simulate > Edit Simulation Cmd to open the Edit Simulation Command window. Click on AC Analysis tab. Select Decade for the Type of Sweep. Type in 1000 for the Number of points, 10MEG for Start Frequency and 10G for Stop Frequency, then press OK.

14) Select Edit > SPICE Directive (or press S on the keyboard) to open the SPICE Directive window. Type in .net I(Rout) V3 and press OK.

15) Select Simulate > Run to simulate the schematic. A blank waveform window opens.

16) Right click on the blank waveform window (or simply Ctrl+A on the computer keyboard) and select Add Trace to open the Add Traces to Plot window, as shown below.

17) In the Expression(s) to add box type in the functions for K and B_1 by using Equations (7-5) and (7-6) in MATLAB scripting format together with the S-parameter at 1 GHz. Press OK to plot K and B_1 values, as shown in Figure 7-15.

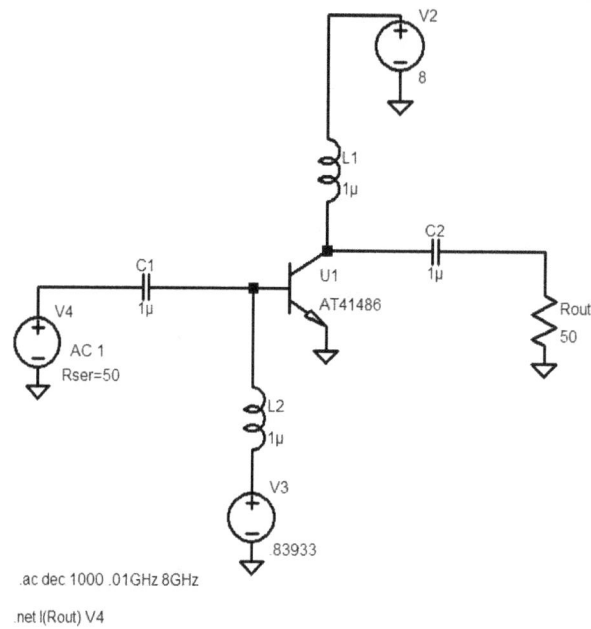

Figure 7-13 Schematic of the AT-41486 device for AC analysis

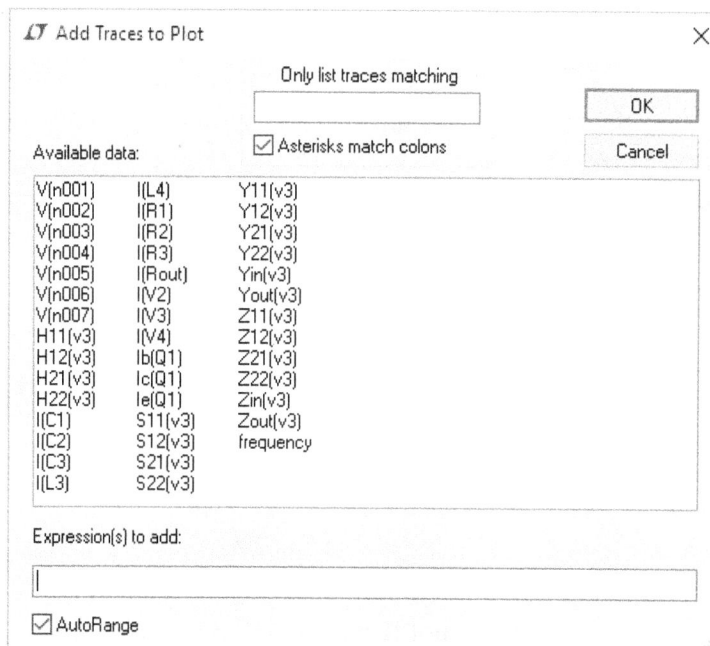

Figure 7-14 Simulation variables for plotting

$$(1-(abs(s11(v4)))^2-(abs(s22(v4)))^2+(abs(S11(v4)*S22(v4)-S21(v4)*S12(v4)))^2)/(2*abs(s21(v4)*s12(v4)))$$
$$1+(abs(s11(v4)))^2-(abs(s22(v4)))^2-(abs(S11(v4)*S22(v4)-S21(v4)*S12(v4)))^2$$

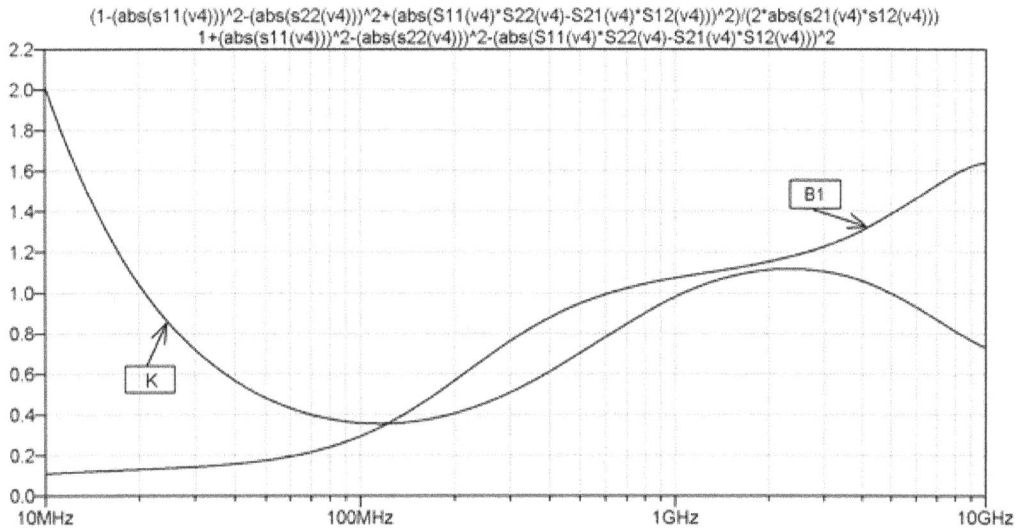

Figure 7-15 Plot of stability parameters for the AT41486 device

Note that stability factor K is less than 1 for most of the frequency band from 10 MHz up to 10 GHz indicating that the transistor is potentially unstable over most of its frequency range.

Solution (b): The effective stabilization network for medium and high power transistors employs a parallel RC circuit at the input of the device, as shown in the Figure 7-16. We keep the resistor value at 50 Ohm to match the source impedance and make the capacitor value C variable {C} and sweep it from 0.1 pF to 0.5 pF in 0.1 pF steps. Then we add .the STEP command to the schematic.

Depending on device characteristics and for higher frequency considerations, a parallel RL network at input or output of the device may be used to improve stability. At lower frequencies, the inductor effectively short circuits the resistor, where the stability is not a concern. As frequency increases, the inductor's impedance increases and the role of the resistor becomes important in improving the stability. The following plot in Figure 7-17 shows the swept K and B_1 parameters over the frequency range from 10 MHz to 10 GHz.

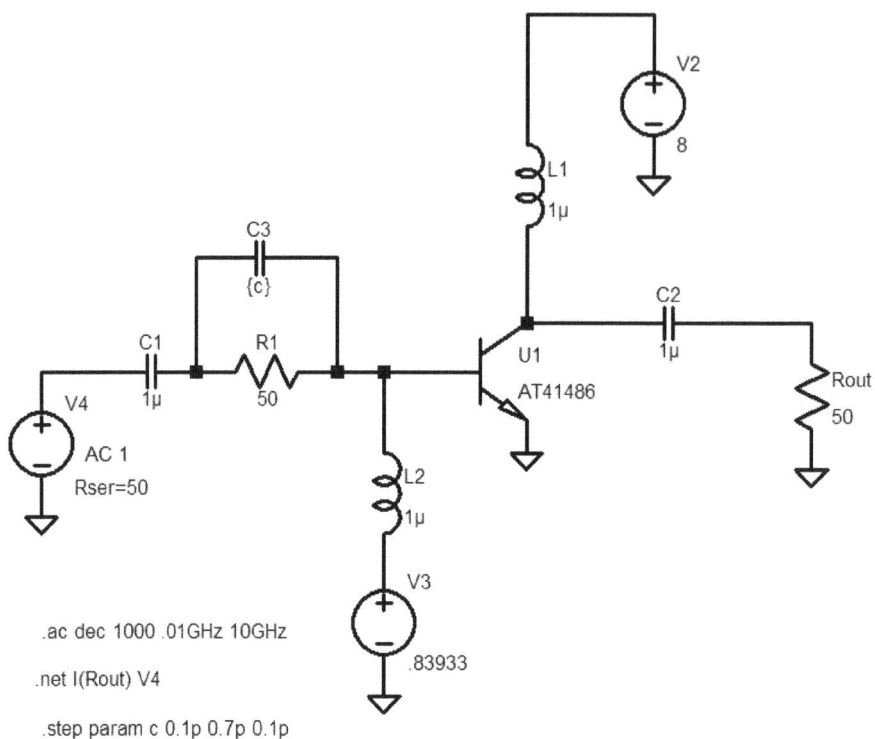

Figure 7-16 Parallel RC stability network

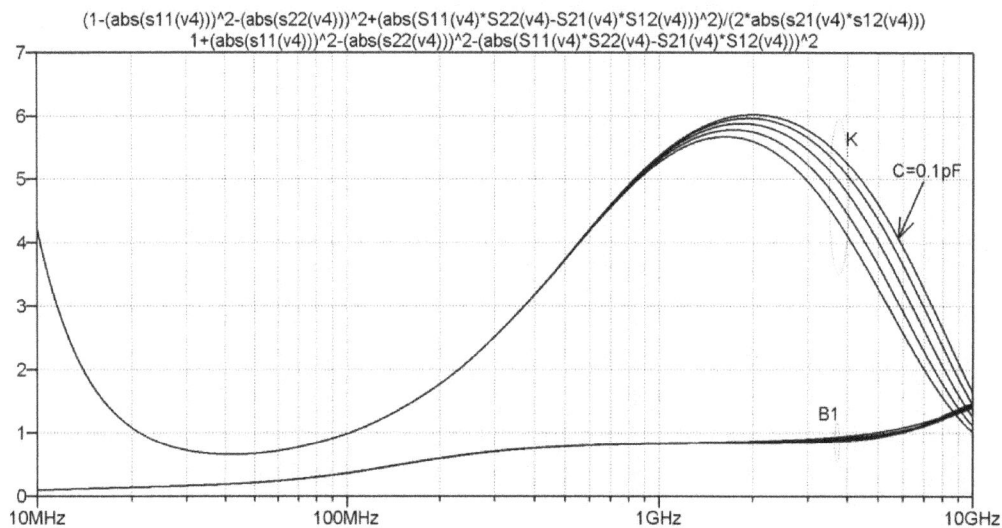

Figure 7-17 Stability parameters with parallel RC network variation

As we can see, the effect of the shunt capacitor is negligible at low frequencies. At frequencies above 1 GHz the stability factor K varies as the capacitor values vary between 0.1 and 0.5 pF Therefore, the capacitor sweep cannot make the transistor unconditionally stable at frequencies below 100 MHz.

To make the device unconditionally stable, we add parallel resistors to base and collector of the device to improve input or output matching. To prevent resistors from drawing any DC current, they are placed in parallel with the biasing inductors. In practice, a small series resistor may be added to the biasing inductor to improve stability at very low frequencies. Practical inductors contain resistive loss in the form of a series resistor that can also be modeled as a parallel resistor. To see the effect of parallel resistors on device stability, we make the resistors variable and sweep them from 200 to 400 ohms in 100 ohm steps as shown in Figure 7-18 and plot the stability factors as shown in Figure 7-19.

Examination of Figure 7-19 shows that device is unconditionally stable across the entire frequency range for the given resistor values (i.e., R=200, 300, 400 ohms). Furthermore, stability increases as parallel resistor value decreases. Here, we use a value of 278 ohms to make GTmax = 13.34 dB at 1 GHz.

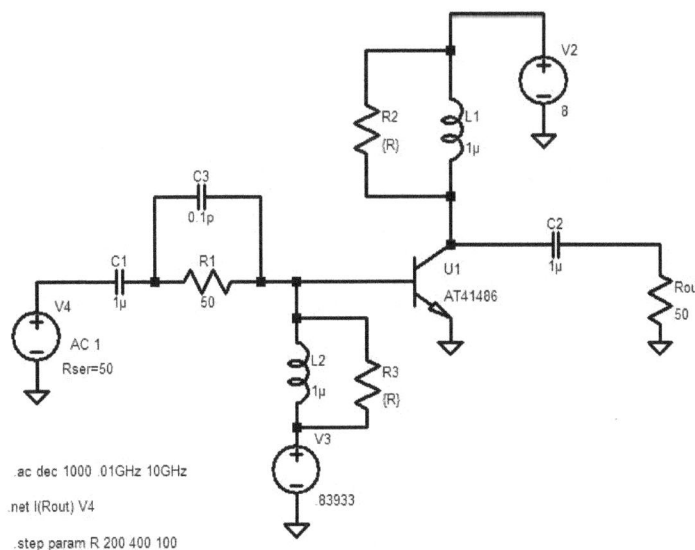

Figure 7-18 Parallel RL variation for unconditional stability

((1-(abs(s11(v4)))^2-(abs(s22(v4)))^2+(abs(S11(v4)*S22(v4)-S21(v4)*S12(v4)))^2)/(2*abs(s21(v4)*s12(v4))))
1+(abs(s11(v4)))^2-(abs(s22(v4)))^2-(abs(S11(v4)*S22(v4)-S21(v4)*S12(v4)))^2

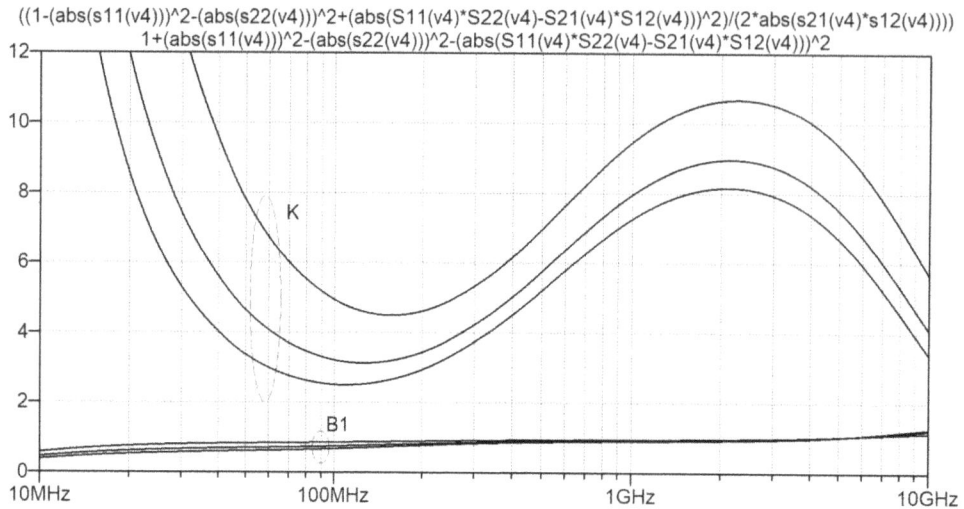

Figure 7-19 Stability parameters of unconditionally stable device

To reduce loss at higher frequencies, the shunt capacitor is changed to 0.5 pF.

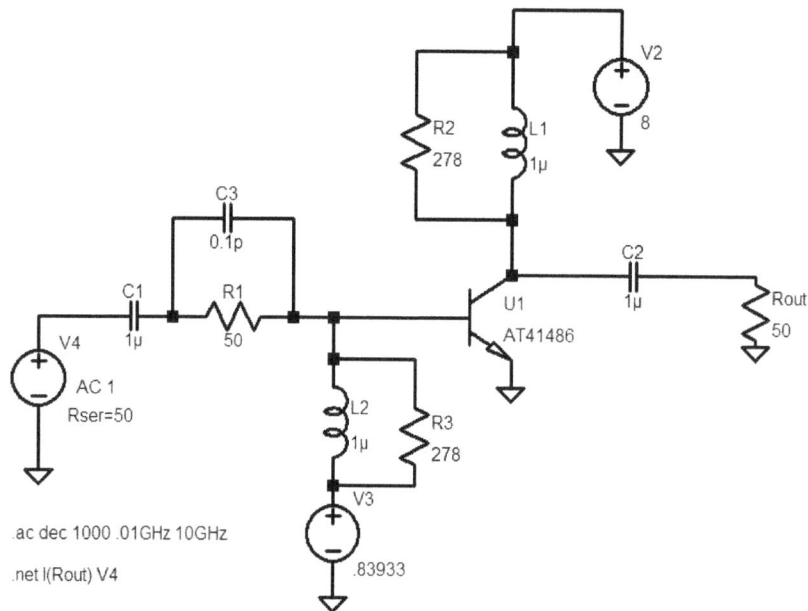

Figure 7-20 Stabilized schematic diagram

Simulated K and B_1 values are plotted in Figure 7-21.

$$\frac{(1-(abs(s11(v4)))^2-(abs(s22(v4)))^2+(abs(S11(v4)*S22(v4)-S21(v4)*S12(v4)))^2)/(2*abs(s21(v4)*s12(v4)))}{1+(abs(s11(v4)))^2-(abs(s22(v4)))^2-(abs(S11(v4)*S22(v4)-S21(v4)*S12(v4)))^2}$$

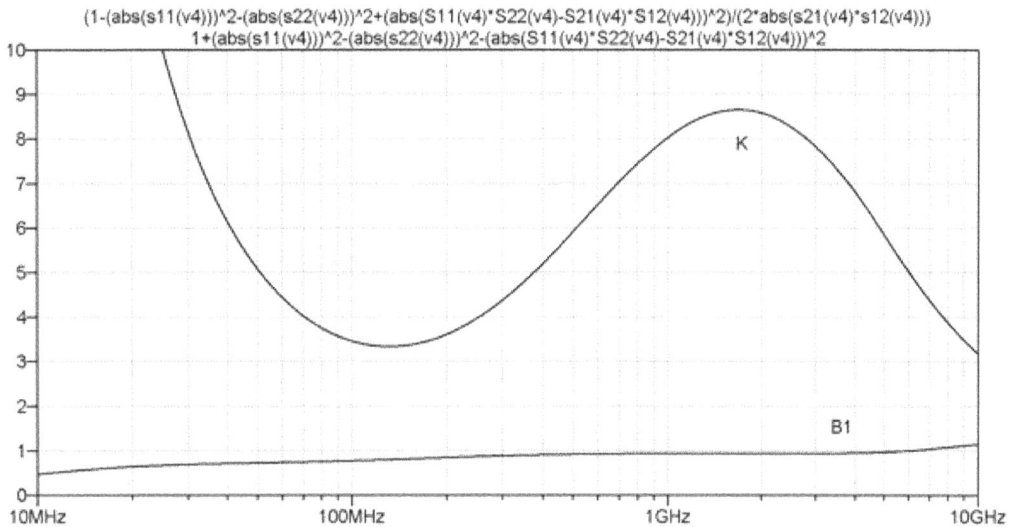

Figure 7-21 Stability parameters of unconditionally stable device

Note that $B_1 > 0$ and $K > 1$ from 10 MHz to 10 GHz.

7.5 Calculation of GTmax and S Parameters at 1 GHz

In this section, maximum transducer power gain, GTmax, of the device under simultaneous conjugate match is calculated. Example 7-3 outlines the details of how this can be accomplished in LTspice.

Example 7-3: (a) Simulate the stabilized schematic of Figure 7-20 and plot the GTmax from 0.5 GHz to 1.5 GHz. Show that GTmax at 1 GHz is equal to 13.34 dB. **(b)** Calculate the device S-parameters at 1 GHz **(c)** Use the S-parameters at 1 GHz to design the simultaneous match input and output reflection coefficients.

Solution (a): To simulate the stabilized schematic of Figure 7-20 and plot the GTmax from 0.5 GHz to 1.5 GHz,

1. Go to Simulate > Edit Simulation Cmd.
2. Change the start and stop frequencies to 0.5 and 1.5 GHz
3. Select Run > Simulation.

4. Right click on the new waveform window and select Add trace.

5. Enter the expression for GTmax in the window under Expression(s).

The result is shown in Figure 7.22. As the Figure 7-22 shows, the value of GTmax at 1 GHz is about 13.34 dB.

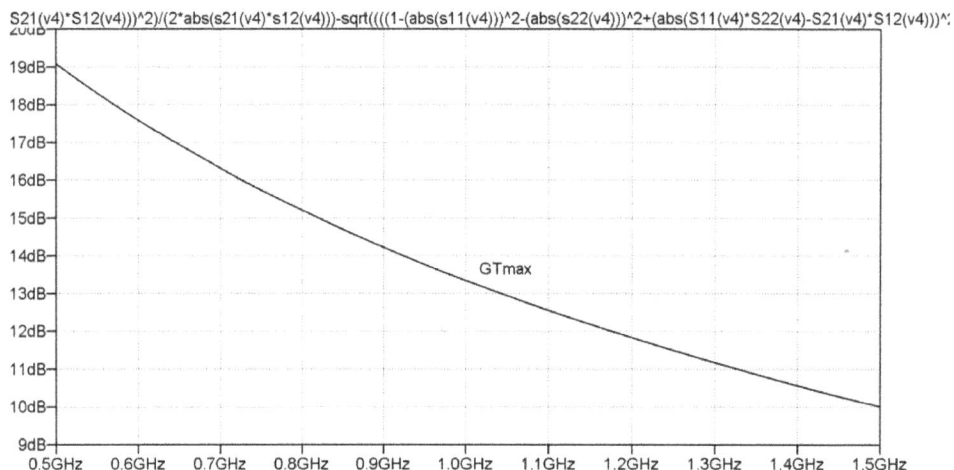

Figure 7.22: GTmax plot for the stabilized device

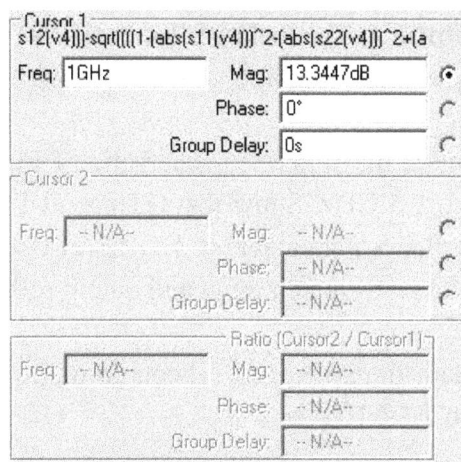

Figure 7.22: GTmax value of 13.34 dB at 1 GHz.

Solution (b): Use the following steps to measure the AT-41486 S-parameters at 1 GHz.

1. Change the start and stop frequencies to 1 GHz and 3 GHz respectively
2. Change the number of points to 3, Type of Sweep to linear and Run simulation
3. Click File > Exports to open the "Select Traces to Export" window
4. Hold the Ctrl key on your keyboard
5. In the Select Traces to Export window, select S11(v4), S12(v4), S21(v4), and S22(v4).
6. In the above window, select the proper format for the tabulated data
7. Use the top-down arrow to select Real and Imaginary > OK.
8. Go to the directory where the generated .txt file is stored. Open the file and organize the S parameters at 1 GHz.

$S11 = 0.1205 - j*0.0559$

$S12 = 0.0062 + j*0.0112$

$S21 = -0.0512 + j*4.4349$

$S22 = 0.2533 - j*0.1115$

Solution (c): Once we have the S parameters at 1 GHz we can use MATLAB script in Equations (7-5) and (7-6) to calculate the stability parameters K and B_1 at 1 GHz and show that GTmax at 1 GHz is 13.34 dB.

If we let N represent the numerator and D the denominator of the stability factor K in Equation (7-5) then:

$$N = (1-(abs(S11))^2 - (abs(S22)^2 + (abs((S11)*(S22) - (S21)*(S12)))*2) \qquad (7-9)$$

$$D = (2*(abs((S12)*(S21)))) \qquad (7-10)$$

$$\text{And } K = N/D \qquad (7-11)$$

Now substituting the S parameter values in Equations (7-9) and (7-10) shows that K = 8.051 at 1 GHz.

Similarly substituting the S parameters values in Equation (7-12) we get $B_1 = 0.933$ at 1 GHz.

B1=1+ (abs (S11))^2 -(abs(S22))^2 -(abs ((S11)*(S22)–(S12)*(S21)))^2   (7-12)

Finally we substitute K and S-parameters in Equation (7-13) to calculate GTmax = 21.601 at 1 GHz.

GTmax = (abs (S21)*(K – sqrt (K^2 – 1)))/ (abs (S12)   (7-13)
Conversion of GTmax to dB is achieved by using Equation (7-14).

(GTmax) $_{dB}$ = 10log (GTmax) = 10 log (21.601) = 13.34 dB.   (7-14)

Using the S-parameters obtained in part **(b)**, we can use the following Equations to compute the simultaneous match input-output reflection coefficients, Gamma1 and Gamma2, and the corresponding input-output impedances Z_{MS} and Z_{ML}.

To achieve this let:

B1=1 + (abs (S11)) ^2 - (abs (S22)) ^2 - (abs ((S11)*(S22)-(S12)*(S21))) ^2

B2=1 + (abs (S22)) ^2 - (abs (S11)) ^2 - (abs ((S11)*(S22)-(S12)*(S21))) ^2

And:

Delta = (S11)*(S22) - (S12)*(S21)

C1=S11-(Delta)*conj (S22)
C2=S22-(Delta)*conj (S11)

Then the simultaneous match input-output reflection coefficients are:

Gama1 = (B1-sqrt ((B1)^2 - 4*(abs(C1)^2)))/(2*C1)

Gama2 = (B2-sqrt ((B2)^2 - 4*(abs(C2)^2)))/(2*C2)

And the simultaneous match input-output impedances Z_{MS} and Z_{ML} are:

Z1sm=50*(1+Gama1)/(1-Gama1)

Z2sm=50*(1+Gama2)/(1-Gama2)

Substituting the values of S-Parameters at 1 GHz we get.
B1 = 0.933, B2 = 1.05, C1 = 0.096-j0.05, C2 = 0.241 – j0.109, and

As a result of calculations we get:

Gamma1 = 0.104 + j0.055, and Gamm2 = 0.247 + j0.111

Finally the source and load impedances for simultaneous match are:

Z_{MS} = 61.184 + j6.789 Ohms and Z_{ML} = 79.887 + j19.21 Ohms.

7.6 Designing the Input-Output Matching Networks

The design of the amplifier can now be started with the design of simultaneous match source and load impedances.

Example 7-4: Use Z_{MS} and Z_{ML} values to design the source and load impedance matching networks at 1 GHz.

Solution: First we use the equations given in Chapter 5 in MATLAB script to design the input impedance matching circuit.

Let Z0=50, RL=61.184, XL=-6.789, f=1000e6, r=RL/Z0, and x=XL/Z0

B3=(x + sqrt(r*(r^2 + x^2-r)))/ (Z0*(r^2 + x^2))

X3=Z0*sqrt ((r^2 + x^2 - r)/r)

LS1=X3/ (2*pi*f)

CP1=B3/ (2*pi*f)

The solution shows that the parallel capacitor CP1=0.97 pF and the series inductor LS1 = 3.888 nH.

Next we use the equations given in Chapter 5 in MATLAB script to design the output impedance matching circuit.
Let Z0=50, RL=79.88, XL=-19.21, f=1000e6, r=RL/Z0, and x=XL/Z0

B3=(x + sqrt(r*(r^2 + x^2 - r)))/ (Z0*(r^2+x^2))

X3=Z0*sqrt ((r^2 + x^2 - r)/r)

LS1=X3/ (2*pi*f)

CP1=B3/ (2*pi*f)

The solution shows that the parallel capacitor CP1 = 1.112 pF and the series inductor LS1 = 6.611 nH.

Example 7-5: Attach the source and load matching networks to the device and simulate the amplifier at 1 GHz. Verify the amplifier gain at 1 GHz and show that the maximum gain is 13.34 dB at 1 GHz.

The maximum gain amplifier with the input-output matching networks is shown in Figure 7-24. Note that inductors L3 and L4 each include a shunt resistor value of 278 ohms (not shown in the schematic).

The simulated S_{21} and GTmax is shown in Figure 7-25. Examination of the plot reveals that S_{21} and GTmax curves cross each other at 1 GHz. The corresponding values for S_{21} and GTmax are shown in Figure 7-26.

Data points in Figure 7-26 shows that both S_{21} and GTmax are equal to 13.34 dB at 1 GHz

The simulated response of the amplifier in Figure 7-27 shows that both S_{11} and S_{22} are perfectly matched at 1 GHz.

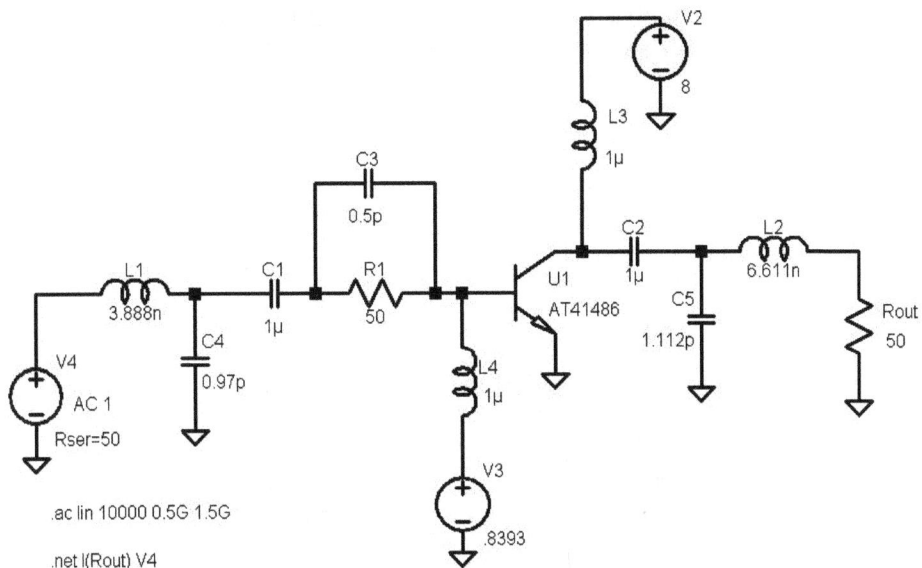

Figure 7-24 Maximum gain amplifier

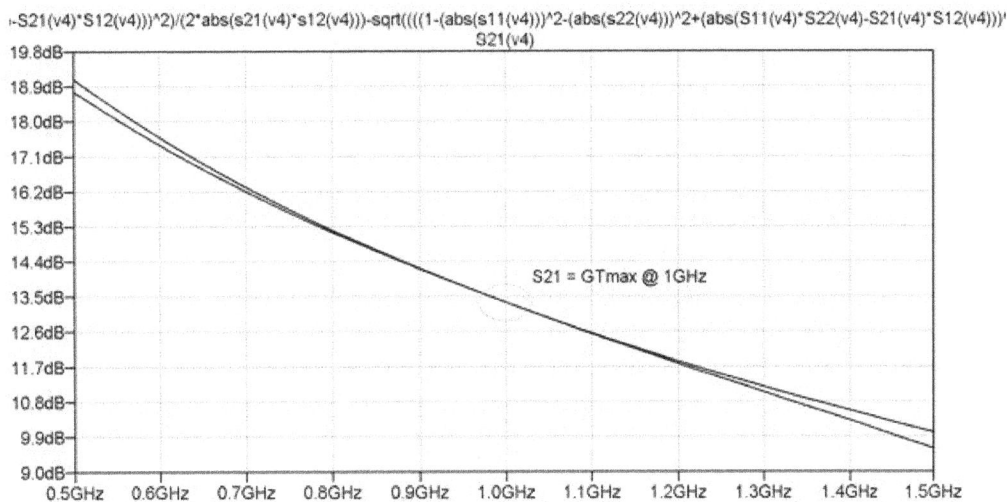

Figure 7-25 Gain plots for S_{21} and GTmax

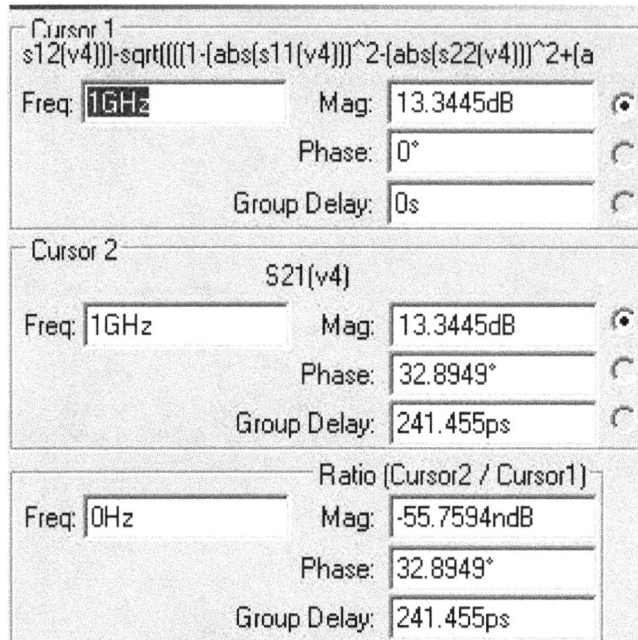

Figure 7-26 S_{21} and GTmax values at 1 GHz

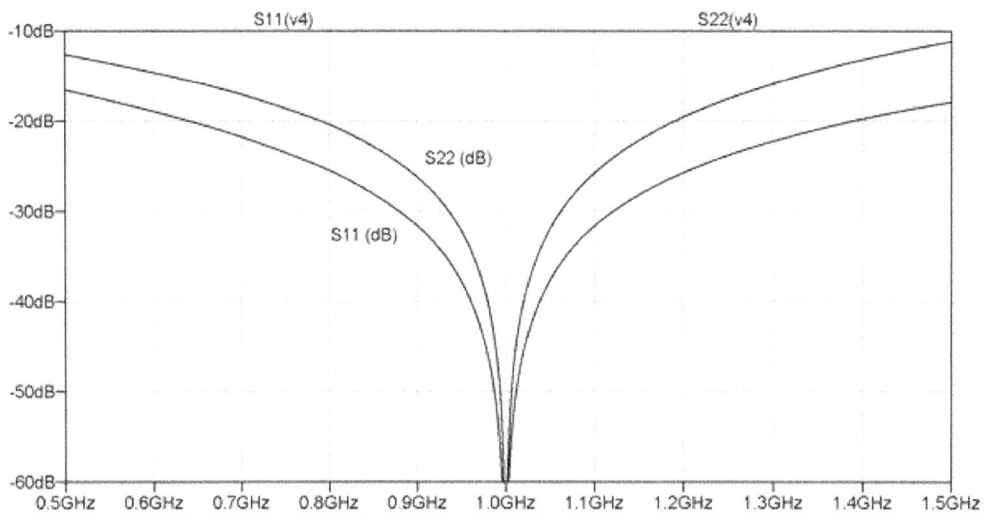

Figure 7-27 Plots of input and output return loss.

7.7 Low Noise Amplifier

One important case of selectively mismatched amplifier design is the Low Noise Amplifier, LNA. In LNA design the input is not matched to the reflection coefficient that results in maximum gain but rather to a reflection coefficient that gives the desired noise figure. Low noise amplifiers are frequently used as the input stage in a radio receiver or satellite down converter to minimize the noise that is added to the amplified signal. Figure 7-28 depicts a signal with noise at the input and output of an LNA.

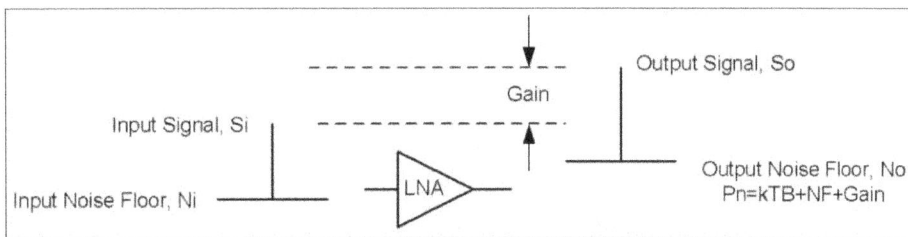

Figure 7-28 Signal and noise through the LNA

All transistors will add some amount of noise to the input signal. Low Noise transistors are optimized so that they will add a minimum amount of noise to the signal. The signal at the output of a linear LNA is simply the signal level at the input plus the gain of the LNA in dB. However the noise at the output will be increased by the gain and the noise figure of the transistor.

The noise figure of the LNA is the degradation of the signal to noise ratio of the input of the LNA to the signal to noise ratio of the output as given by Equation (7-15).

$$NF_{dB} = 10 \cdot \log\left[\frac{\left(S_i/N_i\right)}{\left(S_o/N_o\right)}\right] \qquad (7\text{-}15)$$

The ratio $\left(S_i/S_o\right)$ is the inverse of the LNA power gain, which is a function of frequency. Therefore any measurement of noise must be referred to a specific bandwidth. The thermal noise power is also a function of temperature and is defined by Equation (7-16).

$$P_n = kTB \qquad (7\text{-}16)$$

Where,

k= Boltzmann's constant, 1.374×10^{-23} J/°K
T = Temperature of the input noise source (290 °K = room temperature)
B = Bandwidth of Measurement in Hz

If we normalize the measurement bandwidth to 1 Hz i.e., B=1 the thermal noise can be calculated using Equation (7-16):

$$P_n = \left(1.374 \cdot 10^{-23}\right)(290)(1) = 3.984 \cdot 10^{-21} \; Watts/Hz$$

Or: $3.984 \cdot 10^{-18} \; mW/Hz$

Converting mW to dBm, we can express the thermal noise floor as:

$$P_{no} = 10 \cdot \log\left(3.984 \cdot 10^{-18}\right) = -173.9 \; dBm/Hz$$

Working with the thermal noise floor normalized to a 1 Hz bandwidth it is straightforward to calculate the output noise from the amplifier using the amplifier's noise figure and gain as given by Equation (7-17).

$$P_n(1Hz) = kTB_{1Hz} + Gain_{dB} + NoiseFigure_{dB} \quad \text{dB/Hz} \quad (7\text{-}17)$$

Using Equation (7-18) we can normalize the noise power in a one Hertz bandwidth to any measurement bandwidth.

$$P_n(B) = P_n(1Hz) + 10 \cdot \log(B) \qquad (7\text{-}18)$$

where B = the measurement bandwidth in Hz

Now, suppose that the noisy LNA, having a voltage gain of A_V, is connected to an antenna, as shown in Figure 7-29 [9]. How can we determine the noise figure?

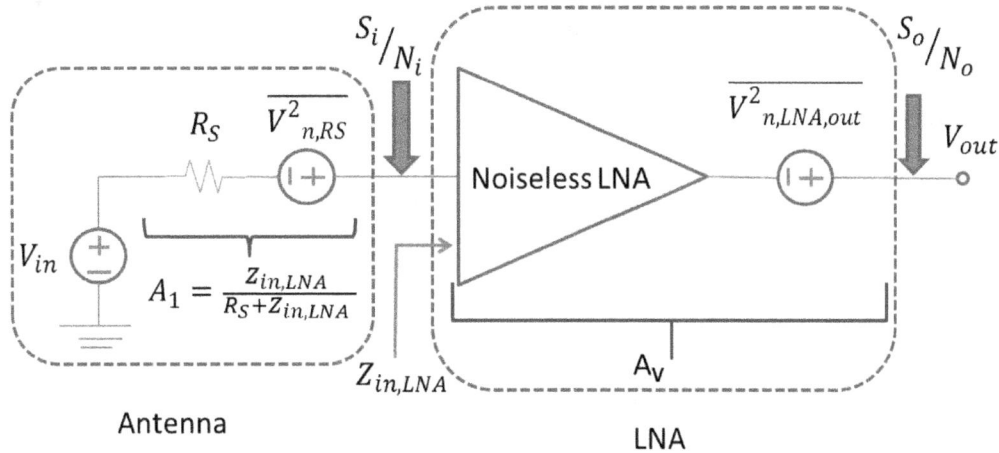

Figure 7-29 Front-end receiver including an antenna and LNA

In Figure 7-29 V_{in} is the rms input voltage source representing the received signal to the antenna. Rs is the source impedance equivalent to antenna radiation resistance. $\overline{V^2_{n,RS}}$ represents antenna's thermal noise and $\overline{V^2_{n,LNA,out}}$ represent the output noise of the LNA. A_V is the LNA's voltage gain and A_1 is the voltage gain between source and antenna input. In the case of infinite LNA input impedance, A_1 reduces to unity. Otherwise, it is less than one and it acts as an attenuator. Therefore, the total voltage gain from input to output is the product of A_1 and A_V. Note that the unit of noise voltage is [V^2/Hz].

For the above configuration, the following can be observed for signal and noise at LNA input:

$$S_i = |A_1|^2\, V^2_{in} \tag{7-19}$$

$$N_i = |A_1|^2\, \overline{V^2_{n,RS}} \tag{7-20}$$

Similarly, the signal and noise at LNA output can be written as:

$$S_o = |A_1|^2\, A^2{}_V\, V^2{}_{in} \tag{7-21}$$

$$N_o = \overline{V^2{}_{n,RS}}\, |A_1|^2\, A^2{}_V + \overline{V^2{}_{n,LNA,out}} = \overline{V^2{}_{n,out}} \tag{7-22}$$

Substituting these equations in Equation (7-15), the noise figure can be written as:

$$NF_{dB} = 10log\left[\frac{\overline{V^2{}_{n,RS}}\,|A_1|^2 A_v{}^2 + \overline{V^2{}_{n,LNA,out}}}{\overline{V^2{}_{n,RS}}\,|A_1|^2 A_v{}^2}\right] \tag{7-23}$$

Or:

$$NF_{dB} = 10log\left[\frac{\overline{V^2{}_{n,out}}}{4KTR_S\,|A_1|^2\,A^2{}_V}\right], \tag{7-24}$$

Where in the above equation,

$\overline{V^2{}_{n,out}}$ is the total noise at LNA output

$4KTR_S = \overline{V^2{}_{n,RS}}$ is the Thermal noise due to source resistance.

From Equation (7-24) it can be seen that noise figure is the ratio of total noise at the LNA output divided by the product of the source impedance thermal noise and the square of the total voltage gain from input to out. As a result, the noise figure can also be expressed as the total noise at the input divided by the source impedance thermal noise.

$$NF_{dB} = 10log\left[\frac{\overline{V^2{}_{n,out}}}{4KTR_S\,|A_1|^2 A^2{}_V}\right] = 10log\left[\frac{\overline{V^2{}_{n,in}}}{4KTR_S}\right], \tag{7-25}$$

where

$$\frac{\overline{V^2{}_{n,out}}}{|A_1|^2 A^2{}_V} = \overline{V^2{}_{n,in}} = \text{the total equivalent noise present at LNA input}$$

$$\tag{7-26}$$

As seen in section 7.9, Equations (7-25) and (7-26) are used in LTspice to calculate LNA's noise figure.

7.8 Noise Figure Analysis

In order to perform noise figure simulation the device S parameter file must include noise parameters. Table 7-3 shows an S parameter file with noise parameters appended to the end of the file. In the S parameter file the noise parameters include the optimum noise figure, NF_{opt}, as a function of frequency. The NF_{opt} is the lowest possible noise figure that can be achieved when the transistor input is matched to optimum reflection coefficient Γ_{opt}. The third and the fourth column give the magnitude and angle of the Γ_{opt}. The final column gives the normalized noise resistance for the device. These noise parameters are usually provided by the device manufacturer on the S parameter data sheet, as shown in Table 7-3 for the AT-41486 device.

Matching the device to Γ_{opt} can potentially lead to very poor input return loss and instability. Just as we have seen in the example of section 7.4.2 the designer must be aware of the location of Γ_{opt} with respect to the input stability circle. In the LNA design we usually do not want to add resistive loading to the input of the device to improve the stability because this would increase the thermal noise power and increase the device noise figure. In certain critical designs the poor input return loss is accepted and an isolator is added to the input of the LNA to provide a good input return loss. This also involves tradeoffs as the losses in the isolator add directly to the noise figure.

As a tradeoff a reflection coefficient can be chosen that corresponds to a noise figure that is slightly greater than minimum noise figure. In this case the resulting noise figure can be calculated for any source reflection coefficient, Γs, by using the following equation [10]. $NF_{min}, R_n,$ and Γ_{opt} are called the *noise parameters* and are provided by transistor manufacturer. The noise resistance R_n can be measured when source reflection Γ_s is made equal to zero. This parameter indicates noise performance of the device away from optimum source reflection coefficient Γ_{opt}.

$$NF = NF_{\min} + \frac{4R_n}{Z_0\left|1+\Gamma_{opt}\right|^2} \cdot \frac{\left|\Gamma_s - \Gamma_{opt}\right|^2}{\left(1-\left|\Gamma_s\right|^2\right)} \qquad (7\text{-}27)$$

```
|!AT-41486
!S AND NOISE PARAMETERS at Vce=8V  Ic=10mA.   LAST UPDATED
07-21-92

# ghz s ma r 50

0.1    .74    -38    25.46    157    .011    68    .94    -12
0.5    .59   -127    12.63    107    .031    47    .60    -29
1.0    .56   -168     6.92     84    .041    46    .49    -29
1.5    .57    169     4.72     69    .049    49    .45    -32
2.0    .62    152     3.61     56    .058    43    .42    -39
2.5    .63    142     2.91     47    .068    52    .40    -42
3.0    .64    130     2.41     37    .078    52    .39    -50
3.5    .68    122     2.06     26    .093    51    .37    -60
4.0    .71    113     1.80     16    .106    48    .35    -70
4.5    .74    105     1.59      7    .125    48    .35    -84
5.0    .77     99     1.42     -4    .139    43    .35    -98
5.5    .79     93     1.27    -13    .153    38    .35   -114
6.0    .81     87     1.13    -22    .170    34    .35   -131

!FREQ   NFopt     GAMMAopt        Rn/Zo
!GHZ     dB      MAG    ANG         -

0.1     1.3     .12      3      0.17
0.5     1.3     .10     16      0.17
1.0     1.4     .04     43      0.16
2.0     1.7     .12   -145      0.16
4.0     3.0     .44    -99      0.40
```

Table 7-3 AT41486 S parameter file with noise parameters included

7.9 Low Noise Amplifier Design

Design of a low noise amplifier starts with establishing its DC operating conditions. To ensure stable operation over all frequencies where there is potential for oscillations, stability analysis is then performed and the device is made stable if necessary. Design for specific parameters such as gain, noise figure, and input / output return loss at a given biasing conditions is then followed. All of the above analyses can be performed in LTspice, as is discussed in following Examples, 7-6, and 7-7.

Example 7-6 Use the Avago NPN transistor AT-41486 to design a low noise amplifier. Perform the following:

(a) Bias the AT41486 transistor spice model for Vce = 8 VDC and Ic = 10 mA.
(b) Perform stability analysis and ensure that device is unconditionally stable.
(c) Determine the stabilized device gain (S_{21}) at 1 GHz.

Solution (a): To bias the AT41486 transistor spice model for Vce = 8 VDC and Ic = 10 mA follow the same steps described Example 7-1. Since the collector current has decreased to 10 mA the analysis shows that base voltage, Vbe, should be decreased to 0.8055 Volt. The biased transistor with 1 µF DC blocking capacitors and 1 µH AC blocking inductors is shown in Figure 7-30.

The operating point data at Vce = 8 VDC, Vbe = 0.8055 VDC, and Ic = 10mA is given in Table 7-4.

Solution (b): To determine device stability, AC analysis is performed and stability parameters are computed as a function of two port S- parameters. Figure 7-31 shows the schematic diagram for .net analysis utilizing LTspice .net command for frequency range of 1 MHz to 10 GHz.

Simulated plots for K and B_1 are shown in Figure 7-32, where K values less than 1 indicate potentially unstable device from a few MHz up to 1 GHz.

```
*SPICE model for AT-41486
*
.SUBCKT AT41486 60    20    40
LL1   20    25    .55NH
* PI NETWORK TO SIMULATE TRANSMISSION LINE T1
C1T1   25    0    .06PF
LT1   25    30    .48NH
C2T1   30    0    .06PF
*
LLB   30    35    .55NH
CCEB   30    50    .02PF
LL2   40    45    .06NH
* PI NETWORK TO SIMULATE TRANSMISSION LINE T2
C1T2   45    0    .04PF
LT2   45    50    .08NH
C2T2   50    0    .04PF
*
LLE   50    55    .1NH
CCEC   50    70    .03PF
LL3   60    65    .25NH
* PI NETWORK TO SIMULATE TRANSMISSION LINE T3
C1T3   65    0    .04PF
LT3   65    70    .30NH
C2T3   70    0    .04PF
*
CCBC   30    70    .03PF
* CALL DIE MODEL
XDIE   70    35    55    AT414
.ENDS

* DIE MODEL (excludes bond wires)
.SUBCKT AT414  75    20    85
CCB   20    60    .032PF
DCD1   20    60    DMOD  572
RRB1   20    25    1.07  TC=0.8E-3
DCD2   25    60    DMOD  680
RRB2   25    30    3.2   TC=1.2E-3
DCD3   30    60    DMOD  340
RRB3   30    35    2.7   TC=1.8E-3
RRC   60    75    5    TC=0.6E-3
RRE   80    85    .24   TC=0.6E-3
CCE   60    85    .032PF
QINT   60    35    80    QDIS  420
.ENDS
*
.MODEL DMOD   D(IS=1E-25, CJO=2.45E-16, VJ=.76, M=.53, BV=45, IBV=1E-9)
.MODEL QDIS   NPN (BF=100, BR=2.5, IS=1.65E-18, VA=20, TF=12PS,
+       CJE=2.4E-15, VJE=1.01, MJE=0.6, PTF=25, XTB=1.818,
+       VTF=6, ITF=3E-4, IKF=1.3E-4, XTF=4, NF=1.03, ISE=5E-15,
+       NE=2.5)
```

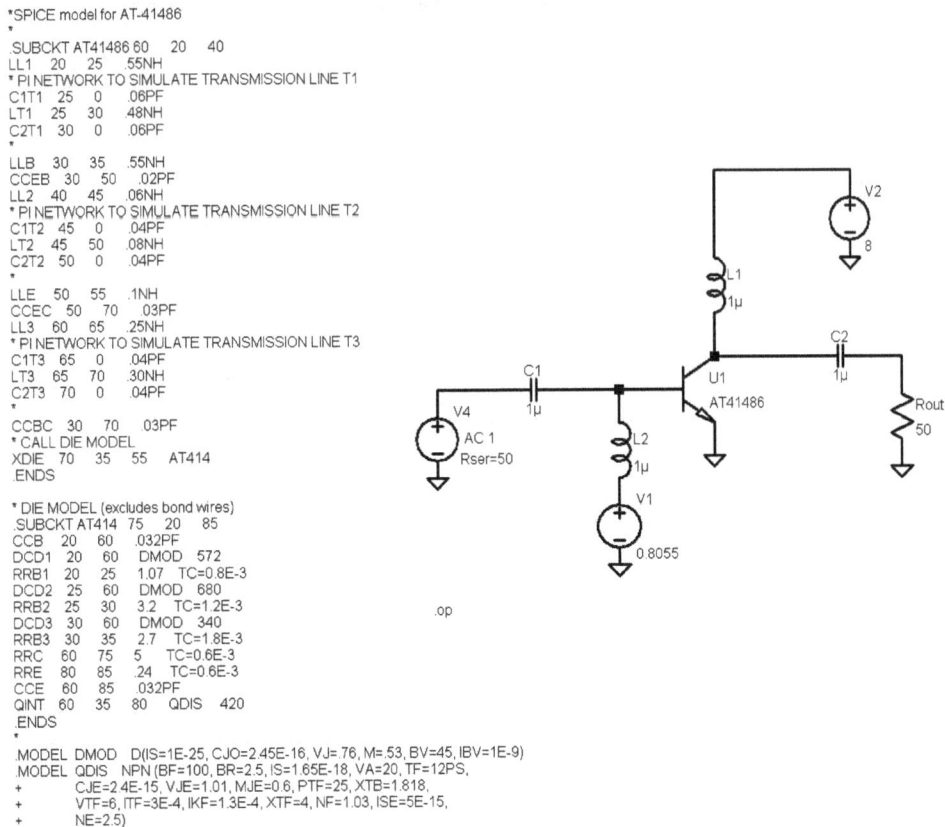

Figure 7-30. LTspice setup for device DC analysis.

```
            --- Operating Point ---

V(n005):        0.8055          voltage
V(n002):        7.99999         voltage
V(n004):        4.0275e-017     voltage
V(n001):        8               voltage
V(n003):        4e-016          voltage
V(n006):        0.8055          voltage
I(C2):          -7.99999e-018   device_current
I(C1):          8.055e-019      device_current
I(L2):          -8.41361e-005   device_current
I(L1):          0.0100001       device_current
I(Rout):        7.99999e-018    device_current
I(V1):          -8.41361e-005   device_current
I(V2):          -0.0100001      device_current
I(V4):          8.055e-019      device_current
Ix(u1:60):      0.0100001       subckt_current
Ix(u1:20):      8.41361e-005    subckt_current
Ix(u1:40):      -0.0100842      subckt_current
```

Table 7-4: Operating point data (Ic = I_{L1} = 10 mA)

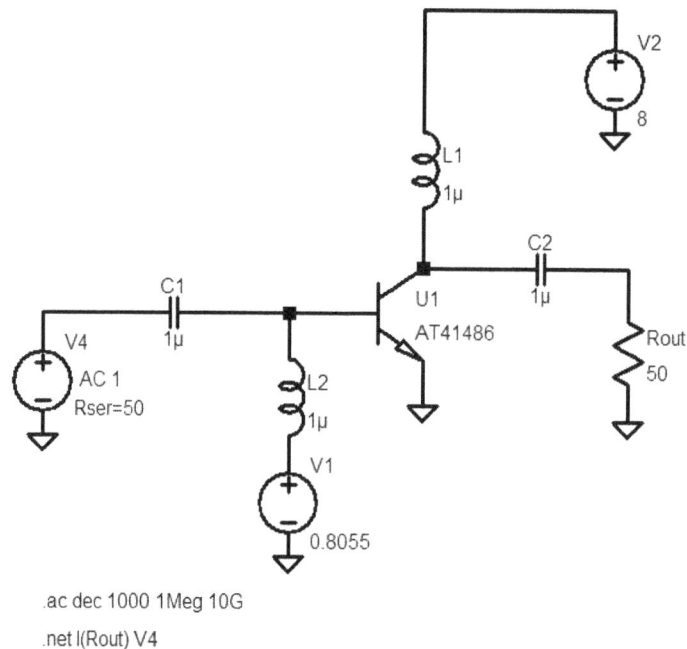

Figure 7-31. AC analysis for AT41486 biased at Vc = 8V, and Ic = 10mA

$$((1-(abs(s11(v4)))^2-(abs(s22(v4)))^2+(abs(S11(v4)*S22(v4)-S21(v4)*S12(v4)))^2)/(2*abs(s21(v4)*s12(v4))))$$
$$1+(abs(s11(v4)))^2-(abs(s22(v4)))^2-(abs(S11(v4)*S22(v4)-S21(v4)*S12(v4)))^2$$

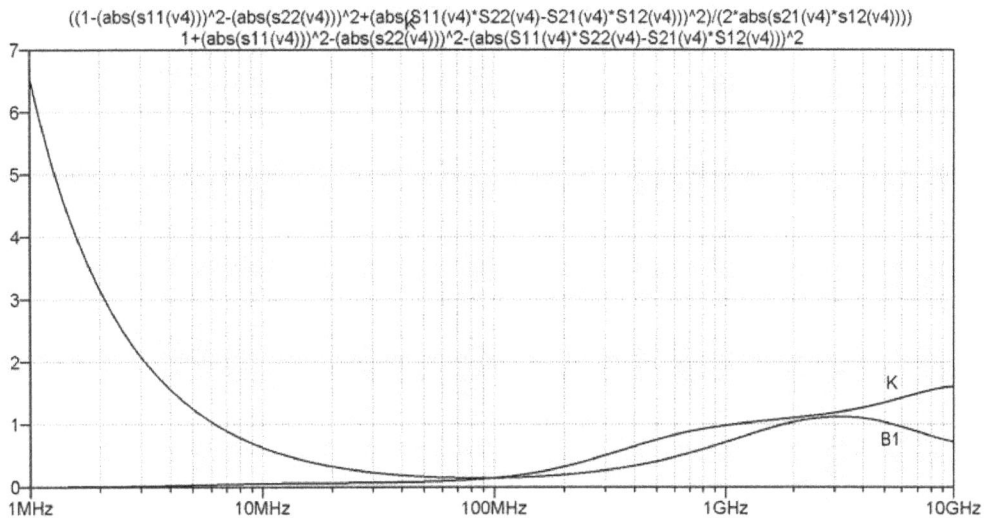

Figure 7-32. Stability parameters K and B1 at 8V, 10mA

To stabilize the device while minimizing noise degradation, a shunt resistor at device output is used. To determine proper resistor value to ensure stability and to meet device gain criterion, R2 value is swept using .step param command, as depicted in Figure 7-33. Using the result in Figure 7-34, a resistor value of 135 ohms for R2 is selected.

Figure 7-35 shows the final schematic diagram for the stabilized device. The resistor R2 is often absorbed as part of the loss associated with L1. However, to meet specific design requirements, an external resistor in parallel to L1 can be used.

Figure 7-36 shows the swept stability factors from 1 MHz to 10 GHz indicating an unconditional stability device. Note that depending on a specific design objective, component values can be adjusted in the final design iteration since the addition of matching components at input or output of the circuit will affect its stability.

Solution (c): Schematic diagram for simulating the gain of the stabilized device is given in Figure 7-37 and the simulated gain plot of the device is shown in Figure 7-38. The value of gain (S_{21}) at 1 GHz is about 16.5 dB.

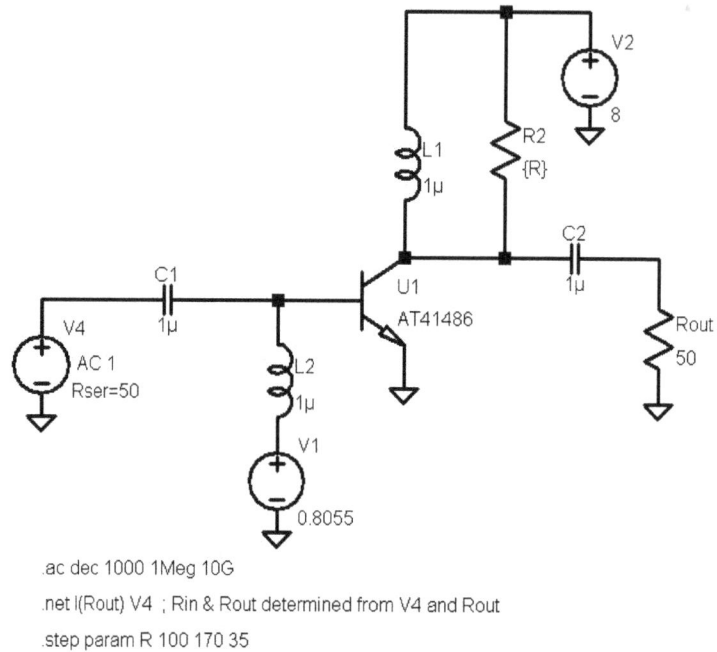

Figure 7-33. Schematic diagram of the LNA for tuning R2

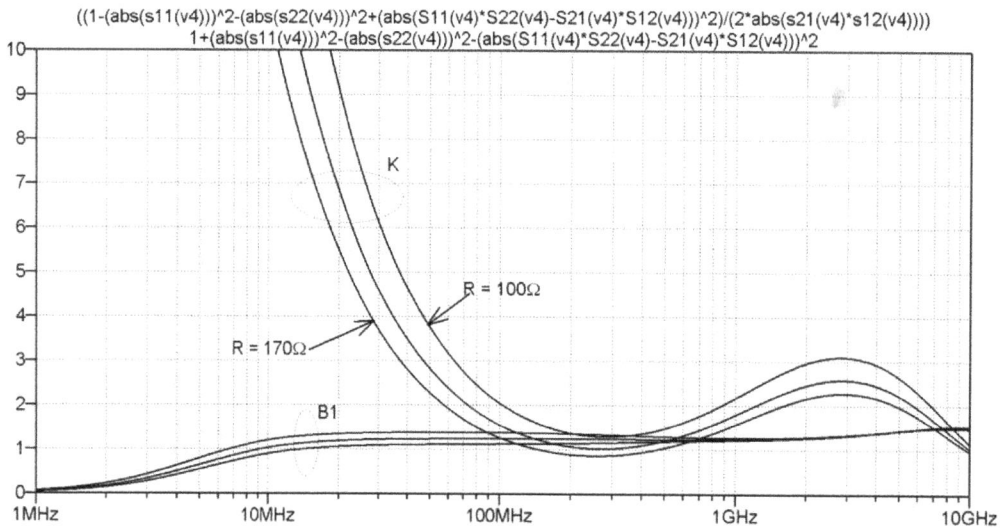

Figure 7-34. K and B_1 plots as a function of R2 tuning.

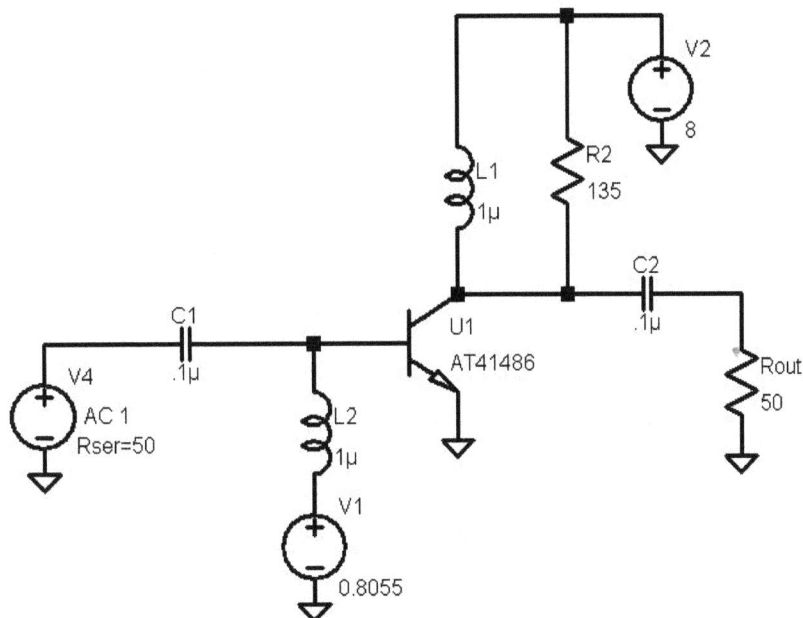

Figure 7-35. Schematic diagram of the unconditionally stable LNA.

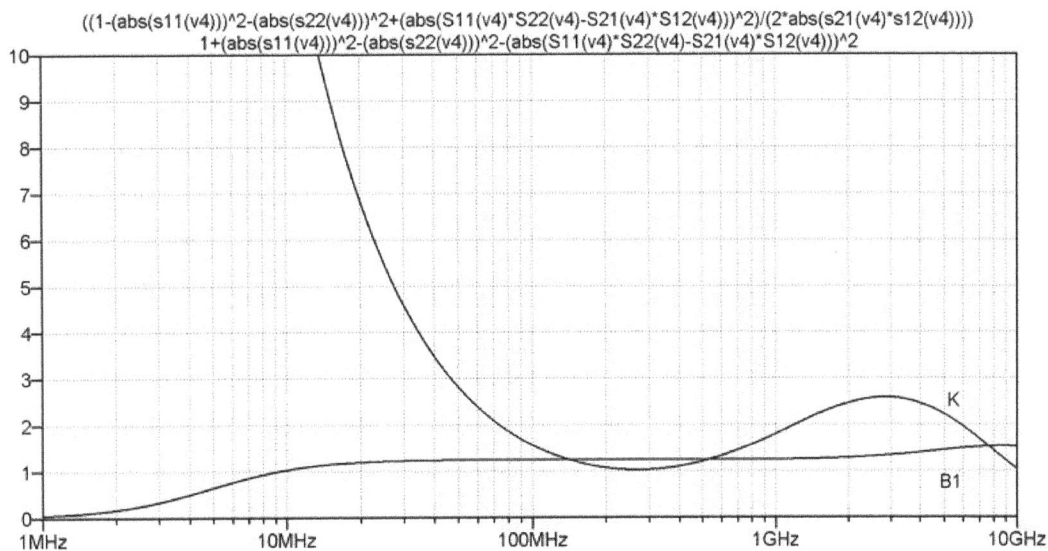

Figure 7-36. K and B1 Plots when R2=135 Ohm.

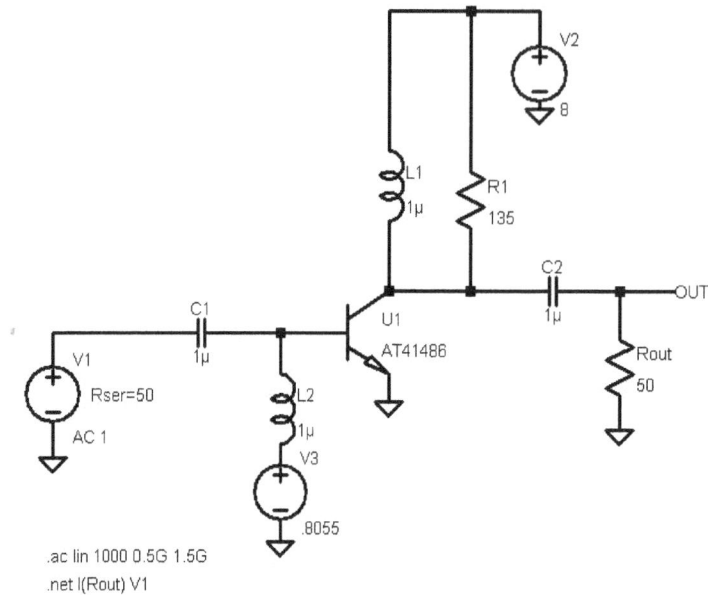

Figure 7-37. LNA simulation from 0.5 to 1.5 GHz

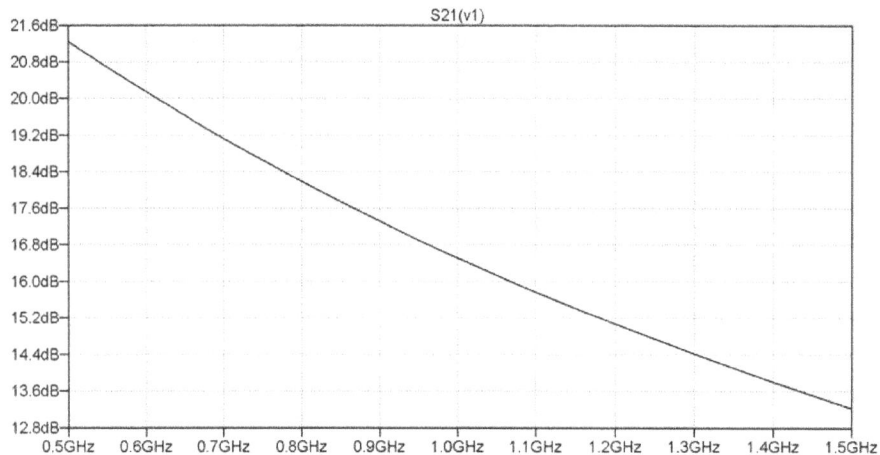

Figure 7-38. LNA gain from 0.5 to 1.5 GHz.

Example 7-7: In this example we examine how to simulate and measure noise figure in LTspice as well as how to design the LNA input-output matching networks to achieve the minimum noise figure.

(a) Determine the noise figure of the stabilized LNA at 1 GHz.

(b) Design the input matching network to achieve minimum noise figure. Use Γ_{opt} = 0.04 < 43° from published data in Table 7-3.

(c) Determine the device S-parameters at 1 GHz.

(d) Design the output matching network.

(e) Assemble the LNA and measure the minimum noise figure at 1 GHz.

Solution (a): LTspice is capable of simulating noise figure of an amplifier versus frequency once the source and load impedances are defined. To simulate noise figure in LTspice, use the following steps:

1) Construct the schematic of the stabilized device including the bias voltages as shown below.

2) Apply an input voltage (V1) with suitable source impedance. The source impedance here is a 50 ohm resistor.

3) Define and construct an output voltage across the load. Here, the load is a 50 ohm resistor:

   a. Attach a wire to the load.

   b. Right click on the wire > Label Net.

   c. Type OUT as the net name.

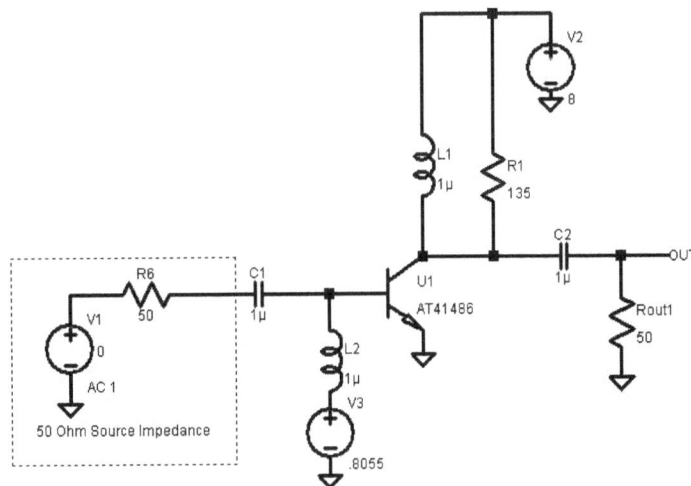

.noise V(OUT) V1 lin 100 500Meg 1.5G

Figure 7-39. LNA Schematic for noise figure calculation

4) Select Simulate > Edit Simulation Cmd

5) Select the Noise tab and complete the Edit Simulation Command window as shown in Figure 7-40:
6) Select Simulate > Run.
7) In Add Traces to Plot window, type in
   10*log10(V(inoise)*V(inoise)/(4*k*300.15*50))

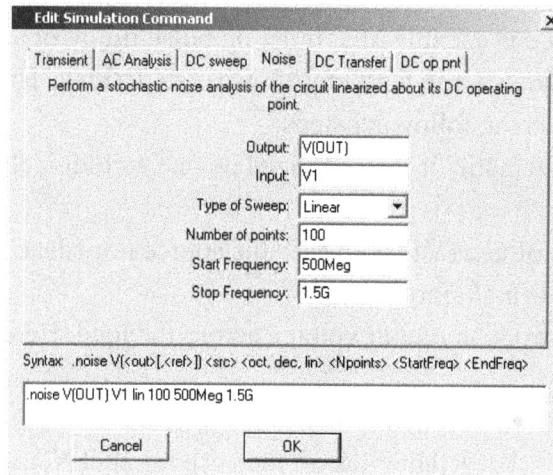

Figure 7-40. Noise measurement in Edit Simulation Command.

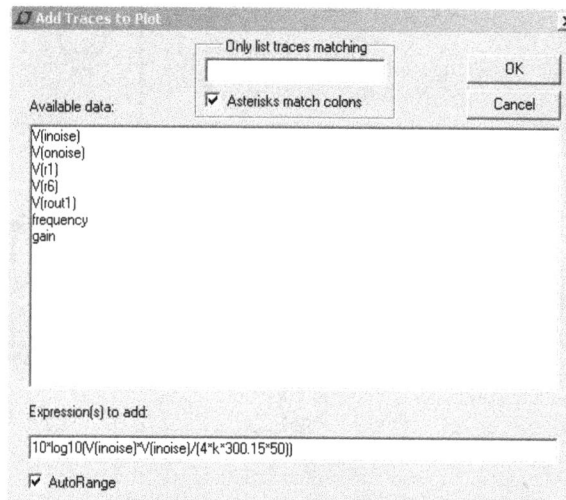

Figure 7-41. Expression for NF plot in dB.

8) Select OK to plot the noise figure versus frequency, as shown in Figure 7-42.

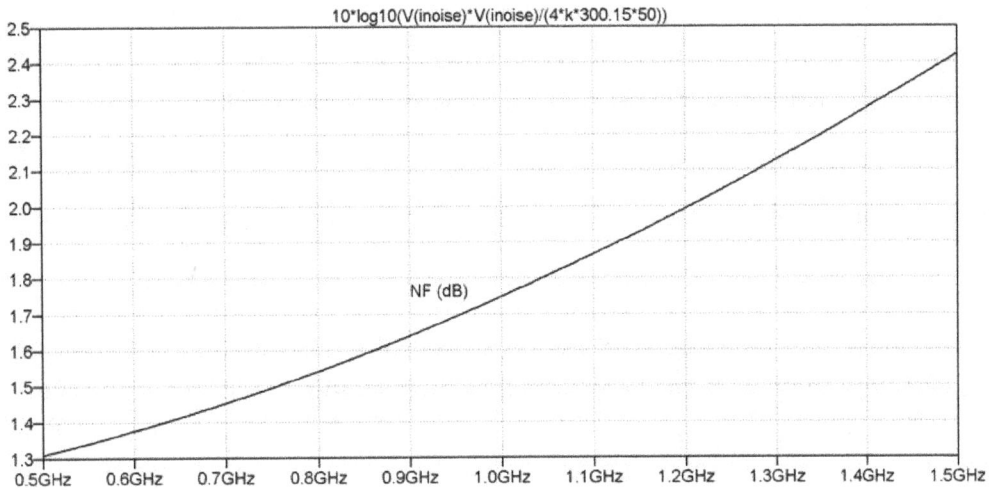

Figure 7-42. Noise Figure in dB as a function of frequency

To measure the noise figure at 1 GHz, move the cursor to the intersection of noise figure plot with the 1GHz line and read the value of noise figure. Notice that the noise figure at 1 GHz is 1.74742 dB. Note that to achieve the minimum noise figure at 1 GHz we must match the transistor input to the optimum input reflection coefficient, Γ_{opt}. or to the optimum input impedance Z_{opt}

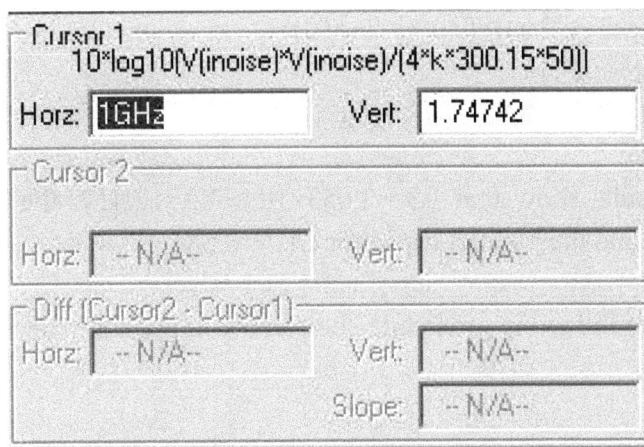

Figure 7-43. Noise Figure in dB at 1GHz.

Solution (b): To determine the optimum source impedance, Z_{opt}, at 1 GHz:
1) Select the optimum source reflection coefficient at 1 GHz from the published noise parameter data in Table 7-3 which is $\Gamma_{opt} = 0.04 < 43°$.

2) Convert the polar to complex form. $\Gamma_{opt} = 0.0292 + 0.0272j$

3) Use Equation $Z_{opt} = (1 + \Gamma_{opt})/(1 - \Gamma_{opt})$ to convert Γ_{opt} to Z_{opt}
$$Z_{opt} = 52.927 + 2.884j \ \Omega$$

The optimum source impedance can be analytically determined by changing the real and imaginary parts of the source impedance and tabulating NF in LTspice. The reader is encouraged to go through this exercise to determine a slightly different optimum source impedance value. Here, we use the typical measured value indicated in Table 7-3 that results in NFmin value of 1.4 dB at 1 GHz. NFmin, however, can range from 1.4 dB to 1.8 dB (see Appendix C).

We use analytical methods in MATLAB script to design the LNA source impedance matching network.

```
%Enter Design Parameters
Z0=50, RL=52.927, XL=-2.88, f=1e9
%Normalize Load Impedance
r=RL/Z0, x=XL/Z0
% Calculate B3, X3, and Element Values
B3=(x + sqrt(r*(r^2+x^2-r)))/(Z0*(r^2+x^2))
X3=Z0*sqrt((r^2+x^2-r)/r)
LS1=X3/(2*pi*f)
CP2=B3/(2*pi*f),
```
The solution results show that B3= 3.653×10^{-3}, X3=12.417, the series inductor LS1= 1.976 nH, and the parallel capacitor CP2= 0.581 pF.

Solution (c): The following procedure shows how to generate the S-Parameters at 1GHz.

1. From the schematic window select Simulate > Edit Simulation Control.
2. Select AC Analysis and complete the simulation box as in Fig. 7-44.
3. Select OK > Simulate > Run.

4. From the waveform window select File > Export to open the Select Traces to Export sown in Figure 7-45.
5. Select the S-Parameter traces and Cartesian re, im format. Use Browse to save the text file in the desired directory.
6. Go to the directory and open the file to access tabulated S-parameters at 1GHz.
7. Organize the S-Parameter in a Table (Table 7-5).

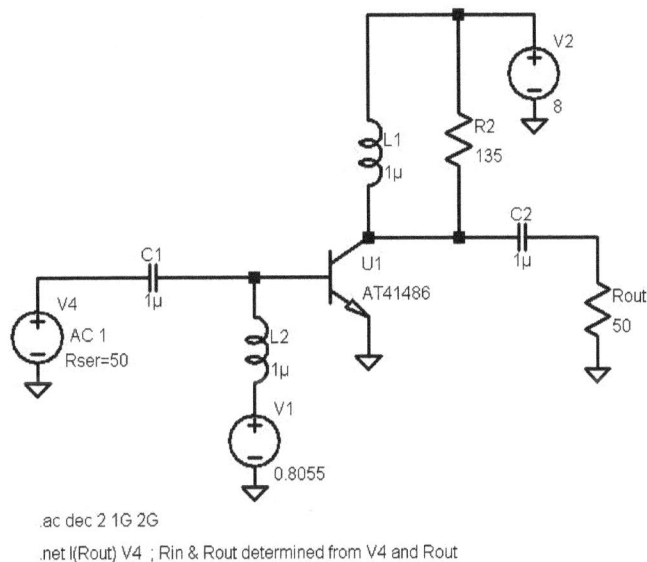

Figure 7-44 Simulation of S parameter measurement

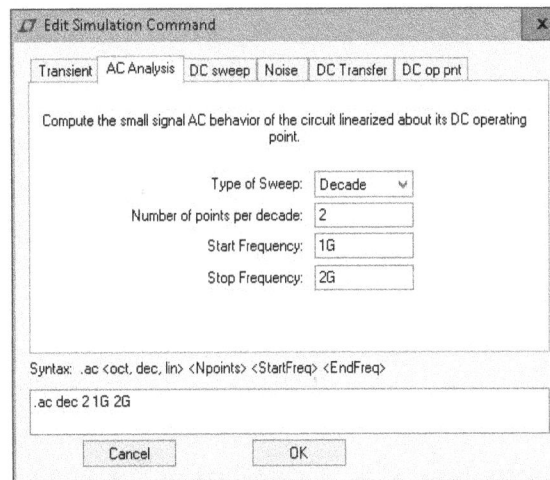

Figure 7-45 S-Parameter measurement at 1 and 2 GHz

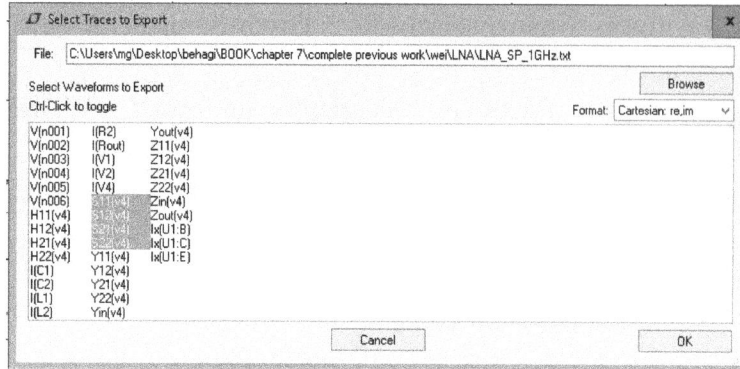

Figure 7-46 Window to Select Traces to Export

$$S_{11} = -0.5058 - j0.2099$$
$$S_{12} = +0.0198 + j0.0188$$
$$S_{21} = -0.4493 + j6.6770$$
$$S_{22} = +0.1739 - j0.1396$$

Table 7-5. Tabulated S parameters at 1 GHz

Solution (d): To achieve maximum power transfer to the load, we use $\Gamma_L = (\Gamma_{out})^*$. Furthermore, to design for minimum noise figure, NFmin, we use $\Gamma_s = \Gamma_{opt}$ in the expression for Γ_{out}. Therefore, Γ_L can be written as:

$$\Gamma_L = \left(S_{22} + \frac{S_{12}S_{21}\Gamma_{opt}}{1 - S_{11}\Gamma_{opt}} \right)^* \tag{7-28}$$

Using Equation 7-28 and the given S-parameters, the corresponding load impedance, Z_L, is determined to be 66.73+19.541j at 1GHz. The output impedance matching network is realized with a series inductor of L=5.332 nH and a shunt capacitance of C=0.828 pF.

Solution (e): The complete LNA circuit is given in Figure 7-47. This circuit can be assembled by adding the input and output LC matching networks to the schematic diagram of Figure 7-44. The simulated NF from 0.5 GHz to 1.5 GHz is depicted in Figure 7-48, where NFmin of 1.72 dB is achieved at 1 GHz.

Note that the above value is within attainable range of NFmin at 1 GHz tabulated in device data sheet in Appendix C. Furthermore, the complete LNA circuit is unconditionally stable at all frequencies. The curious reader is encouraged to verify this fact.

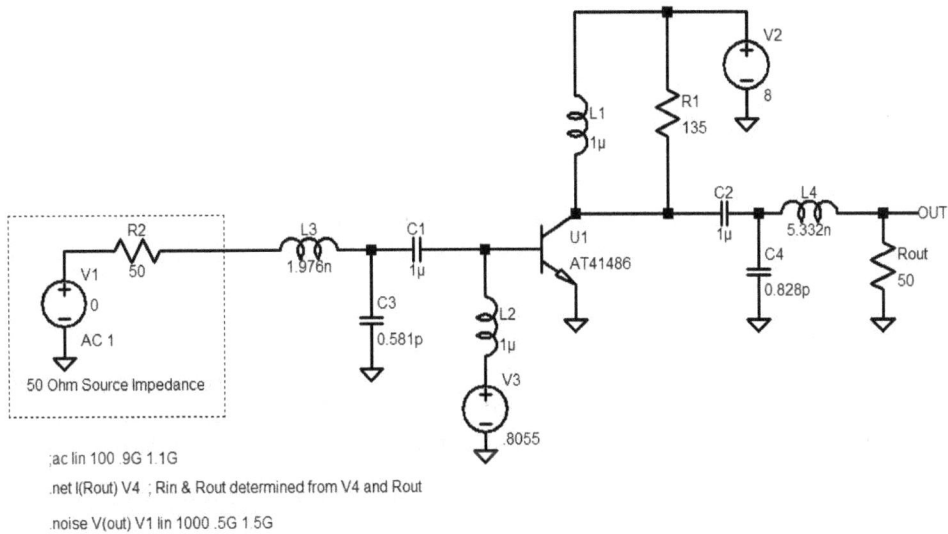

Figure 7-47. Schematic of the LNA to achieve NFmin at 1 GHz

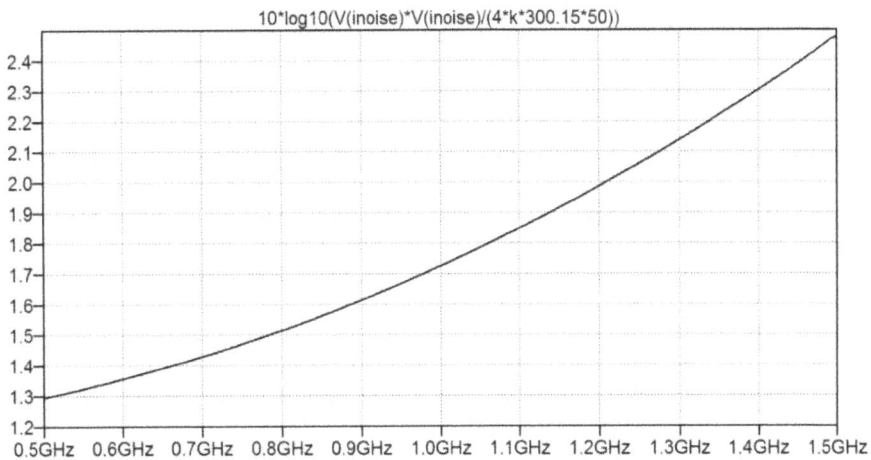

Figure 7-48 Minimum noise figure NFmin =1.72 dB at 1 GHz.

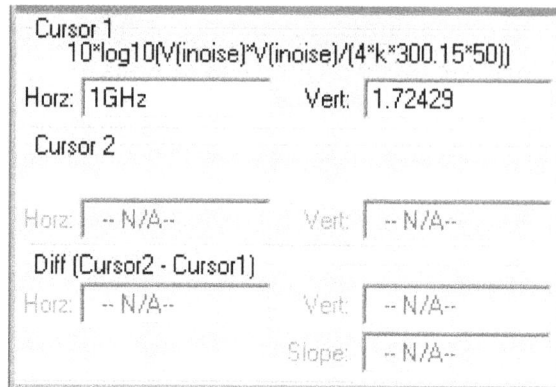

Figure 7-49 Captured NFmin of 1.72dB at 1 GHz

Problems

7-1. In this problem, we are designing a maximum gain amplifier at 800 MHz using the self-biased schematic shown below:

(a) Determine the value of R1 such that collector bias current of 25 mA is achieved.

(b) Sweep stability parameters K and B1 from 10 MHz to 10 GHz to show that the amplifier is potentially unstable.

(c) Use RC stabilizing network at input and shunt resistor across L1 to stabilize the amplifier.

(d) Simulate the stabilize amplifier circuit to achieve GTmax of at least 15 dB at 800 MHz.

(e) Calculate the device S-parameters at 800 MHz.

(f) Use the S-parameters at 800 MHz to design the simultaneous match input and output reflection coefficients.

(g) Simulate the S parameters and plot S_{21}, S_{11}, and S_{22}.

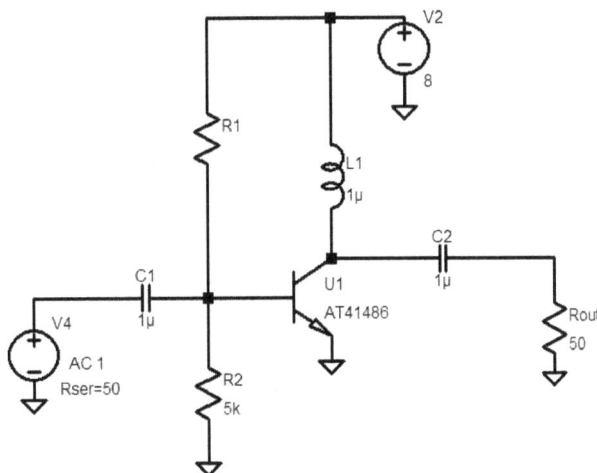

Figure P7.1 Schematic diagram of a self-biased amplifier

7-2. Design a Low Noise Amplifier at 500 MHz using the self-biased circuit of Fig. P7.1 that operates at 8V and 10mA. Make sure that the circuit is unconditionally stable over all frequencies.

(a) Determine the noise figure of the stabilized LNA at 500 MHz.

(b) Design the input matching network to achieve minimum noise figure. Use the tabulated Γ_{opt} at 500 MHz given by the manufacturer in the data sheet in Appendix C.

(c) Determine the device S-parameters at 0.5 GHz.

(d) Design the output matching network.

(e) Assemble the LNA and measure the minimum noise figure at 0.5 GHz.

7-3. Modify the amplifier of P7-2 to achieve input return loss better than 10 dB and gain of 20 dB at 500 MHz, if possible. Determine the NF at 500 MHz. Compare the new NF with the NFmin in problem 7-2.

References and Further Readings

[1]   Ali A. Behagi, *RF and Microwave Circuit Design*, A Design Approach Using (**ADS**), Techno Search, Ladera Ranch, CA 2015.

[2] Ali Behagi and Manou Ghanevati, *Fundamentals of RF and Microwave Circuit Design,* Ladera Ranch, CA 2017

[3] Dale D. Henkes, FAST: *Fast Amplifier Synthesis Tool*, Artech House Publishers, Norwood, MA. 2004

[4] Guillermo Gonzales, *Microwave Transistor Amplifiers – Analysis and Design*, Second Edition, Prentice Hall Inc., Upper Saddle River, NJ.

[5] Randy Rhea, *The Yin-Yang of Matching: Part 1 – Basic Matching Concepts*, High Frequency Electronics, March 2006

[6] Steve C. Cripps, *RF Power Amplifiers for Wireless Communications*, Artech House Publishers, Norwood, MA. 1999

[7] David M. Pozar, *Microwave Engineering*, Third Edition, John Wiley & Sons, New York, 2005

[8] R. Ludwig, P. Bretchko, *RF Circuit Design*, Theory and Applications, Prentice Hall, Upper Saddle River, NJ, 2000

[9] Behzad Razavi, *RF Microelectronics, Second Edition, Prentice Hall*, New York, 2015

[10] Ted Grosch, *Noise Concepts and Design, Noble Publishing*, Atlanta, GA, 2003

Chapter 8

Sinewave Oscillator Circuit Design

8.1 Introduction

Sinewave oscillators [1-8] are extensively used in radio frequency and microwave systems. In transmitter systems, sinewave oscillators provide carrier frequencies whereas in receiver systems, they drive the mixer stages and provide frequency translation of the received signal. This frequency translation from radio frequency (RF) to intermediate frequency (IF) is called down conversion. The process of frequency translation of an IF signal to RF frequency is called up-conversion. Sinewave oscillators are widely used in many up- and down-converters, signal generators, equipment, and instrumentations. In addition to sinewave waveforms, square-waves are extensively used in clock generation and ramp waveforms are used in time-base generation.

In general, a sinewave oscillator circuit can be constructed by means of an amplifier having positive feedback in the form of a frequency selecting block. The amplifier or gain element is often a single active device such as a FET, MOS, or a BJT. By proving positive feedback from output to the input of the amplifier or gain element, device thermal noise is amplified, and when a frequency selecting block is used in the feedback path, a well-defined sinewave waveform can be generated at the output. The feedback block can be constructed in several ways including LC tuned circuits, RC circuits, filters such a BPFs, or delay lines in the form of a long fiber optic cable. Frequency stability of the oscillator is dependent on the quality factor, Q, of the resonant tank. Depending on the desired frequency, a particular device, component, or circuit topology can be utilized. However, in all oscillator circuits, output power, frequency stability (often indicated by phase noise or jitter), and amplitude stability need to be properly addressed.

8.2 The Criteria for Oscillation

To explore the criteria for oscillation in a circuit, let us first consider the general block diagram of an amplifier with positive feedback as depicted in Figure 8-1.

Several fundamental assumptions are made: 1) The block shown as A represents the amplifier having open loop gain A, which in general is frequency dependent, that is A= A(s). 2) The feedback block β represents the frequency selecting block. It is frequency dependent that is β = β(s). 3) The input signal is only transmitted to output through the amplifier A and not through the feedback. 4) The output signal is transmitted from output to input through the feedback only and not the amplifier. 5) Feedback β is independent of loading effects from both input source signal and output load.

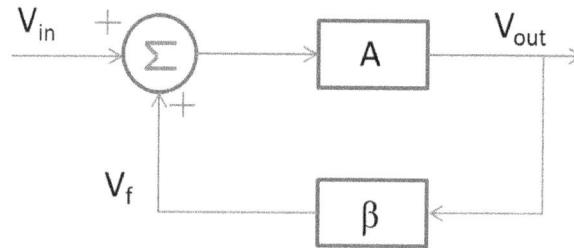

Figure 8-1 General block diagram of an amplifier with feedback

Examination of Figure 8-1 reveals that the output signal is sampled and conditioned by passing through the feedback circuit. The result is the signal V_f which adds to the input signal and is amplified by amplifier A resulting in output signal. Therefore, below equations can be written as

$$(V_{in} + V_f)A = V_{out} \tag{8-1}$$

$$V_{out} \cdot \beta = V_f \tag{8-2}$$

$$A_F = \frac{V_{out}}{V_{in}} = \frac{A}{1 - A\beta} = \frac{A}{1 + T} \tag{8-3}$$

Here, A is the gain of the amplifier independent of feedback circuit β. It is often called the *open-loop gain*. A_F is often called the *closed-loop-gain*. T = -Aβ is called the *loop gain* or return ratio.

When $|A_F| < |A|$, the feedback is *negative* and $|1 - A\beta| > 1$. Negative feedback is often used for amplifier circuit stabilization at the cost of reduced gain. When

$|A_F| > |A|$, the feedback is *positive* or *regenerative* and oscillation can result. Intuitively, A_F can increase when the feedback signal is in phase with the input signal. Therefore, the two signals add constructively. However, certain criteria need to be met for sustained oscillation in the circuit. If the product $A\beta$ approach 1, the denominator can approach 0 indicating infinite closed-loop-gain. However, the gain of the active device is limited due to its inherent nonlinearity and limited supply in the circuit.

The conditions for sustained oscillation for the feedback amplifier of Figure 8-1 is given as Barkhausen Criterion which states that

$$T(j2\pi f_0) = -1$$

The above equation can also be expressed as

$$T(j2\pi f_0) = 1 \quad \angle T(j2\pi f_0) = -180^0$$

Therefore, above equations indicate that for sustained oscillation, below conditions must be met:

1. The phase shift through the amplifier and feedback circuit must be 360°, or $2\pi n$. Phase of $A\beta$ must be 360°, or multiples of 2π.
2. The product of the magnitudes of the amplifier and feedback networks must be unity. Magnitude of $A\beta$ must be equal to unity.

In addition to existence of positive feedback in oscillator circuits, if we were to separate an oscillator circuit into an active device portion and a load, the output impedance of the active device must have a negative real part. Negative resistance concept has traditionally applied in design of one-port oscillator circuits including circuits used in radar and microwave applications. Negative resistance concept is closely related to the ability of an active device to generate or source electrical power. Unlike loads that absorb or sink electrical current or energy, active devices are capable of absorbing direct electrical current and converting part of this energy into a well-defined sinusoidal output waveform, thereby exhibiting negative resistance. A simple dc battery circuit connected to a load, for example, sources current and electrical energy and the existence of negative resistance across the terminals of the battery can be explained by applying voltage polarity convention and the application of Ohm's law.

Figure 8-2 depicts the presence of negative resistance at the output of an active device. While the static resistance of the device, v/i, is always positive, the device can exhibit negative resistance in the region between points a and b due to the negative slope of the i-v curve. Therefore, at the operating point shown as point Q, the device demonstrates incremental or small-signal negative resistance expressed as $r_n = dv/di$. If a tuned LC circuit is connected across the negative resistance device [2], stable oscillations can result under certain conditions.

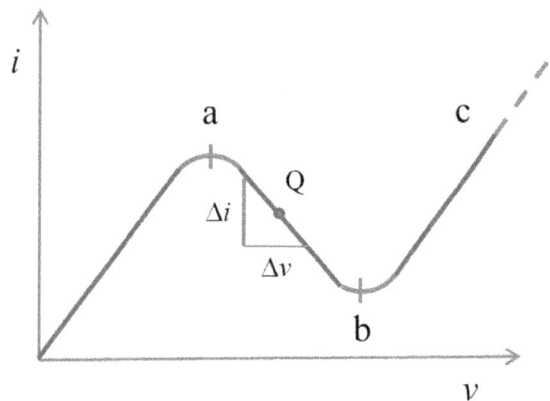

Figure 8-2 Voltage-stable negative resistance characteristics [2]

8.3 General Oscillator Configuration

Many oscillator circuits can be analyzed using the simple configuration shown in Figure 8.3. The active device is an Operation Amplifier (Op-Amp) having extremely high input resistance, small output resistance R_o, and open-circuit negative voltage gain $-A_v$. A network of passive components including Z_1, Z_2, and Z_3 constitutes the load and also provide positive feedback to the input of the active device, the Op-Amp. As the first step, we will find an expression for the *return ratio* $T = -A\beta$.

In figure 8-3, we first observe that $Z_L = Z_2 \parallel (Z_1 + Z_3)$. The feedback voltage V_f can be written as

$$V_f = V_{13} = \frac{V_o Z_1}{Z_1 + Z_3} \qquad (8\text{-}4)$$

where

$$\beta = \frac{Z_1}{Z_1 + Z_3} \qquad (8\text{-}5)$$

The relation between V_o and input voltage to the Op-Amp, V_{in}, is given by

$$V_o = \frac{A_v \, v_{in} \, Z_L}{R_o + Z_L} = \frac{A_v \, v_{in} \, Z_2 \, (Z_1 + Z_3)}{R_o(Z_1 + Z_2 + Z_3) + Z_2 \, (Z_1 + Z_3)} \qquad (8\text{-}6)$$

Or

$$V_o = \frac{A_v \, v_{in} \, Z_2 \, (Z_1 + Z_3)}{R_o(Z_1 + Z_2 + Z_3) + Z_2 \, (Z_1 + Z_3)} = A v_{in} \qquad (8\text{-}7)$$

Therefore, $T = -A\beta$ can be written as

$$T = -A\beta = \frac{-A_v \, Z_2 \, (Z_1 + Z_3)}{R_o(Z_1 + Z_2 + Z_3) + Z_2 \, (Z_1 + Z_3)} \frac{Z_1}{Z_1 + Z_3} \qquad (8\text{-}8)$$

which after simplification reduces to

$$T = -A\beta = \frac{-A_v \, Z_2 \, Z_1}{R_o(Z_1 + Z_2 + Z_3) + Z_2 \, (Z_1 + Z_3)} \qquad (8\text{-}9)$$

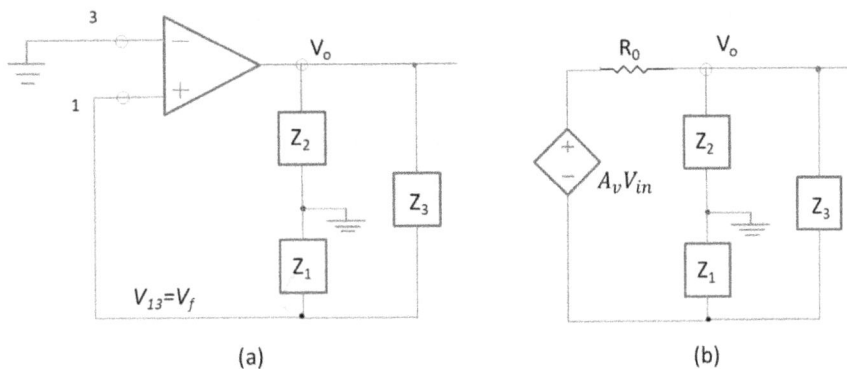

Figure 8-3 (a) General oscillator circuit and (b) The equivalent circuit of (a)

If we let $Z_1 = jX_1$, $Z_2 = jX_2$, and $Z_3 = jX_3$ in equation 8-9, we can write

$$T = -A\beta = \frac{A_v\, X_2\, X_1}{jR_o\,(X_1 + X_2 + X_3) - X_2\,(X_1 + X_3)} \qquad (8\text{-}10)$$

Examination of equation 8-10 reveals that for T to be real, the imaginary part of T has to be zero. Therefore,

$$X_1 + X_2 + X_3 = 0 \qquad (8\text{-}11)$$

Therefore T reduces to

$$T = -A\beta = \frac{-A_v\, X_1}{(X_1 + X_3)} \qquad (8\text{-}12)$$

Since $X_1 + X_3 = -X_2$, equation 8-12 reduces to

$$T = -A\beta = \frac{A_v\, X_1}{X_2} \qquad (8\text{-}13)$$

Equations 8-11 through 8-13 have important applications in oscillator circuit design. In general, the impedance $Z = jX$ can be impedance of an inductor or $Z = j\omega L$. Z can also represents impedance of a capacitor or $Z = (1/\, j\omega C) = -j(1/\omega C)$. In general, X can be equal to ωL if Z is an inductor or X can be equal to $-1/\omega C$ if Z is a capacitor.

From equation 8-11, we can conclude that if Z_1 and Z_2 are capacitors, Z_3 has to be an inductor. Similarly, if Z_1 and Z_2 are inductors, Z_3 has to be a capacitor. This design criterion constitutes the principle design guideline for two very important oscillator circuit topologies, namely *Colpitts* oscillators and *Hartley* oscillators. In a *Colpitts* oscillator circuit, Z_1 and Z_2 are capacitors while Z_3 is an inductor. In a *Hartley* oscillator circuit, Z_1 and Z_2 are inductors while Z_3 is a capacitor.

For a Colpitts oscillator circuit, let $X_1 = -1/\omega C_1$, $X_2 = -1/1/\omega C_2$, and $X_3 = \omega L$. Therefore, from equation 8-11 we can write

$$X_1 + X_2 = -X_3 \qquad (8\text{-}14)$$

Or

$$\left(-\frac{1}{\omega C_1}\right) + \left(-\frac{1}{\omega C_2}\right) = -\omega L \qquad (8\text{-}15)$$

Therefore, the frequency of oscillation ω is given by

$$\omega = 2\pi f = \frac{1}{\sqrt{LC_{eq}}} \qquad (8\text{-}16)$$

where

$$C_{eq} = \frac{C_1 C_2}{C_1 + C_2} \qquad (8\text{-}17)$$

In equation 8-13 we observe that the ratio of impedances Z_1 and Z_2 are proportional to amplifier gain A_v. To satisfy oscillation condition, Av has to be equal to X_2/X_1 so that magnitude of T (= -Aβ) = 1.

Example 8-1 Use the schematic diagram of Fig. 7-37 to design a *Colpitts* oscillator as follows

(a)   Remove the input source and the 50-Ω load. Add a resistor RE = 20 Ω and a CE = 0.5 pF to the basic amplifier. Include a load / feedback network such that Z1 and Z2 are impedance of capacitors with C2 = 2pF and C1 = 10pF. Let Z3 be impedance of an inductor where L3 = 5nH. Use transient analysis: start time = 0, stop time = 500ns, and step size = 0.1ns.

(b)   Compute the expected frequency of oscillation using equations 8-16 and 8-17 assuming low-frequency model Op-Amp analysis hold.

(c)   Simulate the above circuit in time domain and capture the output waveform. Show the start of oscillation and the waveform after about 500ns.

(d)   Insert two cursors to measure the frequency of oscillation when oscillation frequency has reached its steady-state condition. Determine the peak voltage and the output power assuming a load equal to 135. Ignore all the loading effects.

(e)   Perform Fast Fourier Transform (FFT) on the waveform to depict the frequency contents of the waveform. Verify the frequency of operation as obtained in part (c). Compare the magnitude of the second harmonic frequency to the fundamental.

(f)   Comment on the computed versus simulated oscillation frequency.

Solution (a): The schematic diagram of the Colpitts oscillator is depicted in Figure 8-4. The resistor RE is part of the DC biasing network for the transistor, however, it contributes to the stability of the circuit. The DC bias current is reduced due to the presence of RE. The capacitor CE is a bypass capacitor but it can affect the gain of the circuit.

Solution (b): Ceq = 1.67pF, L = 5nH, then f = $(1/(2\pi\sqrt{LC_{eq}}))$ = 1.743GHz.

Solution (c): Output waveform at the oscillator output is shown in Fig. 8-5.

Solution (d): To measure the frequency of oscillation, double-click on V(out) in Fig. 8-5 to create two cursors. Move the cursors to two consecutive peak voltages in the waveform and readout the oscillation frequency at about 1.53 GHz. The peak voltage is about 0.43 V and assuming RL = 135 Ω, the oscillator power of P = $[V^2 / (2\text{x}R_L)]$ = $[0.432^2 / (2\text{x}135)]$ = 0.69 mW = -1.6 dBm is expected.

Solution (e): To perform FFT, click on the waveform plot first, go to *view* and select *FFT*. FFT is performed on the time-domain waveform (Figure 8-6). To measure the frequency of oscillation, insert a cursor and move on the fundamental tone. Note that harmonics of the fundamental tones are also generated by the oscillator. The oscillation frequency is measured at about 1.53 GHz. For FFT to be more accurate, stop time must be large so that many waveform cycles are captured in time domain. In Figure 8-6 (a), the value of the fundamental is about 180 mV while the magnitude of second harmonic is about 9.7 mV. Therefore the voltage ratio of the second harmonic to the fundamental is (9.7/180) = 0.054 or 20*log(0.054) = -25.4 dB. From Figure 8-6 (b), we can see that the second harmonic voltage is 25.4 dB lower than that of the fundamental.

Solution (f): The frequency of oscillation is affected by loading effects of the resonator tank. This effect is called oscillator *load pulling*. The loading effects contain all the passive components in the circuit as well as the active device (the BJT) and its internal parasitic and due to packaging.

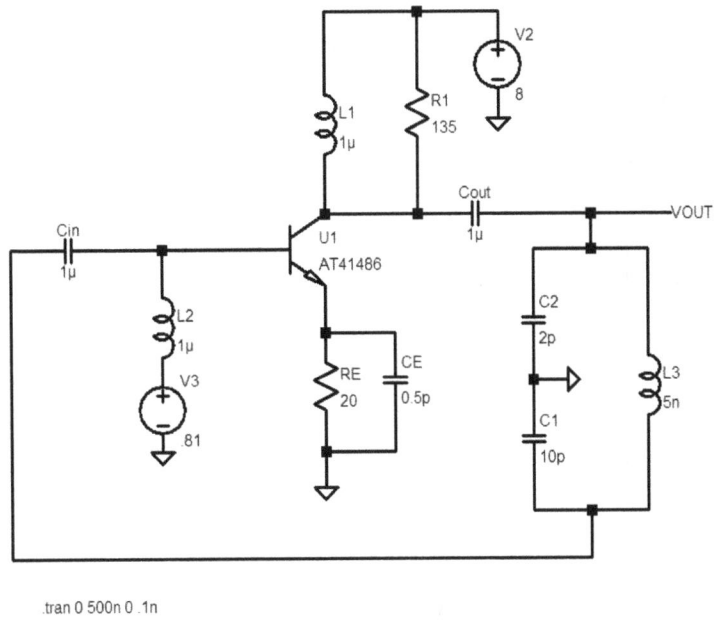

.tran 0 500n 0 .1n

Figure 8-4 Schematic diagram of a simple Colpitts oscillator

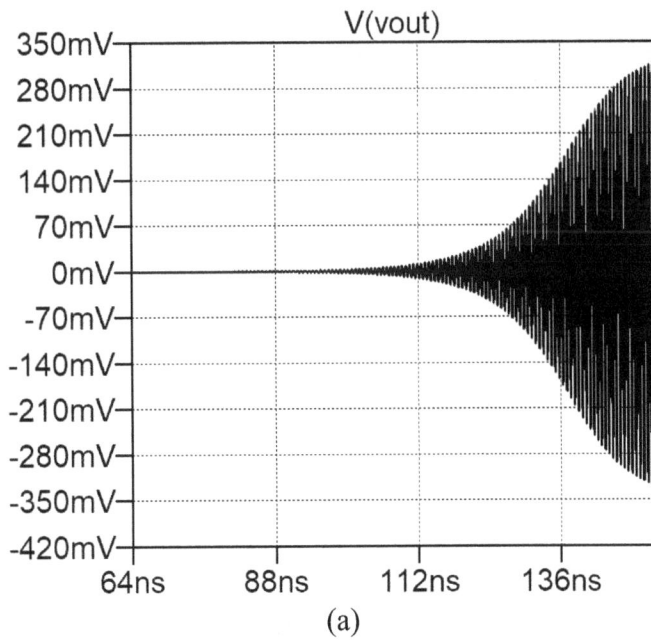

(a)

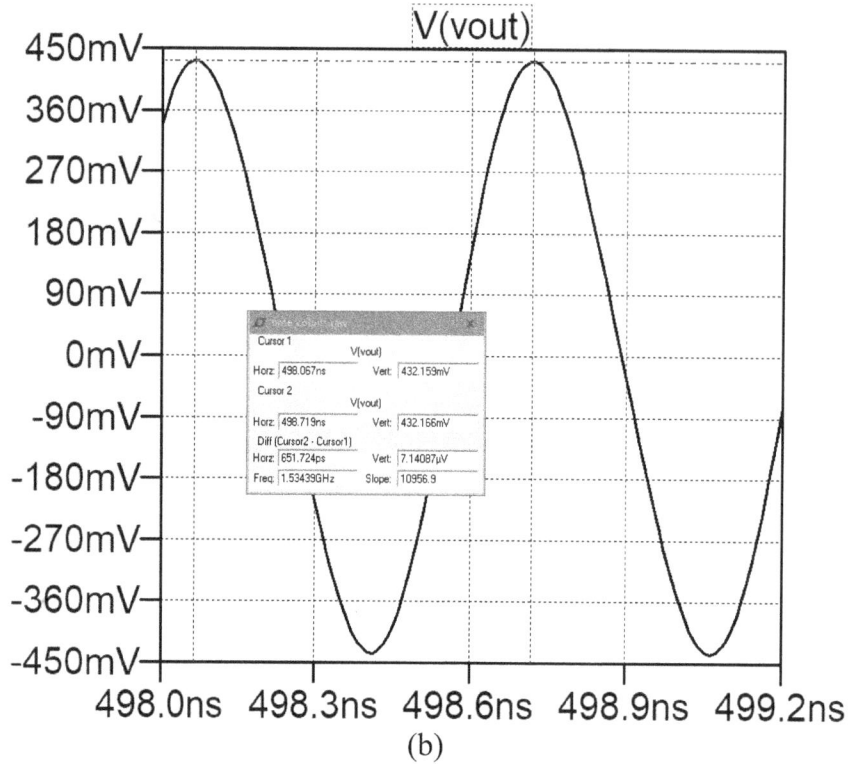

(b)

Figure 8-5 (a) output waveform at the output of the Colpitts oscillator showing start of oscillation, (b) Steady-state output waveform

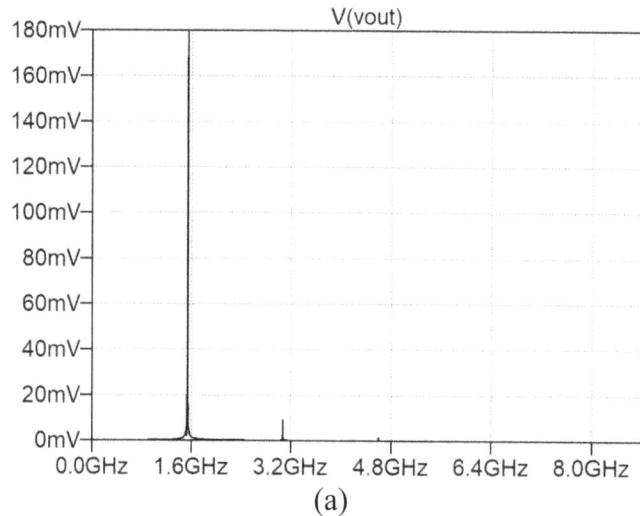

(a)

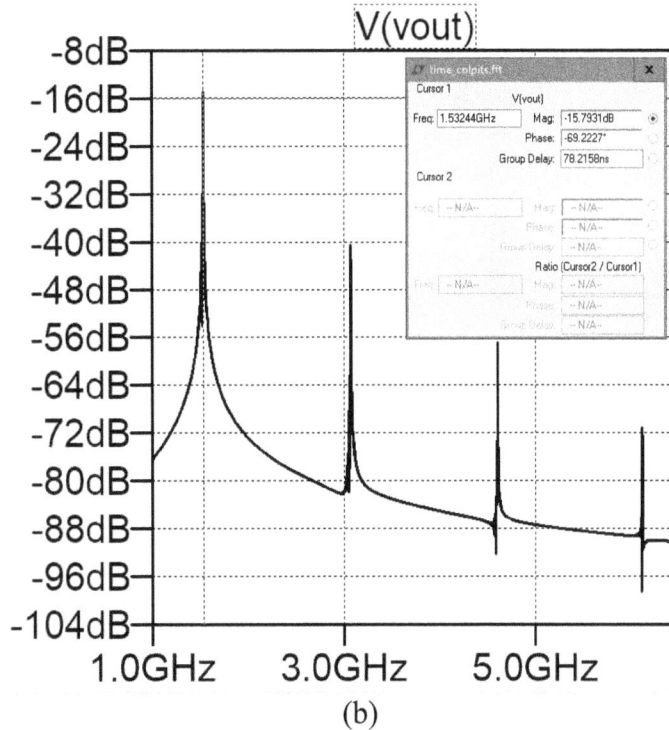

Figure 8-6 Frequency spectrum at oscillator output showing fundamental (at 1.53 GHz) and several of its harmonic frequencies linear (a) and (b) decibel scales

The oscillator circuit in Fig. 8-4 uses two separate voltage supplies to bias the collector and the base of the transistor. When *self-biasing* scheme is implemented, number of supply voltages can be reduced in a circuit. Often a voltage divider resistive network is used to generate a voltage at the gate of the device to ground. To improve stability of the operating condition, gate voltage, V_{BB} of about 10% of V_{CC} is used. Any change in the bias supply affects the oscillation frequency. This phenomenon is called *frequency pushing*.

Example 8-2 In schematic diagram of Fig. 8-4, eliminate the voltage supply biasing the base. Use self-biasing scheme to achieve the same DC voltage at the B-E junction for the transistor. Simulate the circuit to plot and measure the fundamental oscillator frequency.

Solution: A simple voltage divider consisting of 100 kΩ and a 25 kΩ resistors can be connected to the collector voltage supply to generate a voltage equal to 0.810 V at the base of the transistor, as shown in Fig. 8-7. The collector current is about 4.6 mA and the total power consumption is VCC. IC = 36.8 mW. The frequency spectrum of the oscillator in Fig. 8-7 is shown in Fig. 8-8, where the fundamental frequency at 1.54 GHz is shown. Note that the oscillator output power is much higher than the one in Example 8-1 due to increased bias current.

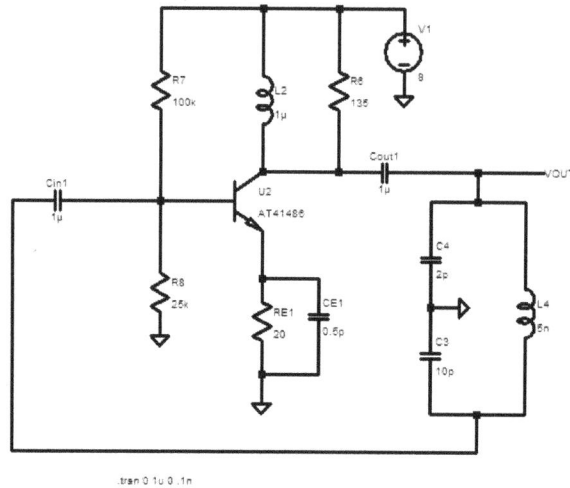

Figure 8-7 Revised schematic diagram of the Colpitts oscillator in Example 8-1

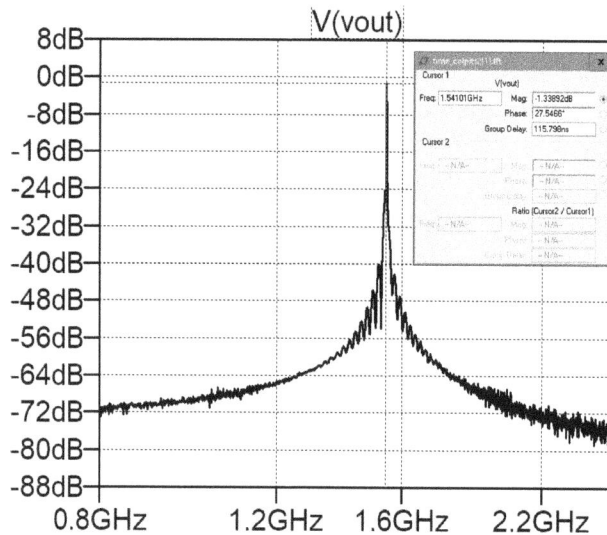

Figure 8-8 Frequency spectra at oscillator output showing fundamental frequency of about 1.54 GHz.

Suppose that the oscillator output is connected to a Spectrum Analyzer (SA) having 50 Ω input impedance. The amplitude of the observed signal is expected to be expressed in dBm (equal to $10log(P_{mW}/1mW)$, power in mW relative to 1 mW). For a high spectral purity signal, the fundamental signal should look like a line with zero bandwidth around the carrier frequency. Practical oscillators exhibit frequency modulated noise (FM noise) around the carrier frequency often called Phase Noise (PN). Spot phase noise for an oscillator can be easily measured using a spectrum analyzer.

Let Fig. 8-9 represent the amplitude spectra of the oscillator in Example 2. Assume the peak amplitude of oscillation is -1.4 dBm and let the frequency equal to 1.540 GHz (Cursor 1). Assume that at the offset frequency of 1.544, a noise power of -28.4 dBm is measured (Cursor 2). Therefore, a delta power equal to 27.1 dB is measured. Assume further that the resolution Bandwidth (RBW) on the spectrum analyzer has been adjusted to 10 kHz for this display. The noise power in 10 kHz bandwidth is equal to 10xlog(10,000) = 40 dB. As a result, the noise power at 1.544 GHz is 67.1 (27.1 + 40) dB below the power at 1.540 GHz in 1 Hz of bandwidth. We can say that the PN for this oscillator is 67.1 dBc/Hz at 4 MHz offset frequency (Cursor 2). This is the single sideband (SSB) PN of the oscillator at fundamental frequency of 1.54 GHz.

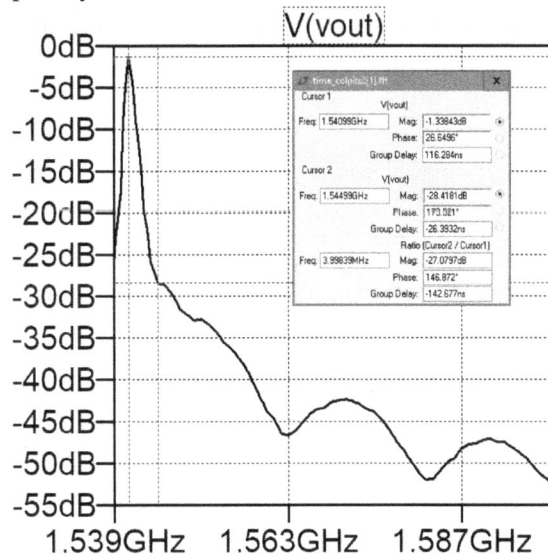

Figure 8-9 Phase noise measurement at 4 MHz offset frequency assuming RBW of 10 kHz.

Several variations of Colpitts oscillator circuits have been introduced in literature. In above example the feedback voltage is applied to the base of the transistor, and the emitter is grounded through the emitter resistor (common emitter topology). However, in some variations of Colpitts oscillator, the feedback voltage is applied to the emitter while transistor gate is grounded. An example of such a circuit topology is depicted in Figure 8-10 and examined at the end of the chapter problems.

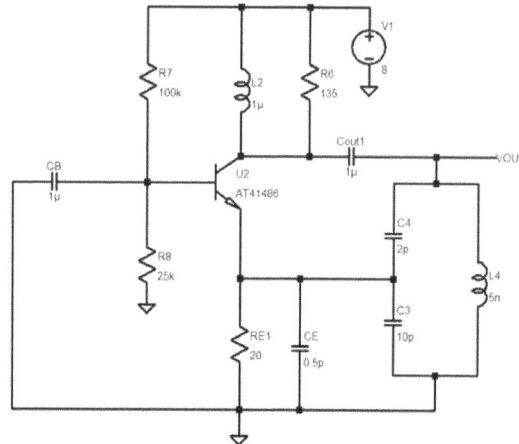

Figure 8-10 Schematic diagram of a common-base topology Colpitts oscillator

In a variation of Colpitts oscillator called Clad oscillator (Figure 8-11), a capacitor is used in series with the inductor in the resonator tank circuit.

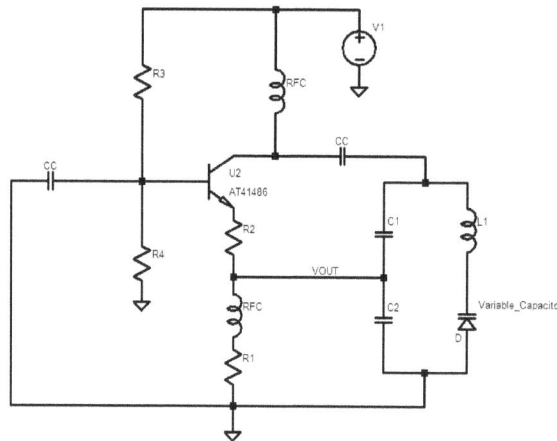

Figure 8-11 Schematic diagram of a Clad oscillator using a varactor diode

If a variable capacitor in a form of a varactor diode is used, the capacitance can be electronically controlled. Therefore, the effective inductance and the inductive impedance of the inductor in the resonator tank circuit can be quickly adjusted. As a result, this oscillator offers higher degree of frequency stability than Colpitts oscillator. Note that C1 and C2 are fixed capacitors and their ratio provided the necessary feedback for oscillation. The resistor R2 (also often used in Colpitts and Hartley oscillators) can further stabilize the circuit by controlling and enhancing the input impedance at the price of some loss in output power. The RFC in the emitter helps with the output swing and reduces RF power dissipation in RE (denoted as R1). The output of the circuit is often taken at the emitter terminal.

8.3.1 Practical Considerations in Oscillator Circuit Design

Two major criteria in oscillator design are the desired oscillation frequency and desired power delivered to a specific load. As a result, technology or device of choice is important. An important device or technology parameter related to frequency of oscillation is called f_{max}, or maximum frequency of operation. At this frequency, maximum available power gain for the transistor is equal to 1. The other important parameter is f_T, transistor's gain-bandwidth frequency. At this frequency, transistor's short circuit current gain h_{FE} is equal to 1 or unity. Examination of Appendix C shows that AT41486 has f_T equal to 6 GHz. Furthermore, the transistor offers $S21_{dB}$ greater than 1 at 6 GHz. However, for practical purposes f_T should be more than twice the oscillation frequency.

Device power handling capability is another important criterion in oscillator design considerations. For many oscillator circuits including most Colpitts oscillators, the transistor should be capable of dissipating 4 times the desired output power across the load. This is analogous to efficiency of different classes of amplifiers. Some techniques to enhance oscillator efficiency are discussed later. Device absolute maximum ratings for current, voltage, and power dissipation are tabulated by manufacturers; all these specifications need to be considered for practical circuit design.

The Colpitts oscillator circuits of figures 8-4 and 8-7 are considered common-emitter circuits as the output is taken from the collector terminal and it is sampled and inputted to the base terminal. Therefore, the base of the transistor can be considered as the input port to the oscillator circuit. For these configurations, it is possible to analyze the oscillator and predict its oscillation frequency using a small-signal hybrid-pi model. Furthermore, for accuracy, parasitic capacitors including Miller capacitors and nonlinear capacitor models can be incorporated. Of course, many of these features have already been devised in the transistor model. For high-frequency design including radio frequency and microwave frequencies, accurate device modeling and design topologies are very important. In reality, base to emitter capacitance, C_π, and base to collector capacitance $C\mu$, will affect the frequency of oscillation at high frequencies. As base to emitter resistance, $r_\pi(=\beta V_T/I_{CQ})$, is dependent on transistor common emitter forward short-circuit current gain , β, thermal temperature, V_T (equal to approximately 26mV), and the operating collector current I_{CQ}, both biasing conditions and frequency of operation affects not only the device input impedance but also the loading of the resonator tank. Also, the Q of the inductor in the resonator tank not only affects the output load but also the frequency of oscillation. As a result, it is mathematically possible to incorporate all these factors and apply a noise pulse (a current source) at the base of the transistor to start and predict the oscillation frequency more accurately. Finally, as the output power depends on device transconductance, g_m, higher oscillator output power is achieved at higher DC power consumption as $g_m = I_{CQ}/V_T$.

The oscillator circuits in Figures 8-10 (Common-base Colpitts oscillator) and 8-11 (Clad oscillator) are some of the widely used oscillator circuits. For applications requiring large output load resistor, Figure 8-10 may be more appropriate, whereas for smaller load resistors, Figure 8-11 may be more appropriate. In this configuration, usually no impedance matching is required and an output in a 50-Ω or 75-Ω system can be readily available. A more detailed analysis of common-base Colpitts oscillator is presented by Krauss [2], where resonator tank loading effects due to transistor and non-ideal components and their impact on the frequency of oscillation is discussed.

8.3.2 Application of Monte Carlo Simulation in Oscillator Circuit Design

So far, in all previous analyses, we assumed that all lumped components are ideal. Therefore, we were able to accurately predict the oscillation frequency and amplitude of the output signal. In practice, manufacturers design and categorize different components with certain tolerance in values. For example, a 100-Ω resistor can have values as low as 80 Ω and as high as 120 Ω having a tolerance of $\pm$ 20%. A common simulation method in many disciplines including engineering applications to predict the outcome of a specified result is Monte Carlo simulation. We apply Monte Carlo method in analyzing the oscillator circuit of Figure 8-7. Suppose that the most sensitive components in determining the frequency of oscillation in Figure 8-7 are the LC components used in the resonator tank circuit. Assume further that each inductor and capacitor has a 5% tolerance in regards to its nominal value. Monte Carlo method is capable of generating random numbers (e.g., component values) within the specified tolerances and specified number of trials to predict the expected range of amplitude and frequency of oscillation for this circuit.

The Monte Carlo analysis setup for Figure 8-7 is depicted in Figure 8-12. Its expected output waveform and frequency spectrum are shown in Figure 8-13 (a) and (b) respectively. As a result of 5% tolerance in component values and given only 5 trials, the frequency of oscillation is expected to vary from 1.51 GHz to 1.59 GHz. Note also that the amplitude of the waveform is impacted as a result of given tolerance values for each component. It can be shown (P. 7-4) that changes in frequency of oscillation can be minimized (to 10 MHz) by reducing the tolerance in component values to 1%.

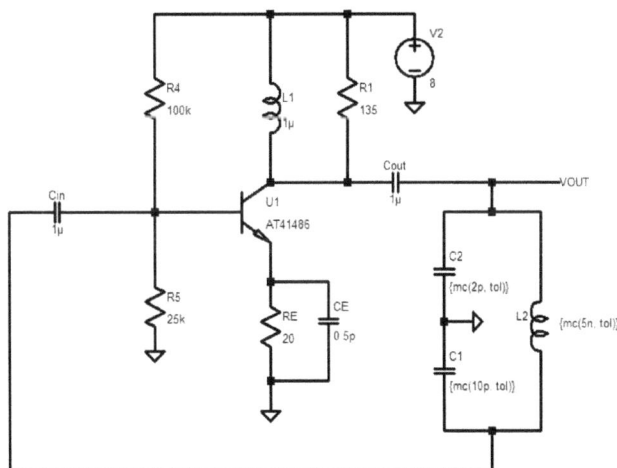

Figure 8-12 Application of Monte Carlo analysis to Figure 8-7

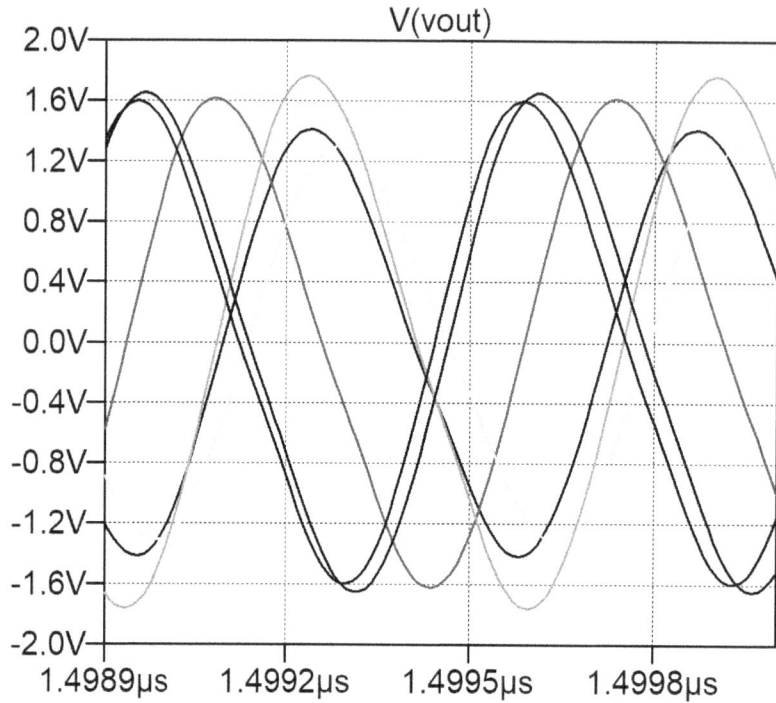

(a)

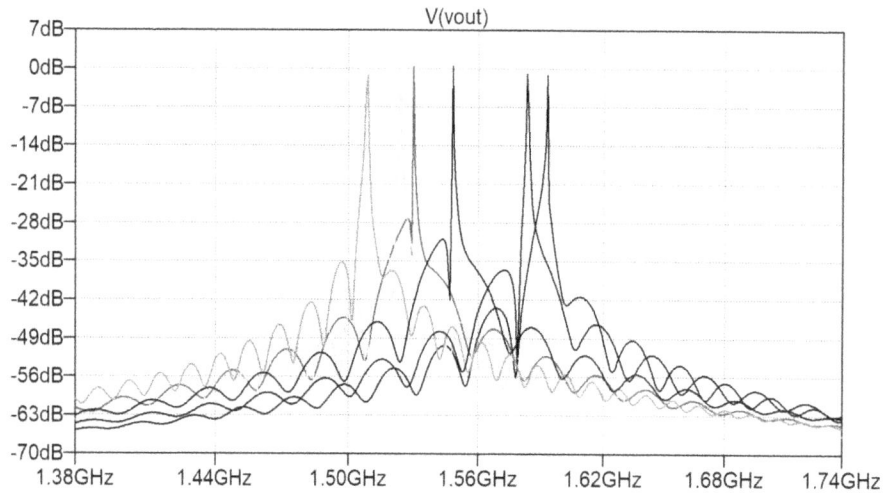

(b)

Figure 8-13 Output waveform (a) and frequency spectrum (b) for Fig. 8-12

8.4 Gunn Diode Oscillators

In section 8.2 we briefly discussed the phenomenon of negative incremental resistance observed in some devices including Gunn diodes. This is attributed to energy band gap properties of certain semiconductor materials such as gallium arsenide (GaAs), indium phosphide (InP), and cadmium telluride (CdTe) These materials have two closely spaced energy bands in the conduction band. Depending on the electric field strength in the material, electrons are either placed in the low-energy band (high-mobility band) or in the high-energy band (low-mobility band). Since the conductivity of the material is proportional to its mobility, the transition of electrons from low to high-energy band as a result of certain electric field strength causes a decrease in conductivity and hence current in the material. That is the current decreases with increased electric field or voltage resulting in incremental negative resistance. The application of this observation in realization of negative resistance oscillator is attributed to J. B. Gunn. Two major modes of operation in a Gunn diode device result in oscillations. These modes are called transit-time and limited-space-charge (LSA) [5]. Gunn oscillators operated in LSA mode can provide several watts of power with 20% efficiency. They can be operated at millimeter wave frequencies. At 100 GHz, several milliwatts of power can be achieved.

The general schematic diagram of a Gunn oscillator operating in the LSA mode is shown in Figure 8-14. The equivalent circuit of the Gunn device is shown in Figure 8-15. The negative resistance device is connected to a resonant LC tank circuit. An external voltage is applied across the Gunn device and output signal is selected across a load resistance R. The parallel combination of device resistance and the load resistance is –R Rd/(R – Rd). Since the device exhibits resistance in the range of -5 Ω to -20 Ω, the load resistance R should be slightly higher than – Rd for the parallel combination to remain negative. In practice, the Gunn device is positioned inside a rectangular cavity and under a post, as shown in Figure 8-16. Typical voltage of 12 V is applied to the Gunn device. The resonant frequency of the cavity is determined by adjusting its length using a sliding short. The sliding short acts as a short circuit for electric fields inside the cavity. The amount of coupling to the external waveguide is determined by adjusting the opening is the diaphragm which is inductive in nature. The post is usually inserted at the middle of the waveguide in the y-direction where the RF current is minimum, providing a low-impedance loading for the Gunn device. The resonance frequency inside the cavity can be fine-tuned by means of a tuning. Other cavity arrangements are possible. In another arrangement, a two-section quarter-wave transformer inside the cavity is used to transform high cavity impedance to low impedance at the Gunn device.

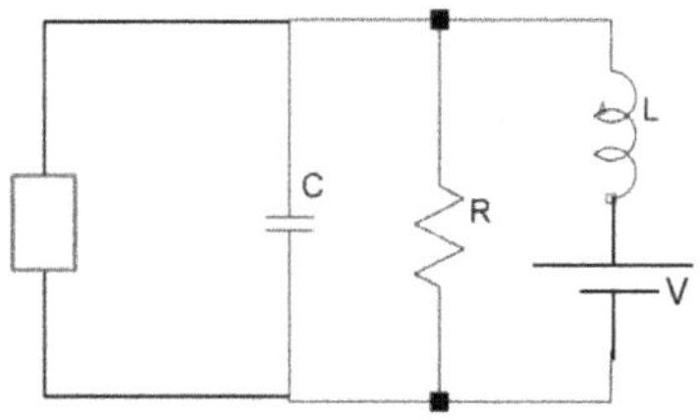

Figure 8-14 A basic Gunn oscillator operating in LSA mode [5]

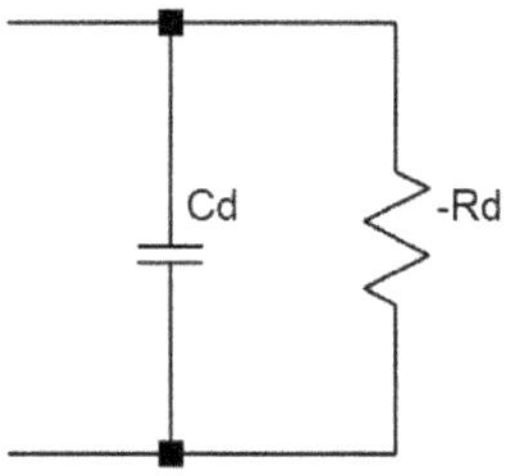

Figure 8-15 Equivalent circuit of a Gunn device oscillator operating in the LSA mode showing negative resistance [5]

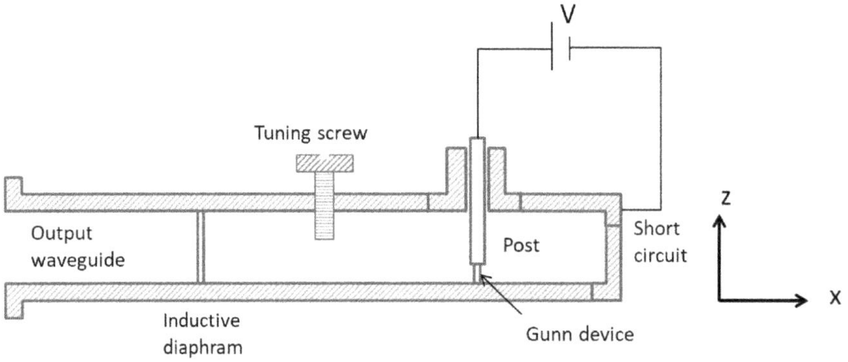

Figure 8-16 A simple waveguide cavity for a Gunn oscillator [5]

8.5 Tuned RC Oscillation Circuits

Tuned RC oscillators using Op-Amps are widely used for generation of low frequency oscillators up to several hundred kilohertz. These oscillators are in generation of audio frequency signals. Wien Bridge oscillator is a very popular low frequency oscillator in which a network of RC circuits along with an Op-Amp is used, as shown in Fig. 8-17. The RC network provides a positive feedback from output to input of the Op-Amp. It is responsible for determination of oscillation frequency. However, oscillation frequency depends on the slew rate of the Op-Amp. A network of resistor R1 and R2 provide the necessary gain by means of negative feedback. To determine the frequency of operation, we need to examine $T = -A\beta$, where $A = 1 + R_2/R_1$ is the gain of the Op-Amp and β is determined from feedback voltage, V_f.

The feedback voltage can be obtained from

$$V_f(s) = \frac{V_o . Z_2(s)}{Z_1 + Z_2(s)} \tag{8-18}$$

where $Z_1(s) = R + 1/SC$ and $Z_2 = R/(1 + RCS)$.

$$V_f(s = j\omega) = \frac{V_o . Z_2(j\omega)}{Z_1 + Z_2(j\omega)} = \frac{V_o . j\omega RCS}{1 - \omega^2 R^2 C^2 + 3j\omega RC} = V_o\beta \tag{8-19}$$

Therefore,

$$T(s = j\omega) = -A\beta = -(1 + \frac{R_2}{R_1})\frac{j\omega RCS}{1 - \omega^2 R^2 C^2 + 3j\omega RC} \tag{8-20}$$

As a required condition for sustained oscillation, T has to be real. Therefore, $1 - \omega^2 R^2 C^2 = 0$. As a result,

$$\omega = 2\pi f = \frac{1}{RC} \tag{8-21}$$

Furthermore,

$$T(s = j\omega) = A\beta = (1 + \frac{R_2}{R_1})\frac{1}{3} \qquad\qquad (8\text{-}22)$$

Equation 8-22 states that $\beta = 1/3$ and that the gain of oscillator has to be equal to 3 for $T = 1$. This requires $(R_2/R_1) = 2$. However, to sustain stable oscillation, we select the ratio (R_2/R_1) to be slightly greater than 2.

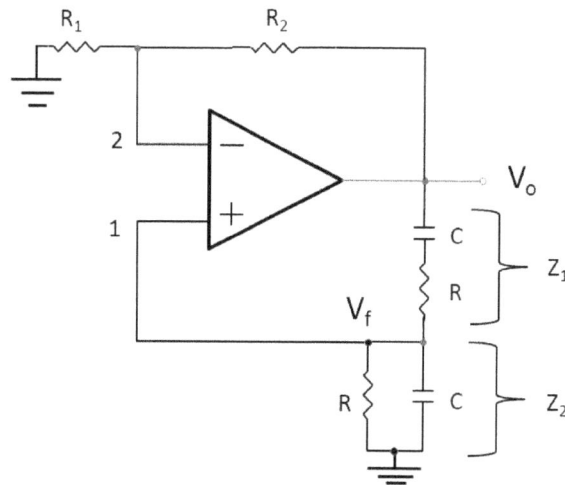

Figure 8-17 A Wien bridge oscillator circuit

Example 8-3 For the Wien bridge oscillator in Fig. 8-18, select LT1638 as the Op-Amp from LTspice component library. Let R_2= 10.1 kΩ, Let R_1= 5 kΩ, Let R= 2 kΩ, and C = C_1 = C_2 = 3 nF.

(a) Compute the expected gain and frequency of oscillation.
(b) Construct the schematic in LTspice. Use ± 5 V voltage supplies and simulate the circuit. Capture the output waveform.
(c) Adjust the value of C to achieve computed oscillation frequency if necessary.
(d) Comment on the time for start of oscillation. How can you speed up the time for start of oscillation?

Solution (a): Gain = A = $1 + \frac{10.1}{5}$ = 3.2 and f = $\dfrac{1}{2\pi(2\text{x}10^3)(3\text{x}10^{-9})}$ =26.53 kHz.

Solution (b): Figures 8-18 (a) and (b) show the schematic diagram of the Wien bridge oscillator and its output waveform where 21.3 kHz oscillation frequency is captured.

Solution (c): Changing the value of C1 and C2 to 2.31 nF result in oscillation frequency equal to 26.5 kHz.

Solution (d): Start of oscillation can become faster by adjusting the gain. If the gain is increased significantly, oscillation frequency and its shape including its harmonics can be affected.

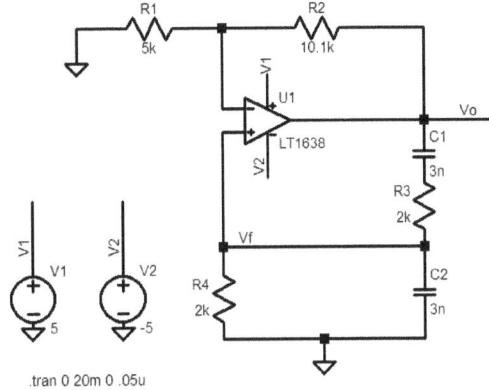

(a)

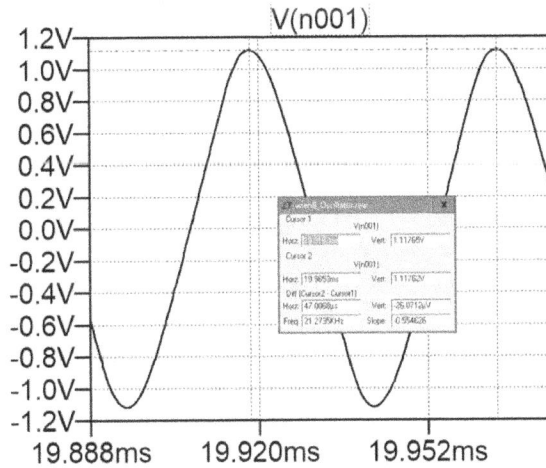

(b)

Figure 8-18 (a) The Wien bridge oscillator circuit for Example 8-3 and (b) Its output waveform showing oscillation frequency of 21.3 kHz

8.6 Special Oscillator Circuits

One of the most important and widely used class of oscillator circuits is crystal oscillators. Crystal (Xtal) oscillators offer highly stable signals from several tens of kHz to about 100 MHz or slightly higher. They can also be used as building blocks in equipment such as frequency synthesizers to generate microwave and millimeter-wave frequency signals through frequency multiplication using frequency multipliers. Many RF and microwave test equipment, for example, use a reference stable signal at 10 MHz. A crystalline such as quartz display piezoelectric effect in which mechanical deformation along one crystal axis results in electric potential along another crystal axis. Mechanical vibrations take place in a number of modes. The mode resulting in lowest-resonance frequency is called the *fundamental* mode. For this mode, the crystal acts like a series-tuned circuit having a large inductance with extremely high quality factor, Q (in the order of several tens of thousands). Higher order mode called *overtones* also exist. The symbol and the equivalent circuit for a crystal is shown in Figure 8-19.

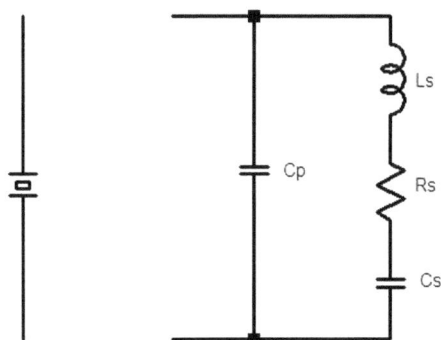

Figure 8-19 Crystal symbol and its equivalent circuit [2]

Two resonance frequencies can be associated with the above figure: a series resonance f_s and a parallel resonance frequency, f_p (see problem 8-5). Typically, f_p is about 1 percent higher than f_s. Furthermore, the crystal experiences extremely high impedance region between these two frequencies. As a result, when a crystal is used as part of a frequency selective network in an oscillator circuit, frequency stability is quickly achieved due to change in crystal's impedance as a result of changes in the frequency. Figure 8-20 shows a Colpitts oscillator utilizing a crystal at the base of its transistor. The crystal is used in its series-resonance mode, therefore, transistor base is effectively grounded as series resonance acts as a short circuit at the resonance frequency. The feedback again is proportional to the capacitor ratio C_1/C_2. The capacitance values for C_1 and C_2 must be higher than device input capacitance by several orders of magnitude to be effective.

While crystal oscillators can be *oven-controlled* to achieve much higher frequency stability over temperature, they exhibit excellent phase noise behavior. A crystal oscillator operating at 10 MHz can offer PN of about -165dBc/Hz at 1 kHz offset. If we were going to use this oscillator to generate a signal at 20 MHz, we would expect a 6-dB degradation in PN performance. In general, PN degrades according to $20*\log (f_{final}/f_{initial})$. When frequency doubles, PN degrades by $20*\log(2) = 6$ dB. Suppose that the 10-MHz source is frequency multiplied to generate a 10-GHz signal, we therefore, expect to achieve a PN degradation equal to $20*\log(10000/10) = 60$ dB. The 10-GHz source now has a PN equal to -105 dBc/Hz at 1 kHz offset. This is still a very good performance. However, several classes of microwave oscillators including *YIG* oscillators[4], *Dielectric Resonator* oscillators [3-4], and optoelectronic microwave oscillators [8] attempt at achieving high spectral purity signals at microwave frequencies.

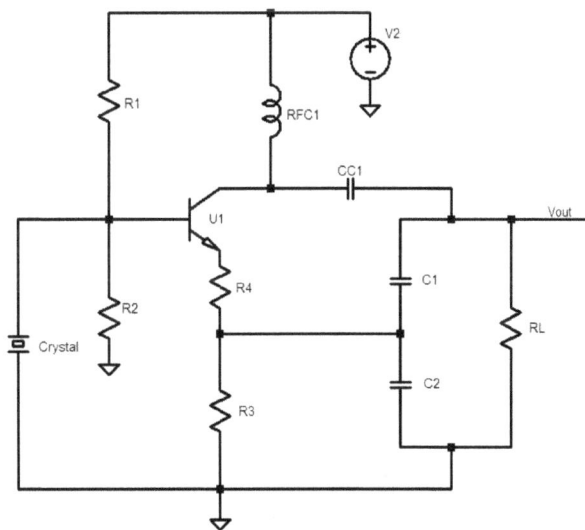

Figure 8-20 A Colpitts oscillator utilizing a series-resonance crystal at the base of its transistor

A unique class of efficient oscillators utilizes two transistors in a push-pull configuration. With this configuration, even harmonics in the output signal can be significantly reduced depending on class A or class AB operation. Some of these oscillators can operate as multifunction circuits providing both frequency oscillation and frequency mixing actions. They are called *Self-Oscillating Mixers (SOM)* and have found applications in frequency convertors. An example of this circuit suitable for operation at 900 MHz is depicted in Figure 8-21. Here, the two transistors are biased in class AB for efficient operation. A series LC circuit from the base of one transistor to the collector of the transistor provides the frequency selective circuit for oscillation. Once oscillation is achieved, an IF/RF signal can be applied to the base of each transistor using a *balun* (balanced input balanced output) since the transistor bases are out of phase by 180°. The resultant RF/IF signal after mixing with the LO (local oscillator) signal can then be sampled at the output of another balun. Therefore, this circuit plays the role of an oscillator and a mixer in one circuit.

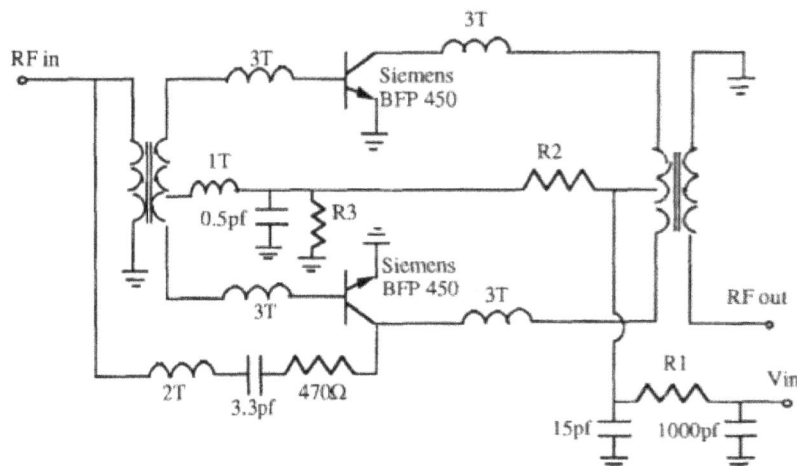

Figure 8-21 A Self-Oscillating Mixer (SOM) for operation at 900 MHz [5]

Another SOM circuit suitable for operation in modern Integrated Circuits (ICs) is shown in Figure 8-22. This circuit is based on the push-pull differential amplifier (after Gilbert cell) operating in calss AB as a gain block. A push-pull concept is effective when matched pair transistors and components are used in the circuit, which makes this design most suitable for MMIC and IC realization. Matched pair Si BJTs in SOIC-8 plastic package have been used in this circuit. Here, the feedback resistors, R_f, sample the collector current in transistor Q1 (Q2) and provide a positive feedback voltage to the base of transistors Q2 (Q1). Thus, the bases of both transistors are 180° out of phase due to the coupling between the collector of Q1 (Q2) with the base of Q2 (Q1). The class AB push-pull amplifier provides high gain as long as this $180°$ phase condition is satisfied. The phase condition necessary for push-pull amplification is destroyed at all frequencies except at the resonant frequency of the resonant tank circuit. Therefore, oscillation frequency is established by the parallel resonator circuit, which acts as an open circuit at the frequency of oscillation. This of course depends on the resonator tank Q factor and resonator loading effects. The capacitor in the resonant tank circuit is realized using a varactor diode. As a result, the oscillator circuit becomes a voltage-controlled oscillator (VCO), capable of achieving 90 MHz of tuning for the oscillating frequency signal. By applying an IF/RF to the base of transistor Q3, the circuit becomes a self-oscillating mixer (SOM). The RF/IF signals after mixing action with the oscillating signal can be samples at the output using a balun. Stable oscillation (5 dBm output) and good up-conversion gain (20 dB) has been reported with very low mixer NF. The circuit can also be injection-locked and phased-locked to an external signal (ILPL) for certain applications.

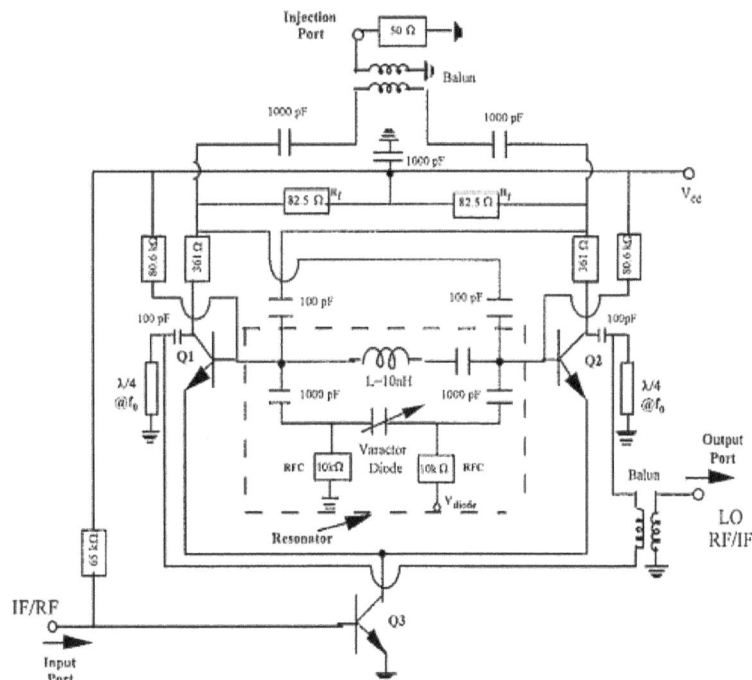

Figure 8-22 A multi-function Self-Oscillating Mixer (SOM) for operation at UHF frequencies [6]

Modern technologies such as 40nm CMOS technology allow for realization of oscillator/VCO circuits suitable for microwave and millimeter-wave frequency operation in highly integrated circuits. All active devices including transistors and diodes and passive components including resistors, capacitors, and inductors are realized using the same technology. An example of such circuits is depicted in Figure 8-23. The core of the circuit is designed based on Gilbert cell that is widely used in modern IC technology. Here, both PMOS and NMOS are used in a differential configuration to achieve low current consumption. Stable oscillation from about 22 GHz to 30 GHz is achieved with this extremely low DC. By using MOS varactors in the resonant tank circuit, the oscillation frequency can be tuned resulting in a VCO. Addition of electronically controlled group of parallel capacitors called cBank across the resonator circuit makes wideband electronic tuning of oscillation frequency possible.

The advantage of high-frequency oscillator design is that modern ICs and technology makes realization of such circuits possible. Furthermore, by method of frequency division to lower frequency, sources with much lower PN are possible. For example, if the oscillation frequency of 25 GHz is divided down to about 2 GHz, a PN improvement of about 22 dB can be achieved (20*log(2/25) = -21.94 dB) at the same carrier offset frequency.

(a)

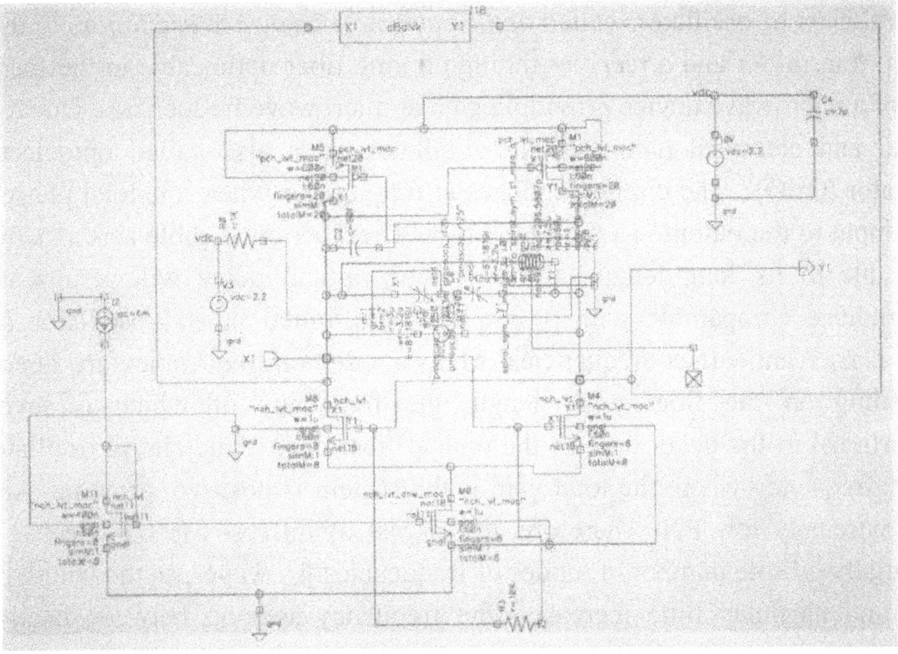

(b)

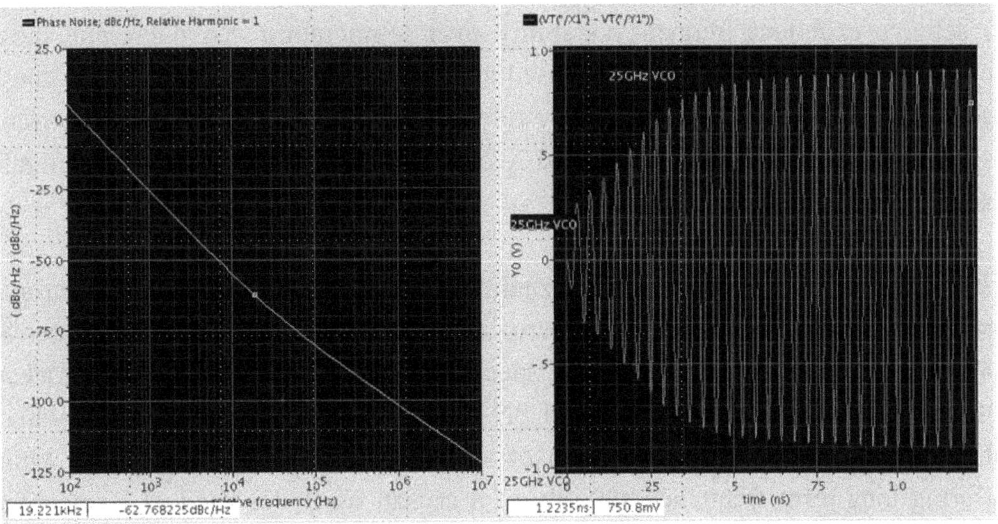

(c)

Figure 8-23 An oscillator (and VCO) circuit for operation at ≥25 GHz using 40nm CMOS technology [Courtesy of RFPTA, USA]: (a) Its simplified schematic diagram, (b) its realization in Cadence and (c) its output waveform showing oscillating frequency at 25 GHz and its corresponding PN plot, where PN of -62.8 dBc/Hz at 19.2 kHz offset is shown.

A novel class of oscillators called *optoelectronic microwave oscillators* [7] uses an optical transmitter and a receiver through a long fiber optic cable in the feedback path of a microwave device providing gain at microwave frequencies. Due to both optical and electrical nature of the oscillator, it is also called optoelectronic oscillator (OEO). The circuit oscillates at frequencies where the total phase shift from input to the output is a multiple of 2π. The fiber optic cable acts as a high Q filter due to its long length. Using a long optical delay will results in PN performance comparable with a frequency-multiplied crystal oscillator at the same carrier and offset frequencies. Many modes or frequencies are generated depending on the fiber optic length the frequency of which is inversely proportional to the delay time of the optical fiber cable. The circuit oscillates up to any frequency where the total gain is the system is positive, providing signals with extremely low PN. Therefore, any signal of interest can be selected from potentially infinite number of modes or frequencies. However, as the length of the fiber and its delay time increases, the frequency spacing between oscillating frequencies decreases. As a result, a very high-Q BPF is used in the feedback path for selection of desired signal. Nevertheless, the realization of such filters at high frequencies becomes a big challenge. Another challenge with the entire system is to achieve excellent frequency stability over temperature. To overcome these challenges, Ghanevati et al. [8] proposed and implemented phase-locking of these oscillators to an external highly stable reference oscillator. Several optoelectronic oscillators at L band (1 - 2 GHz) and X band (7 - 11.2 GHz) were designed and tested.

The schematic diagram for a free-running and a phase-locked optoelectronic oscillator suitable for operation at microwave frequencies are shown in Figures 8-24 and Figures 8-25. The frequency spectrum of free-running and phase-locked OEO at L band with their associated PN plots are shown in Figure 8-26, where phase-locking results in suppression of spurious modes by about 20 dB while offering long term frequency stability of a crystal oscillator used as a reference external source [8]. Excellent PN (better than -124dBc/Hz at 10 kHz offset frequency) is achieved by the phased-locked OEO, however, the performance shown here is limited by the capability of the equipment for lower offset frequencies within the *loop filter bandwidth* of the phase-locked oscillator. Accurate PN measurements at low offset frequencies can be accomplished by using a novel method called *frequency discriminator* method.

While phase-locking to an external source can provide both short-term and long-term frequency stability, there has been a great deal of research in transforming these oscillators to small form factor circuits.

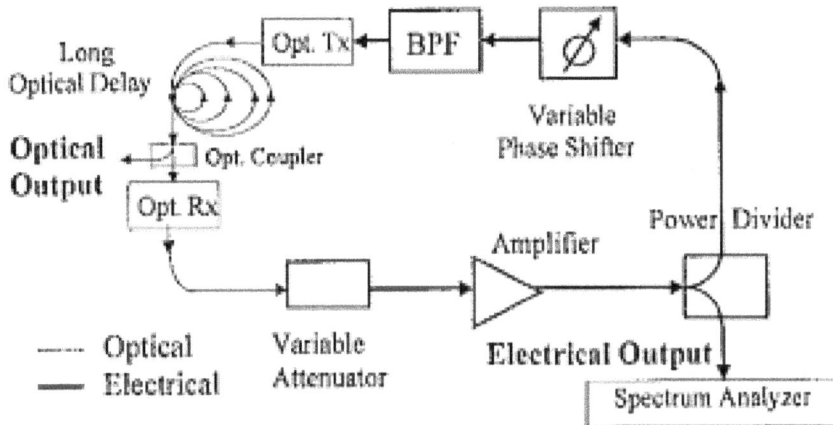

Figure 8-24 A free-running optoelectronic oscillator for microwave applications [9]

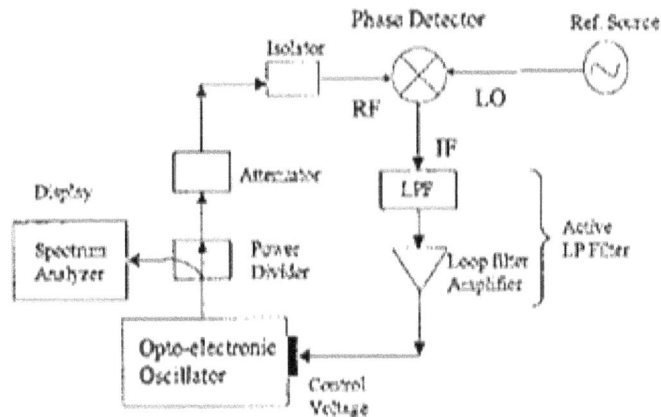

Figure 8-25 The block diagram of the phase-locked optoelectronic oscillator for microwave applications [8]

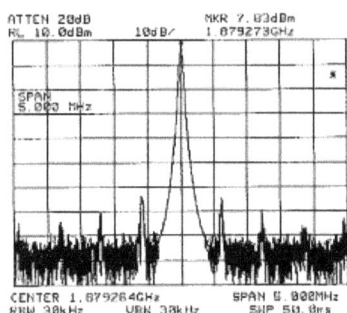

Fig. 2   Power spectrum of the free-running OEO at L band showing mode spacing of about 670 kHz, span of 5 MHz and RBW of 30 kHz.

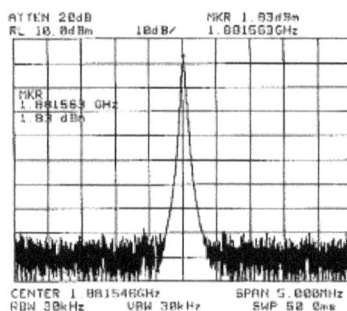

Fig. 5   Power spectrum of the phased-locked OEO at L band showing single-mode operation.

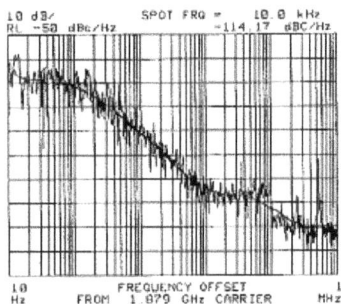

Fig. 3   Phase noise plot of the OEO at L band using HP 8562E spectrum analyzer.

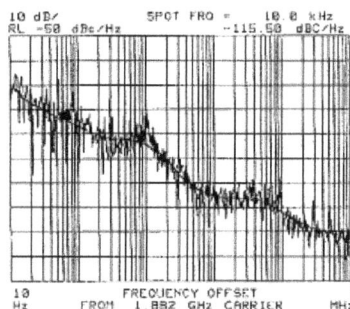

Fig. 6   Plot of phase noise for the phased-locked OEO at L band.

Figure 8-26 Frequency spectrum of free-running and phase-locked OEO at L band with their associated PN plots [8]

8.7 Dielectric Resonator Oscillators

Finally, we briefly discuss one of the widely used class of microwave oscillators know as Dielectric Resonator Oscillators. These oscillators can be designed at microwave frequencies between a few GHz to well over 20 GHz. They use a high-Q, compact, and temperature stable dielectric resonator (DR) as part of frequency stabilizing element to achieve spectrally clean signal sources. Q factor of several thousand (9000 or higher at 10 GHz) can be easily achieved with these ceramic resonators. Therefore, when used as the primary resonator element in variety of circuits including FET and bipolar-based transistor circuits, they determine the frequency of oscillation. These resonators can be used in a variety of oscillator configurations. They can be used as frequency stabilizing elements in an oscillator circuit, or as the primary resonator element. Figure 8-27 shows as example of a DR-based oscillator circuit where the resonator is used in a positive feedback configuration to provide coupling between the drain and gate of a FET device. The DR disk is situated at a distance λ /4 from the open-circuited ends of the coupling line where the standing wave and magnetic field intensity is high and coupling between the lines is the strongest. The length of the line segments l_1 and l_2 can be optimized so that correct phase for the feedback voltage can be achieved. The amount of coupling can also be adjusted by adjusting the position of the resonator. Sometimes, the frequency of oscillation can be adjusted mechanically by inserting a screw in the DR disk. Electronic control of oscillation frequency can be achieved by using varactor diodes (typically at the source of the FET device). In Fig. 8-27, biasing networks for the drain (positive voltage) and gate (negative voltage) of the device is shown. The load at the output (50 Ω resistor) is ac-coupled to the drain of the transistor using a dc blocking capacitor.

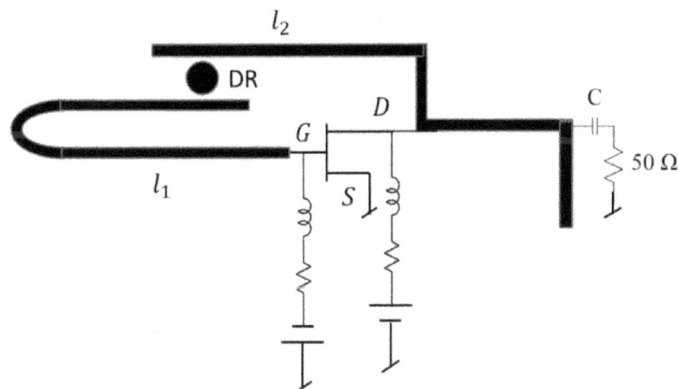

Figure 8-27 An FET oscillator using a dielectric resonator in the feedback path from drain to the gate [3]

Problems

8-1. Construct the schematic diagram of the Colpitts oscillator below.

(a) Let V1 = 5 v, R6 = 500 Ω, and RE = 20 Ω. Simulate the circuit and comment on the output waveform.
(b) Let V1 = 12 v, R6 = 500 Ω, and RE = 20 Ω. Simulate the circuit and comment on the output waveform.
(c) Let V1 = 12 v, R6 = 500 Ω, and RE = 100 Ω. Simulate the circuit and comment on the output waveform. Perform FFT on the output waveform. Comment on the output power by examining the frequency spectra.

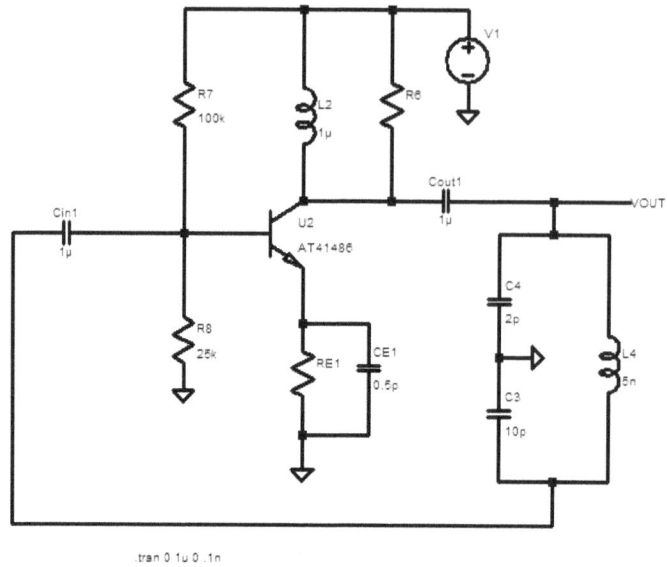

Figure P. 8-1 Schematic diagram of a Colpitts oscillator

8-2. Construct the schematic diagram of the Colpitts oscillator below in which the transistor base is grounded. Simulate the circuit and observe the time domain output waveform and its frequency spectra. Compare the result with that of Example 8-2.

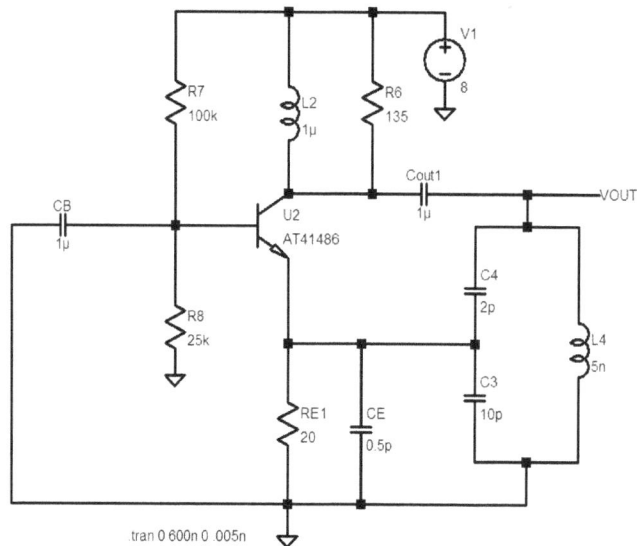

Figure P. 8-2 Schematic diagram of a Colpitts oscillator with base shorted out.

8-3. Construct the schematic diagram of the Wien bridge oscillator in Example 8-2. Vary the negative supply voltage from 0 to -5 V in 1 V steps. Simulate the circuit and comment on output waveform for each voltage.

8-4. Assume 1% tolerance in component values for the resonator tank circuit of Figure 8-12. Perform Monte Carlo analysis using 5 trials, Determine the expected change in frequency of oscillation.

8-5. For the equivalent circuit of a crystal shown in Figure 8-19, show that the difference between f_s and f_p is given by

$$\Delta f = f_p - f_s = f_s \left[1 - \left(1 + \frac{C_s}{C_p} \right)^{1/2} \right], \text{ where}$$

$$f_s = \frac{1}{2\pi\sqrt{L_s C_s}}$$

References and Further Readings

[1]   Jacob Millman & Arvin Grabel, *Microelectronics*, Second Edition, McGraw-Hill, Inc., New York, 1987.

[2]   Herbert L. Kraus, Charles W. Bostian, and Fredrick H. Raab, *Solid State Radio Engineering*, John Wiley & Sons, Inc., New York, 1980.

[3]   Robert E. Collin, *Foundations for Microwave*, Second Edition, McGraw-Hill, Inc., New York, 1996.

[4]   Guillermo Gonzales, *Microwave Transistor Amplifiers – Analysis and Design*, Second Edition, Prentice Hall Inc., Upper Saddle River, NJ.

[5]   V. Thangavelu, H. P. Moyer, M. Ghanevati, A. S. Daryoush, and R. Gutierrez, " Push-Pull Frequency Converter for Mobile Communication, " Microwave Symposium Digest, 1997, IEEE MTT-S International, Volume: 2, 1997, Page(s): 661 –664.

[6]   M. Ghanevati and A. S. Daryoush, "A low-power-consuming SOM for wireless communications," Microwave Theory and Techniques, IEEE Transactions on, Volume: 49 Issue: 7, July 2001, Page(s): 1348 -1351.

[7]   XS Yao and L Maleki , "Optoelectronic Microwave Oscillator," JOSA B, 1996.

[8]   M. Ghanevati[+], R. Saedi*, and R. Gandham, "Phase Locking of Optoelectronic Oscillators," Princeton Optronics, Princeton, NJ 08550, U.S.A., June 2000. Submitted to PTL, 2000.

Appendix A

Transmission Line Equations and Parameters

We start our analysis by first considering a lumped element representation of a two-wire transmission line segment of length Δz, as shown in Fig. 2-1 and Fig. A-1. In the figure b, R is the per unit length series resistance of both lines in Ω/m, L is the per unit length series inductance of both lines in Henries/meter, G is the per unit length shunt conductance of the line in Siemens/meter, and C is the per unit length capacitance of the line in Farads/meter.

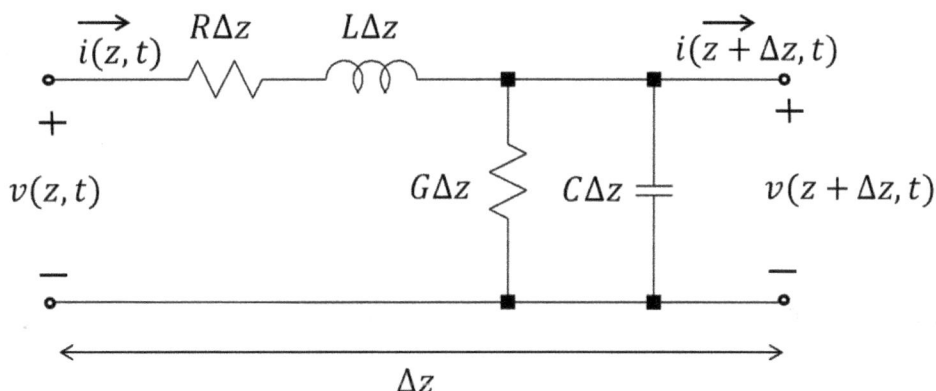

Figure A-1 Lumped element representation of a transmission line segment

Using Kirchhoff's voltage law, KVL, we can write

$$v(z, t) - i(z, t)R\Delta z - L\Delta z \frac{\partial i(z, t)}{\partial t} - v(z + \Delta z, t) = 0 \qquad \text{(A-1)}$$

$$-\frac{v(z + \Delta z, t) - v(z, t)}{\Delta z} = i(z, t)R + L \frac{\partial i(z, t)}{\partial t} \qquad \text{(A-2)}$$

In the limit as Δz goes to zero, the above equation reduces to

$$-\frac{\partial v(z, t)}{\partial z} = i(z, t)R + L \frac{\partial i(z, t)}{\partial t} \qquad \text{(A-3)}$$

Using Kirchhoff's current law, KCL, we can write

$$i(z,t) - v(z + \Delta z, t)G\Delta z - C\Delta z \frac{\partial v(z + \Delta z, t)}{\partial t} - i(z + \Delta z, t) = 0 \qquad \text{(A-4)}$$

$$-\frac{i(z + \Delta z, t) - i(z,t)}{\Delta z} = v(z + \Delta z, t)G + C \frac{\partial v(z + \Delta z, t)}{\partial t} \qquad \text{(A-5)}$$

In the limit as Δz goes to zero, the above equation reduces to

$$-\frac{\partial i(z,t)}{\partial z} = v(z,t)G + C \frac{\partial v(z,t)}{\partial t} \qquad \text{(A-6)}$$

Let

$$v(z,t) = Re\{V(z)e^{j\omega t}\} \qquad \text{(A-7)}$$

$$i(z,t) = Re\{V(z)e^{j\omega t}\} \qquad \text{(A-8)}$$

Using the above relations in (A-3) and (A-6), result in

$$-\frac{dV(z)}{dz} = (R + j\omega L)I(z) \qquad \text{(A-9)}$$

$$-\frac{dI(z)}{dz} = (G + j\omega C)V(z) \qquad \text{(A-10)}$$

Since the voltage and current in above equations are coupled together, a second derivative can be performed to decouple the quantities. Taking derivative with respect to z on both sides of (A-9) results in

$$-\frac{d^2V(z)}{dz^2} = (R + j\omega L) \frac{dI(z)}{dz} \qquad \text{(A-11)}$$

Substituting Eq. (A-10) in (A-11), results in

$$-\frac{d^2V(z)}{dz^2} = -(R + j\omega L)\,(G + j\omega C)V(z) \tag{A-12}$$

or

$$\frac{d^2V(z)}{dz^2} - \gamma^2 V(z) = 0 \tag{A-13}$$

where

$$\gamma = \sqrt{(R + j\omega L)\,(G + j\omega C)} \quad m^{-1} \tag{A-14}$$

Following the same procedure as above, we can write

$$\frac{d^2I(z)}{dz^2} - \gamma^2 I(z) = 0 \tag{A-15}$$

To satisfy equations (A.13) through (A.15), the voltage and current wave equations can be given as

$$V(z) = A\,e^{-\gamma z} + B\,e^{\gamma z} \tag{A-16}$$

$$I(z) = \bar{A}\,e^{-\gamma z} + \bar{B}\,e^{\gamma z} \tag{A-17}$$

Therefore,

$$\frac{dI(z)}{dz} = -\gamma \bar{A}\,e^{-\gamma z} + \gamma\,\bar{B}e^{\gamma z} \tag{A-18}$$

Also,

$$\frac{dI(z)}{dz} + (G + j\omega C)(R + j\omega L)V(z) = 0 \tag{A-19}$$

Therefore,

$$-\gamma \bar{A}\,e^{-\gamma z} + \gamma\bar{B}e^{\gamma z} + (G + j\omega C)(Ae^{-\gamma z} + Be^{\gamma z}) = 0 \tag{A-20}$$

From which we can obtain the relationship between A and $\bar{A}$:

$$\bar{A} = A\frac{(G+j\omega C)}{\gamma} = A\frac{\gamma}{(R+j\omega L)} = A\frac{\sqrt{(G+j\omega C)(R+j\omega L)}}{(R+j\omega L)} = A\frac{1}{Z_0} \tag{A-21}$$

Following the same procedure as above, the relationship between B and $\bar{B}$ can be determined resulting in

$$I(z) = \frac{A}{Z_0} e^{-\gamma z} - \frac{B}{Z_0} e^{\gamma z} \tag{A-22}$$

where,

$$Z_0 = \sqrt{\frac{R + j\omega L}{G + j\omega C}}$$

The quantity Z_0 is related to the voltage and current ratio of incident or reflected waves as if the medium is unbounded, that is the wave travels in one direction only.

Now, consider the finite transmission line shown below, where it is connected to a load impedance Z_L at one end and a voltage source having internal source impedance Z_S at the other end. We note that $V_{in} = I_{in} Z_{in}$ and $V_L = I_L Z_L$. The voltage and current at any point on the transmission line can be determined if A and B are known. To determine A and B, we can use the boundary condition for current and voltage at the load where z = d and ź = 0. Therefore, the expressions for current and voltage reduce to

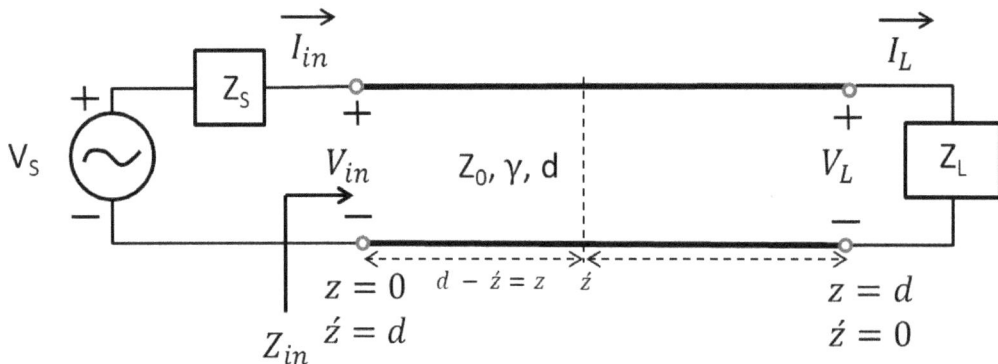

Figure A-2 A finite length transmission line with source and load arrangement

$$I(z = d) = \frac{A}{Z_0} e^{-\gamma d} - \frac{B}{Z_0} e^{\gamma d} = I_L \qquad \text{(A-23)}$$

and

$$V(z = d) = A e^{-\gamma d} + B e^{\gamma d} = V_L \qquad \text{(A-24)}$$

Solving the above equations simultaneously for A and B and noting that voltage at the load $V_L = I_L Z_L$, we obtain

$$A = \frac{e^{\gamma d}}{2} I_L (Z_L + Z_0) \qquad \text{(A-25)}$$

$$B = \frac{e^{-\gamma d}}{2} I_L (Z_L - Z_0) \qquad \text{(A-26)}$$

For a lossless transmission line, $\alpha = 0$, and $\gamma = j\beta$. Therefore, A and B become

$$A = \frac{e^{j\beta d}}{2} I_L (Z_L + Z_0) \qquad \text{(A-27)}$$

$$B = \frac{e^{-j\beta d}}{2} I_L (Z_L - Z_0) \qquad \text{(A-28)}$$

The ratio of the reflected to incident voltage for a lossy line is then

$$\Gamma_{in}(z) = \frac{B e^{\gamma z}}{A e^{-\gamma z}} = \frac{\dfrac{e^{-\gamma d}}{2} I_L (Z_L - Z_0) e^{\gamma z}}{\dfrac{e^{\gamma d}}{2} I_L (Z_L + Z_0) e^{-\gamma z}}$$

$$= \frac{Z_L - Z_0}{Z_L + Z_0} e^{-2\gamma d} e^{2\gamma z} \qquad \text{(A-29)}$$

At the load end $z = d$, and

$$\Gamma_{in}(z = d) = \frac{Z_L - Z_0}{Z_L + Z_0} = \Gamma_L = |\Gamma_L| e^{j\theta_L} \qquad \text{(A-30)}$$

At the source end $z = 0$, and

$$\Gamma_{in}(z=0) = \frac{Z_L - Z_0}{Z_L + Z_0}\, e^{-2\gamma d} = |\Gamma_L| e^{j\theta_L}\, e^{-2\gamma d} \tag{A-31}$$

For a lossless line

$$\Gamma_{in}(z=0) = \frac{Z_L - Z_0}{Z_L + Z_0}\, e^{-2j\beta d} = |\Gamma_L| e^{j\theta_L}\, e^{-2j\beta d} \tag{A-32}$$

The expressions for current and voltage on a lossy line after substituting for A and B then reduce to

$$V(z) = \frac{e^{\gamma d}}{2} I_L\, (Z_L + Z_0)\, e^{-\gamma z} + \frac{e^{-\gamma d}}{2} I_L\, (Z_L - Z_0)\, e^{\gamma z} \tag{A-33}$$

$$I(z) = \frac{e^{\gamma d}}{2} I_L\, (Z_L + Z_0)\frac{1}{Z_0}\, e^{-\gamma z} - \frac{e^{-\gamma d}}{2} I_L\, (Z_L - Z_0)\frac{1}{Z_0}\, e^{\gamma z} \tag{A-34}$$

After simplification and using a new variable $\acute{z} = d - z$, we can write

$$\begin{aligned} V(\acute{z}) &= \frac{I_L}{2}\left[(Z_L + Z_0)\, e^{\gamma \acute{z}} + (Z_L - Z_0)\, e^{-\gamma \acute{z}}\right] \\ &= \frac{I_L}{2}\, (Z_L + Z_0)\, e^{\gamma \acute{z}}\left[1 + |\Gamma_L| e^{j\theta_L}\, e^{-2\gamma \acute{z}}\right] \end{aligned} \tag{A-35}$$

$$I(\acute{z}) = \frac{I_L}{2Z_0}\, (Z_L + Z_0)\, e^{\gamma \acute{z}}\left[1 - |\Gamma_L| e^{j\theta_L}\, e^{-2\gamma \acute{z}}\right] \tag{A-36}$$

The summation of incident and reflected waves along the transmission line for current and voltage result in standing waves. For a lossless line, these waves are sinusoidal in nature along the spatial z axis with their maxima and minima separated by $\lambda/2$. The voltage in Eq. (A-35) is a maximum when $e^{-j(2\beta\,\acute{z}-\theta)} = 1$ and a minimum when $e^{-j(2\beta\,\acute{z}-\theta)} = -1$. The ratio of maximum to minimum voltage is defined as standing wave ratio (SWR) and is given by

$$SWR = \frac{|V(\acute{z})_{max}|}{|V(\acute{z})_{min}|} = \frac{1 + |\Gamma_L|}{1 - |\Gamma_L|} \tag{A-37}$$

When the line is lossy, standing waves undergo attenuation as they bounce back and forth between the voltage source end and the load end. Therefore, Equation (A-37) is applicable to a lossless line.

The use of new variable $\acute{z}$ is convenient in calculation of the input impedance to the transmission line at a distance $\acute{z} = d$.

The input impedance at any point $\acute{z}$ away from the load can be determined from the ratio $V(\acute{z})/I(\acute{z})$. Therefore, we can write

$$Z(\acute{z}) = \frac{V(\acute{z})}{I(\acute{z})} = \frac{\frac{I_L}{2}\left[(Z_L+Z_0)\,e^{\gamma\acute{z}}+(Z_L-Z_0)\,e^{-\gamma\acute{z}}\right]}{\frac{I_L}{2Z_0}\left[(Z_L+Z_0)\,e^{\gamma\acute{z}}-(Z_L-Z_0)\,e^{-\gamma\acute{z}}\right]} \tag{A-38}$$

Simplify and dividing both the numerator and denominator by $Z_L + Z_0$, we obtain

$$Z(\acute{z}) = Z_0\frac{\left[1+\frac{(Z_L-Z_0)}{(Z_L+Z_0)}e^{-2\gamma\acute{z}}\right]}{\left[1-\frac{(Z_L-Z_0)}{(Z_L+Z_0)}e^{-2\gamma\acute{z}}\right]} = Z_0\frac{1+\Gamma_L\,e^{-2\gamma\acute{z}}}{1-\Gamma_L\,e^{-2\gamma\acute{z}}} \tag{A-39}$$

where,

$$\Gamma_L = \frac{(Z_L - Z_0)}{(Z_L + Z_0)} = |\Gamma_L|\,e^{j\theta_L} \tag{A-40}$$

Γ_L is called the load reflection coefficient, which is a complex number. Note that at the load, $\acute{z} = 0$, and $Z(\acute{z} = 0)$ reduces to Z_L.

Expressions for voltage, current, and their ratio can also be defined in terms of $\sinh(\gamma\acute{z})$, $\cosh(\gamma\acute{z})$, and $\tanh(\gamma\acute{z})$, using $\cosh(\gamma\acute{z}) = (e^{\gamma\acute{z}} + e^{-\gamma\acute{z}})/2$, $\sinh(\gamma\acute{z}) = (e^{\gamma\acute{z}} - e^{-\gamma\acute{z}})/2$, and $\tanh(\gamma\acute{z}) = \sinh(\gamma\acute{z}))/\cosh(\gamma\acute{z})$.

As a result,

$$Z(\acute{z}) = \frac{V(\acute{z})}{I(\acute{z})} = \frac{I_L\left[Z_L\cosh(\gamma\acute{z}) + Z_0\sinh(\gamma\acute{z})\right]}{\frac{I_L}{Z_0}\left[Z_0\cosh(\gamma\acute{z}) + Z_L\sinh(\gamma\acute{z})\right]}$$
$$= Z_0\frac{Z_L + Z_0\tanh(\gamma\acute{z})}{Z_0 + Z_L\tanh(\gamma\acute{z})} \tag{A-41}$$

Therefore, the input impedance at a distance d away from the load is

$$Z_{in} = Z(\acute{z} = d) = Z_0\frac{Z_L + Z_0\tanh(\gamma d)}{Z_0 + Z_L\tanh(\gamma d)} \tag{A-42}$$

For a lossless transmission line ($\alpha = 0$), $\tanh(\gamma d) = \tanh(j\beta d) = j\tan(\beta d)$. As a result, we have the following

$$Z_{in} = Z(\acute{z} = d, \alpha = 0) = Z_0 \frac{Z_L + j Z_0 \, \tan(\beta d)}{Z_0 + j Z_L \tan(\beta d)} \tag{A-43}$$

Now that the voltage, current, and impedance equations along the transmission line have been determined, the average power delivered by the voltage source to the input of the line, $P_{\text{ave-in}}$, and the power delivered to the load, $P_{\text{ave-load}}$, can be calculated by first calculating V_{in} and I_{in}:

$$V_{in} = I_{in}Z_{in} \tag{A-44}$$

where,

$$I_{in} = \frac{V_s}{Z_s + Z_{in}} \tag{A-45}$$

$$P_{ave-in} = \frac{1}{2} Re \, [V_{in}I_{in}^*] \tag{A-46}$$

$$P_{ave-L} = \frac{1}{2} Re \, [V_L I_L^*] = \frac{1}{2} Re \, [Z_L I_L I_L^*] = \frac{1}{2} |I_L|^2 Re[Z_L]$$
$$= \frac{1}{2} |I_L|^2 Re[R_L + X_L] = \frac{1}{2} |I_L|^2 R_L \tag{A-47}$$

When the load impedance $Z_L = Z_0$, V_{in} can be determined by substituting $\acute{z} = d$ in voltage expression in Equation (A-35). As a result, we have

$$V_{in} = V(\acute{z} = d) = \frac{I_L}{2} \left[(Z_0 + Z_0) \, e^{\gamma 0} + (Z_0 - Z_0) \, e^{-\gamma \acute{z}} \right] = I_L Z_0 e^{\gamma d} \tag{A-48}$$

and

$$I_{in} = I(\acute{z} = d) = I_L e^{\gamma d} \tag{A-49}$$

Since for power calculation considerations it is more convenient to use the variable z instead of $\acute{z}$, voltage and current equations (with $Z_L = Z_0$) simplify to

$$V(z) = [I_L \, Z_0 \, e^{\gamma d}]e^{-\gamma z} \tag{A-50}$$

$$I(z) = [I_L \ e^{\gamma d}]e^{-\gamma z} \tag{A-51}$$

or

$$V(z) = V_{in}e^{-\gamma z} \tag{A-52}$$

$$I(z) = I_{in}e^{-\gamma z} \tag{A-53}$$

Since the transmission line is terminated in its own characteristics impedance, that is since $Z_L = Z_0$, there are no reflected waves. As a result, the wave is only moving in the +Z direction. As $\gamma = \alpha + j\beta$, the wave undergoes phase variation (it is a moving wave in the +z direction) while it also undergoes attenuation due to attenuation constant α. In summary, Equations (A-52) to (A-53) can be used in computing current, voltage, and power at any point in the +z direction.

A generalized equation for current and voltage at any point on the transmission line involving load reflection coefficient Γ_L and source reflection coefficient Γ_S (to be defined) can be constructed as follows.

First, by examining Figure A-2, we observe that

$$V_S - I_{in}Z_S = V_{in} \tag{A-54}$$

V_{in} and I_{in} can be determined from equations for $V(\acute{z})$ and $I(\acute{z})$ by setting $\acute{z} = d$ and substituting back in the above equation.

Therefore,

$$V(\acute{z} = d) = \frac{I_L}{2} \ (Z_L + Z_0) \ e^{\gamma d}[\, 1 + \Gamma_L \ e^{-2\gamma d}] \tag{A-55}$$

$$I(\acute{z}) = \frac{I_L}{2Z_0} \ (Z_L + Z_0) \ e^{\gamma d}[\, 1 - \Gamma_L \ e^{-2\gamma d}] \tag{A-56}$$

After substitution, we obtain

$$\frac{I_L}{2}(Z_L + Z_0)\, e^{\gamma d} = \frac{Z_0 V_S}{Z_S + Z_0}\, \frac{1}{1 - \Gamma_S\, \Gamma_L\, e^{-2\gamma d}} \tag{A-57}$$

where,

$$\Gamma_S = \frac{(Z_S - Z_0)}{(Z_S + Z_0)} = |\Gamma_S|\, e^{j\theta_S} \tag{A-58}$$

is the voltage reflection coefficient at the voltage source. Using the result obtained in Eq. (A-58) in (A-35) and (A-36), we arrive at

$$V(\acute{z}) = \frac{V_S Z_0}{Z_S + Z_0}\, e^{-\gamma z} \left(\frac{1 + \Gamma_L\, e^{-2\gamma \acute{z}}}{1 - \Gamma_S\, \Gamma_L\, e^{-2\gamma d}} \right) \tag{A-59}$$

$$I(\acute{z}) = \frac{V_S}{Z_S + Z_0}\, e^{-\gamma z} \left(\frac{1 - \Gamma_L\, e^{-2\gamma \acute{z}}}{1 - \Gamma_S\, \Gamma_L\, e^{-2\gamma d}} \right) \tag{A-60}$$

The above phasor equations not only can be used to determine voltage and current at any point on a transmission line segment, it can show the build-up of infinite reflection at load and source interfaces due to mismatch. All the forward and backward moving voltage waves, have $\frac{V_S Z_0}{Z_S + Z_0}$ in common. When the incident wave is initially launched by the voltage source, it has not sensed the mismatch at the load. Therefore, the incident wave "sees" the characteristic impedance of the line as if the line is infinitely long or matched. This quantity is analogous to the first incident wave A_1 moving toward the load.

How can we compute α and β? These values can be obtained by performing two separate measurements on identical finite transmission lines, namely input impedance of a short-circuited and an open-circuited transmission line. Using Equation (A-41), we can write

$$Z_{insc} = Z(Z_L = 0) = Z_0 \frac{0 + Z_0 \tanh(\gamma d)}{Z_0 + (0)\ \tanh(\gamma d)} = Z_0 \tanh(\gamma d) \tag{A-61}$$

$$Z_{inoc} = Z(Z_L = \infty) = Z_0 \coth(\gamma d) \tag{A-62}$$

Therefore, Z_0 can be determined by multiplying Z_{insc} by Z_{inoc}:

$$Z_0 = \sqrt{Z_{insc}Z_{inoc}} \tag{A-63}$$

On the other hand, γ can be calculated from the ratio Z_{insc}/Z_{inoc}:

$$\gamma = \frac{1}{d} tanh^{-1}\left[\sqrt{\frac{Z_{insc}}{Z_{inoc}}}\right] \tag{A-64}$$

Finally, since a practical transmission line characteristic γ, contains both real and imaginary parts, one should be able to define a quality factor Q for the line. To do this, input impedance of a short-circuited finite transmission line can be determined and the Q factor can be obtained from its 3-dB or half-power frequency bandwidth. Q factor determination using the fractional bandwidth concept for a half-wavelength line is illustrated in Figure A-3. Similar procedure has been illustrated for a quarter-wavelength line in second edition of *Field and Waves Electromagnetics* by D.K. Cheng.

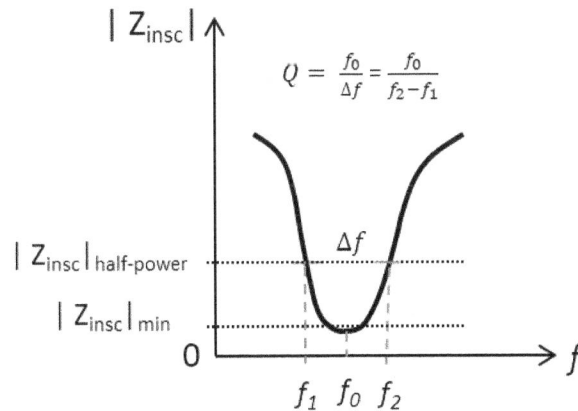

Figure A-3 Q factor calculation method of a short-circuited half-wavelength line

The input impedance to a short-circuited lossless transmission line, $Z_{insc} = j\,Z_0\,tan(\beta d)$. If the length of the line is a multiple of half-wavelength at the center frequency of operation, the input impedance is zero since $\beta d = (2\pi/\lambda)(\lambda/2) = \pi$.

When, the line is lossy (i.e., $\alpha \neq 0$), the input impedance is small but not equal to zero.

This is analogous to the series RLC circuit connected to a sinusoidal voltage source having peak amplitude V. At resonance, the input impedance Z to the circuit is purely real and equal to R. The current delivered to the load is then maximum and equal to I_{max} = V/R. The power delivered to R is therefore equal to (1/2) $(I_{max})^2$ R = $V^2/(2R)$. It can be shown that at half-power frequencies, the magnitude of reactive part of the input impedance is equal to its real part (i.e., |X| = R). As a result, at half-power frequencies, |Z| = $\sqrt{2}$ R and the current is equal to $I_{max}/\sqrt{2}$.

Using the above analogy, an expression for Z_{insc} (of the short-circuited half-wavelength line) at the center frequency can be obtained. Z_{in} is a minimum at the center frequency, and its value at f_1 and f_2 should be 3 dB higher (i.e., 20 log($\sqrt{2}$)). The Q factor can then be determined by examining their ratio. The details of this analysis are given below.

$$Z_{insc} = Z(\acute{z} = d, Z_L = 0)$$
$$= Z_0 \tanh(\gamma d) = Z_0 \frac{\sinh(\alpha + j\beta)}{\cosh(\alpha + j\beta)}$$
$$= Z_0 \frac{\sinh(\alpha d)\cos(\beta d) + j\cosh(\alpha d)\sin(\beta d)}{\cosh(\alpha d)\cos(\beta d) + j\sinh(\alpha d)\sin(\beta d)} \qquad \text{(A-65)}$$

Let d = n(λ/2), therefore βd = nπ and sin(βd) =0 at the center frequency. At a frequency $f = f_0 + \delta f$, we can write

$$\beta d = \frac{\omega}{v_p}d = \frac{2\pi (f_0 + \delta f)}{v_p}d = \frac{2\pi f_0 d}{v_p} + \frac{2\pi \delta f d}{v_p}$$
$$= \frac{2\pi f_0 \left(\frac{n\lambda}{2}\right)}{\lambda f_0} + \frac{2\pi \delta f \left(\frac{n\lambda}{2}\right)}{\lambda f_0}$$
$$= n\pi + n\pi \left(\frac{\delta f}{f_0}\right) \qquad \text{(A-66)}$$

$$\sin(\beta d) = \sin\left(n\pi + n\pi\left(\frac{\delta f}{f_0}\right)\right) \begin{cases} = -n\pi\left(\frac{\delta f}{f_0}\right) & for\ n\ odd \\ = n\pi\left(\frac{\delta f}{f_0}\right) & for\ n\ even \end{cases} \quad \text{(A-67)}$$

$$\cos(\beta d) = \cos\left(n\pi + n\pi\left(\frac{\delta f}{f_0}\right)\right) = \begin{cases} -1\ for\ n\ odd \\ 1\ for\ n\ even \end{cases} \quad \text{(A-68)}$$

Consider the case where n = odd and assume that $\alpha d \ll 1$. We have

$$\sin(\beta d) = -n\pi\left(\frac{\delta f}{f_0}\right)$$
$$\cos(\beta d) = -1$$
$$\tanh(\alpha d) \approx \alpha d$$

The expression for Z_{insc} then becomes

$$Z_{insc} = Z_0\frac{-\alpha d - jn\pi\ (\delta f/f_0)}{-1 - j(\alpha d)n\pi\ (\delta f/f_0)} \approx Z_0\big(\alpha d + jn\pi\ (\delta f/f_0)\big) \quad \text{(A-69)}$$

since $\alpha d \ll 1$.

At $f = f_0$, $\delta f = 0$, we expect Z_{insc} to be a minimum.

$$(Z_{insc})_{f_0} = Z_0(\alpha d) \quad \text{(A-70)}$$

At the frequency $\delta f = \Delta f/2$, Z_{insc} for 3-dB frequencies can be determined. We now make the observation that

$$(Z_{insc})_{f_0} = 0.707\ (Z_{insc})_{f_{1,2}} \quad \text{(A-71)}$$

$$\frac{\left|(Z_{insc})_{f_0}\right|^2}{\left|(Z_{insc})_{f_{1,f2}}\right|^2} = \frac{|Z_0(\alpha d)|^2}{\left|Z_0\big(\alpha d + jn\pi\ (\Delta f/2f_0)\big)\right|^2}$$

$$= \frac{(Z_0\alpha d)^2}{(Z_0\alpha d)^2 + (n\pi\Delta f/2f_0)^2} = \frac{1}{2} \quad \text{(A-72)}$$

After simplification and substituting $(\beta d) = n\pi$, we obtain

$$\frac{\beta}{2\alpha}\frac{\Delta f}{f_0} = \frac{\beta}{2\alpha}\frac{1}{Q} = 1 \tag{A-73}$$

or

$$Q = \frac{f_0}{\Delta f} = \frac{\beta}{2\alpha} \tag{A-74}$$

Appendix B

Two-Port Network Parameters

Two-port networks are widely used in communication systems and in RF and microwave circuits including filters, impedance matching circuits and power distribution networks, to name a few. Furthermore, different two-port networks require different network parameters. Also, the knowledge about these networks and their corresponding parameters is very useful in circuits and system analysis without knowledge of the details of a given network or networks. The various configuration and arrangement of networks also demands for use of certain parameters over others. We therefore provide a brief description of some of the most widely used networks and parameters.

We start our introduction by referring to figure B-1, which depicts a linear two-port network identified by currents I_1, I_2, and voltages V_1 and V_2 at its input and output port. We then define several important two-port parameters.

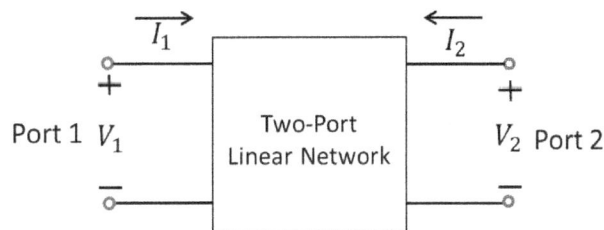

Figure B-1. A linear two-port network

Impedance parameters or *open-circuit parameters* relate the terminal voltages to terminal currents as

$$V_1 = Z_{11}I_1 + Z_{12}I_2 \tag{B-1a}$$

$$V_2 = Z_{21}I_1 + Z_{22}I_2 \tag{B-1b}$$

In matrix form, we can write

$$\begin{bmatrix} V_1 \\ V_2 \end{bmatrix} = \begin{bmatrix} Z_{11} & Z_{12} \\ Z_{21} & Z_{22} \end{bmatrix} \begin{bmatrix} I_1 \\ I_2 \end{bmatrix} \tag{B-2}$$

or

$$[V] = [Z][I] \tag{B-3}$$

Impedance parameters can be determined by open-circuiting port 1 or port 2, one port at a time:

$$Z_{11} = \frac{V_1}{I_1}\Big|_{I_2=0}, \qquad Z_{12} = \frac{V_1}{I_2}\Big|_{I_1=0}$$

$$Z_{21} = \frac{V_2}{I_1}\Big|_{I_2=0}, \qquad Z_{22} = \frac{V_2}{I_2}\Big|_{I_1=0} \tag{B-4}$$

Impedance parameters are useful when analyzing networks used in series.

Admittance parameters or *short-circuit parameters* are used to express terminal currents in terms of terminal voltages as

$$I_1 = Y_{11}V_1 + Y_{12}V_2 \tag{B-5a}$$

$$I_2 = Y_{21}V_1 + Y_{22}V_2 \tag{B-5b}$$

In matrix form, we can write

$$\begin{bmatrix} I_1 \\ I_2 \end{bmatrix} = \begin{bmatrix} Y_{11} & Y_{12} \\ Y_{21} & Y_{22} \end{bmatrix} \begin{bmatrix} V_1 \\ V_2 \end{bmatrix} \tag{B-6}$$

or

$$[I] = [Y][V] \tag{B-7}$$

Admittance parameters can be determined by short-circuiting port 1 or port 2, one port at a time:

$$Y_{11} = \frac{I_1}{V_1}\Big|_{V_2=0}, \qquad Y_{12} = \frac{I_1}{V_2}\Big|_{V_1=0}$$

$$Y_{21} = \frac{I_2}{V_1}\Big|_{V_2=0}, \qquad Y_{22} = \frac{I_2}{V_2}\Big|_{V_1=0} \tag{B-8}$$

Admittance parameters are useful when analyzing networks used in parallel.

Hybrid parameters or *h parameters* are widely used in defining certain linear transistor parameters tabulated by manufacturers, and defined as

$$V_1 = h_{11}I_1 + h_{12}V_2$$ (B-9a)

$$I_2 = h_{21}I_1 + h_{22}V_2$$ (B-9b)

In matrix form, we can write

$$\begin{bmatrix} V_1 \\ I_2 \end{bmatrix} = \begin{bmatrix} h_{11} & h_{12} \\ h_{21} & h_{22} \end{bmatrix} \begin{bmatrix} I_1 \\ V_2 \end{bmatrix}$$ (B-10)

Hybrid parameters can be determined by short-circuiting port 2 or by open-circuiting port 1:

$$h_{11} = \left.\frac{V_1}{I_1}\right|_{V_2=0}, \qquad h_{12} = \left.\frac{V_1}{V_2}\right|_{I_1=0}$$

$$h_{21} = \left.\frac{I_2}{I_1}\right|_{V_2=0}, \qquad h_{22} = \left.\frac{I_2}{V_2}\right|_{I_1=0}$$ (B-11)

Example B-1: Compute h21 for the 10-dB attenuator circuit below designed in a 50-Ω system. Verify your result in LTspice.

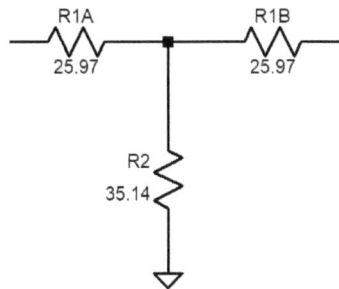

Figure B-2. The 10-dB attenuator network realized in 50-Ω system

Solution: To compute h21, port 2 is short-circuited. To simplify the analysis, it is convenient to apply an independent current source at port 1.

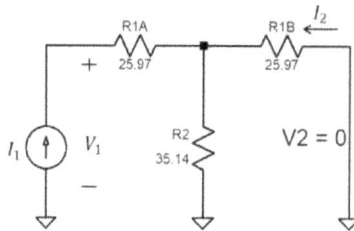

Figure B-3. Schematic diagram for computing h21 in Example B.1

$-I_2 = I_1 (35.14/(25.97 + 35.14)) = I_1 (0.575)$

Therefore, $h_{21} = \frac{I_2}{I_1}\Big|_{V_2=0} = -0.575.$

To simulate h-parameters, we construct the schematic diagram of Figure B-4. We then simulate and plot all h-parameters. Note that the circuit can be simulated at any arbitrary single frequency since the circuit is frequency independent.

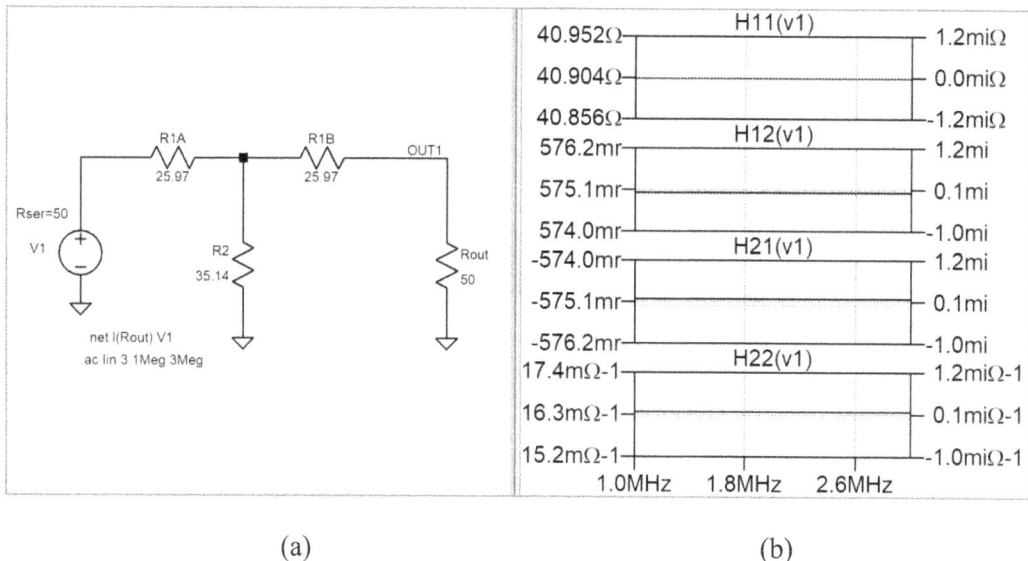

(a)   (b)

Figure B-4. Schematic diagram for simulating hybrid parameters and its value, (a) Schematic diagram and (b) Simulated h parameters showing h_{11} = 40.90 Ω, h_{12} = 0.575, h_{21} = -0.575, and h_{22} = 0.0164 Ω^{-1}

Inverse Hybrid parameters or *g parameters* are closely related to h parameters and defined as

$$I_1 = g_{11}V_1 + g_{12}I_2 \tag{B-12a}$$

$$V_2 = g_{21}V_1 + g_{22}I_2 \tag{B-12b}$$

In matrix form, we can write

$$\begin{bmatrix} I_1 \\ V_2 \end{bmatrix} = \begin{bmatrix} g_{11} & g_{12} \\ g_{21} & g_{22} \end{bmatrix} \begin{bmatrix} V_1 \\ I_2 \end{bmatrix} \tag{B-13}$$

Transmission parameters or *ABCD* parameters relate the input variables at the input to the variables at the output, and they are defined as

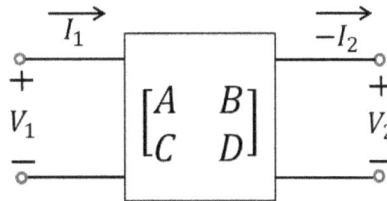

Figure B-5. A linear two-port network representing ABCD parameters

$$V_1 = AV_2 - BI_2 \tag{B-14a}$$

$$I_1 = CV_2 - DI_2 \tag{B-14b}$$

In matrix form, we can write

$$\begin{bmatrix} V_1 \\ I_1 \end{bmatrix} = \begin{bmatrix} A & B \\ C & D \end{bmatrix} \begin{bmatrix} V_2 \\ -I_2 \end{bmatrix} \tag{B-15}$$

or

$$\begin{bmatrix} V_1 \\ I_1 \end{bmatrix} = [T] \begin{bmatrix} V_2 \\ -I_2 \end{bmatrix} \tag{B-16}$$

ABCD parameters can be determined by open-circuiting or short-circuiting port 2.

$$A = \frac{V_1}{V_2}\bigg|_{I_2=0}, \qquad B = -\frac{V_1}{I_2}\bigg|_{V_2=0}$$

$$C = \frac{I_1}{V_2}\bigg|_{I_2=0}, \qquad D = -\frac{I_1}{I_2}\bigg|_{V_2=0} \tag{B-17}$$

For a reciprocal network $Z_{12} = Z_{21}$, therefore below is true:

$$AD - BC = 1 \qquad\qquad (B\text{-}18)$$

ABCD parameters are extremely important in analysis of cascaded networks, as will be discussed later.

Example B-2: Determine the ABCD parameters for the attenuator circuit below. Is this network reciprocal? Compute its values and verify your result in LTspice.

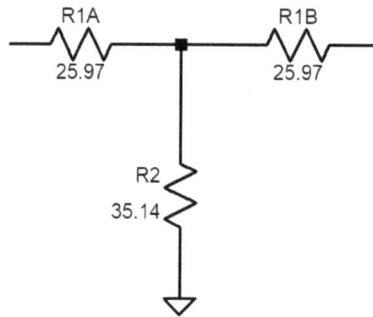

Fig. B-6. The 10-dB attenuator network realized in 50-Ω system

Solution: To determine A, we open circuit port 2 and apply a voltage source V_1 at port 1.

Therefore, $V_2 = [V_1/(R1A + R2)]\, R2$ resulting in $\mathbf{A} = (V_1/V_2) = 1 + (R1A/R2)$.

To determine C, with port 2 open-circuited, we apply a current source at port 1. Therefore, $V_2 = (I_1)\, R2$ resulting in $\mathbf{C} = I_1/V_2 = 1/R2$.

To determine B, port 2 is short-circuited and voltage source V_1 can be applied at port 1. Therefore, we can write

$$-I_2 = I_1\left(\frac{R2}{R2 + R1B}\right)$$

$$I_1 = \frac{V_1}{R1A + R2||R1B} = \frac{V_1}{R1A + \frac{(R2)(R1B)}{R2 + R1B}}$$

$$-I_2 = \left[\frac{V_1}{R1A + \frac{(R2)(R1B)}{R2 + R1B}}\right]\left(\frac{R2}{R2 + R1B}\right)$$

$$B = -\frac{V_1}{I_2} = R1A + R1B + \frac{(R1A)(R1B)}{R2}$$

To determine D, we short circuit port 2 and apply a current source at port 1. We can write

$$-I_2 = I_1\left(\frac{R2}{R2 + R1B}\right)$$

Therefore,

$$D = \frac{I_1}{-I_2} = \frac{R2 + R1B}{R2} = 1 + \frac{R1B}{R2}$$

Using the values for R1A, R1B, and R2 we can compute the ABCD parameters as

A = 1.739, B= 71.133, C = 0.02846, D = 1.739

Since AD − BC = (1.739)(1.739) − (71.133)(0.02846) = 1, the network is reciprocal.

Example B-3. Determine the ABCD parameters of a transmission line having length d, characteristic impedance Z0, and propagation constant β, as shown in Figure B-7 below.

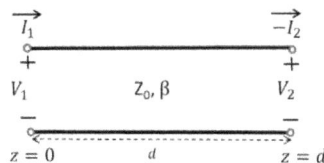

Figure B-7. Phasor voltage and current at input and output of a lossless line

Solution: We first write general equations for voltage and current on the transmission line in terms of incident and reflected voltage waves. We then use boundary conditions for voltage and current to find the relation between incident and reflected waves.

$$V(z) = A\,e^{-j\beta z} + B\,e^{j\beta z}$$

$$I(z) = \frac{A}{Z_0}\,e^{-j\beta z} - \frac{B}{Z_0}\,e^{j\beta z}$$

To determine **A** and **C**, port 2 is first open-circuited since $I_2{=}0$, resulting in

$$I(z = d) = \frac{A}{Z_0}\,e^{-j\beta d} - \frac{B}{Z_0}\,e^{j\beta d} = 0$$

Therefore, $B = Ae^{-2j\beta d}$ and

$$V(z) = A\,e^{-j\beta z} + A\,e^{-2j\beta d}\,e^{j\beta z}$$

$$I(z) = \frac{A}{Z_0}\,e^{-j\beta z} - \frac{A}{Z_0}\,e^{-2j\beta d}e^{j\beta z}$$

From which we can write

$$A = \frac{V_1}{V_2}\bigg|_{I_2=0} = \frac{V(z=0)}{V(z=d)} = \frac{A\,e^{-j\beta 0} + A\,e^{-2j\beta d}\,e^{j\beta 0}}{A\,e^{-j\beta d} + A\,e^{-2j\beta d}\,e^{j\beta d}} = \frac{1 + e^{-2j\beta d}}{2e^{-j\beta d}} = cos\beta d$$

$$C = \frac{I_1}{V_2}\bigg|_{I_2=0} = \frac{I(z=0)}{V(z=d)} = \frac{\dfrac{A}{Z_0} - \dfrac{A}{Z_0}\,e^{-2j\beta d}}{A\,e^{-j\beta d} + A\,e^{-2j\beta d}\,e^{j\beta d}} = \frac{\dfrac{1}{Z_0}\left(1 - e^{-2j\beta d}\right)}{2e^{-j\beta d}} = j\frac{1}{Z_0}sin\beta d$$

To determine **B** and **D**, port 2 is first short-circuited since $V_2{=}0$, resulting in

$$V(z = d) = A\,e^{-j\beta d} + B\,e^{j\beta d} = 0$$

Therefore, $B = -Ae^{-2j\beta d}$ and

$$V(z) = A\,e^{-j\beta z} - A\,e^{-2j\beta d}\,e^{j\beta z}$$

$$I(z) = \frac{A}{Z_0}\,e^{-j\beta z} + \frac{A}{Z_0}\,e^{-2j\beta d}e^{j\beta z}$$

Now, we can write

$$B = -\frac{V_1}{I_2}\Big|_{V_2=0} = \frac{V(z=0)}{I(z=d)} = \frac{A\,(1 - e^{-2j\beta d})}{\dfrac{A}{Z_0}e^{-j\beta d} + \dfrac{A}{Z_0}e^{-2j\beta d}e^{j\beta d}} = jZ_0 sin\beta d$$

$$D = \frac{I_1}{I_2}\Big|_{V_2=0} = \frac{I(z=0)}{I(z=d)} = \frac{\dfrac{A}{Z_0}e^{-j\beta 0} + \dfrac{A}{Z_0}e^{-2j\beta d}e^{j\beta 0}}{\dfrac{A}{Z_0}e^{-j\beta d} + \dfrac{A}{Z_0}e^{-2j\beta d}e^{j\beta d}} = \frac{\dfrac{A}{Z_0}(1 + e^{-2j\beta d})}{2\dfrac{A}{Z_0}e^{-j\beta d}} = cos\beta d$$

Note that AD – BC = $(cos\beta d)^2 - (jZ_0 sin\beta d)\left(j\frac{1}{Z_0}sin\beta d\right) = 1$. This is true because the network is reciprocal, that is $Z_{12} = Z_{21}$.

Example B-4. Determine the ABCD parameters for the transformer shown below.

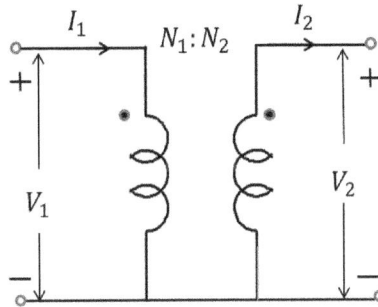

Figure B-8. An ideal transformer with primary and secondary windings having N_1 and N_2 turns respectively

Solution: Let $(V_2/V_1) = (N_2/N_1)$ and $(I_2/I_1) = (N_1/N_2)$. With port two open-circuited we can write

A = $(V_1/V_2) = [(V_2 N_1/N_2)/V_2] = N_1/N_2$

C = $(I_1/V_2) = (0/V_2) = 0$ since $I_2 = 0$.

With port two short-circuited we can write

B = $(V_1/I_2) = (0/I_2) = 0$ since $V_2 = 0$.

D = $(I_1/I_2) = (N_2/N_1)$.

ABCD parameters are extremely important in many RF and microwave applications where networks are being cascaded. In that case, the overall ABCD parameters can be computed by matrix multiplication of individual matrices. This is not true of scattering parameters. To compute the S matrix of the overall cascaded network, T (or ABCD) matrix of the overall matrix needs to computed first and then be converted to S matrix.

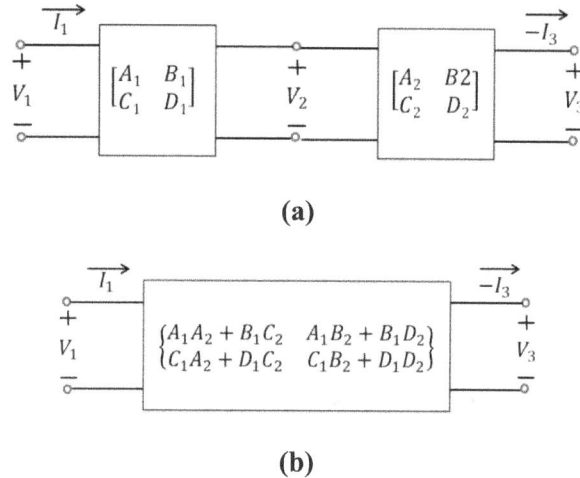

(a)

(b)

Figure B-8. ABCD matrix of two-port networks, (a) Cascaded two-port networks, and (b) Its overall ABCD or T matrix

The ABCD parameters for a two-port network can be easily expressed in terms of other parameters such as Z or Y parameters. For Z parameters given in Equations (B-1a) and (B-1b), we can perform the following to arrive at ABCD parameters in terms of Z parameters:

Adding Equations (B-1a) and (B-1b) together and substituting for I_1 using (B-1b) on one hand and also rearranging (B-1b) on the other hand result in

$$V_1 = (Z_{11}/Z_{21})V_2 - [(Z_{11}Z_{22} - Z_{21}Z_{12})/Z_{21}]I_2 \qquad \text{(B-19a)}$$

$$I_1 = (1/Z_{21})V_2 - (Z_{22}/Z_{21})I_2 \qquad \text{(B-19b)}$$

or

$$\begin{bmatrix} V_1 \\ I_1 \end{bmatrix} = \begin{bmatrix} (Z_{11}/Z_{21}) & [(Z_{11}Z_{22} - Z_{21}Z_{12})/Z_{21}] \\ 1/Z_{21} & Z_{22}/Z_{21} \end{bmatrix} \begin{bmatrix} V_2 \\ -I_2 \end{bmatrix} \qquad \text{(B-20)}$$

The ability to convert two-port parameters from one form to another useful form is very convenient. Some two-port networks lack certain parameters. For example, a transformer network lack Z parameters since voltage and current cannot be expressed in terms of one another. On the other hand, determination of certain network parameters and then conversion to other forms may be more advantages for some networks. As a result, a table containing conversion between various parameters is developed and tabulated in some text books. Here we show how Z parameters can be used to obtain the corresponding Y and S parameters.

To obtain Y parameters we start with Equation (B-2)

$$\begin{bmatrix} V_1 \\ V_2 \end{bmatrix} = \begin{bmatrix} Z_{11} & Z_{12} \\ Z_{21} & Z_{22} \end{bmatrix} \begin{bmatrix} I_1 \\ I_2 \end{bmatrix} = [Z] \begin{bmatrix} I_1 \\ I_2 \end{bmatrix}$$

or

$$[Z]^{-1} \begin{bmatrix} V_1 \\ V_2 \end{bmatrix} = [Z]^{-1}[Z] \begin{bmatrix} I_1 \\ I_2 \end{bmatrix}$$

$$\begin{bmatrix} I_1 \\ I_2 \end{bmatrix} = [Z]^{-1} \begin{bmatrix} V_1 \\ V_2 \end{bmatrix} = [Y] \begin{bmatrix} V_1 \\ V_2 \end{bmatrix}$$

It can be shown that

$$\begin{bmatrix} Y_{11} & Y_{12} \\ Y_{21} & Y_{22} \end{bmatrix} = \begin{vmatrix} \dfrac{Z_{22}}{\Delta_Z} & -\dfrac{Z_{12}}{\Delta_Z} \\ -\dfrac{Z_{21}}{\Delta_Z} & \dfrac{Z_{11}}{\Delta_Z} \end{vmatrix} \qquad (B-21)$$

where

$$\Delta_Z = Z_{11}Z_{22} - Z_{12}Z_{21}$$

To obtain S parameters we start with Equation (B-3) and expand in terms of incident and reflected current and voltage waves:

$$[V] = [V^+] + [V^-] = [Z][I] = [Z]\{[I^+] - [I^-]\} \qquad (B-22)$$

$$[Z][I^-] + [V^-] = [Z][I^+] - [V^+]$$

$$[Z][I^-] + [Z_0][I^-] = [Z][I^+] - [Z_0][I^+]$$

$$\{[Z] + [Z_0]\}[I^-] = \{[Z] - [Z_0]\}[I^+]$$

Therefore,

$$[S] = \frac{I^-}{I^+} = \{[Z] + [Z_0]\}^{-1}\{[Z] - [Z_0]\} \qquad \text{(B-23)}$$

where

$$[Z_0] = \begin{bmatrix} Z_0 & 0 \\ 0 & Z_0 \end{bmatrix}$$

is the two-dimensional diagonal matrix having identical values in both ports. Equation (B-17) can be solved for [Z] in terms of [S] resulting in

$$[Z] = [Z_0]\{[U] + [S]\}\{[U] - [S]\}^{-1} \qquad \text{(B-24)}$$

where

$$[U] = \begin{bmatrix} 1 & 0 \\ 0 & 1 \end{bmatrix}$$

is the two-dimensional unit matrix.

The scattering parameters were discussed in detail in Chapter 3.

Appendix C

AT-41486 Data Sheet

AT-41486
Up to 6 GHz Low Noise Silicon Bipolar Transistor

Avago
TECHNOLOGIES

Data Sheet

Description

Avago's AT-41486 is a general purpose NPN bipolar transistor that offers excellent high frequency performance. The AT-41486 is housed in a low cost surface mount .085" diameter plastic package. The 4 micron emitter-to-emitter pitch enables this transistor to be used in many different functions. The 14 emitter finger interdigitated geometry yields an intermediate sized transistor with impedances that are easy to match for low noise and moderate power applications. Applications include use in wireless systems as an LNA, gain stage, buffer, oscillator, and mixer. An optimum noise match near $50\,\Omega$ at 900 MHz, makes this device easy to use as a low noise amplifier.

The AT-41486 bipolar transistor is fabricated using Avago's 10 GHz f_T Self-Aligned-Transistor (SAT) process. The die is nitride passivated for surface protection. Excellent device uniformity, performance and reliability are produced by the use of ion-implantation, self-alignment techniques, and gold metalization in the fabrication of this device.

Features

- Low Noise Figure:
  1.4 dB Typical at 1.0 GHz
  1.7 dB Typical at 2.0 GHz

- High Associated Gain:
  18.0 dB Typical at 1.0 GHz
  13.0 dB Typical at 2.0 GHz

- High Gain-Bandwidth Product: 8.0 GHz Typical f_T

- Surface Mount Plastic Package

- Tape-and-Reel Packaging Option Available

- Lead-free Option Available

86 Plastic Package

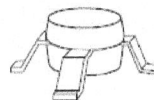

Pin Connections

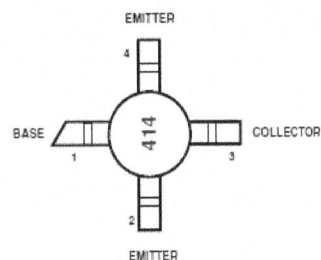

2

AT-41486 Absolute Maximum Ratings

Symbol	Parameter	Units	Absolute Maximum[1]
V_{EBO}	Emitter-Base Voltage	V	1.5
V_{CBO}	Collector-Base Voltage	V	20
V_{CEO}	Collector-Emitter Voltage	V	12
I_C	Collector Current	mA	60
P_T	Power Dissipation [2,3]	mW	500
T_j	Junction Temperature	°C	150
T_{STG}	Storage Temperature	°C	-65 to 150

Thermal Resistance[2,4]:
θ_{jc} = 165°C/W

Notes:
1. Permanent damage may occur if any of these limits are exceeded.
2. T_{CASE} = 25°C.
3. Derate at 6 mW/°C for T_C > 68°C.
4. See MEASUREMENTS section "Thermal Resistance" for more information.

Ordering Information

Part Numbers	No. of Devices	Comments
AT-41486-BLK	100	Bulk
AT-41486-BLKG	100	Bulk
AT-41486-TR1	1000	7" Reel
AT-41486-TR1G	1000	7" Reel
AT-41486-TR2	4000	13" Reel
AT-41486-TR2G	4000	13" Reel

Note: Order part number with a "G" suffix if lead-free option is desired.

Electrical Specifications, T_A = 25°C

Symbol	Parameters and Test Conditions		Units	Min.	Typ.	Max.
$|S_{21E}|^2$	Insertion Power Gain; V_{CE} = 8 V, I_C = 25 mA	f = 1.0 GHz f = 2.0 GHz	dB		17.5 11.5	
$P_{1\,dB}$	Power Output @ 1 dB Gain Compression V_{CE} = 8 V, I_C = 25 mA	f = 2.0 GHz	dBm		18.0	
$G_{1\,dB}$	1 dB Compressed Gain; V_{CE} = 8 V, I_C = 25 mA	f = 2.0 GHz	dB		13.5	
NF_O	Optimum Noise Figure: V_{CE} = 8 V, I_C = 10 mA	f = 1.0 GHz f = 2.0 GHz f = 4.0 GHz	dB		1.4 1.7 3.0	1.8
G_A	Gain @ NF_O; V_{CE} = 8 V, I_C = 10 mA	f = 1.0 GHz f = 2.0 GHz f = 4.0 GHz	dB	17.0	18.0 13.0 9.0	
f_T	Gain Bandwidth Product: V_{CE} = 8 V, I_C = 25 mA		GHz		8.0	
h_{FE}	Forward Current Transfer Ratio; V_{CE} = 8 V, I_C = 10 mA		—	30	150	270
I_{CBO}	Collector Cutoff Current; V_{CB} = 8 V		μA			0.2
I_{EBO}	Emitter Cutoff Current; V_{EB} = 1 V		μA			1.0
C_{CB}	Collector Base Capacitance[1]; V_{CB} = 8 V, f = 1 MHz		pF		0.25	

Note:
1. For this test, the emitter is grounded.

3

AT-41486 Typical Performance, $T_A = 25°C$

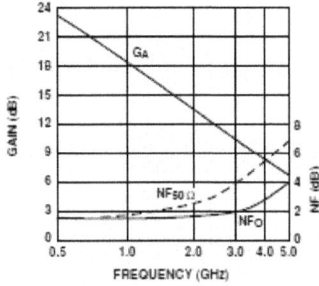

Figure 1. Noise Figure and Associated Gain vs. Frequency. $V_{CE} = 8$ V, $I_C = 10$ mA.

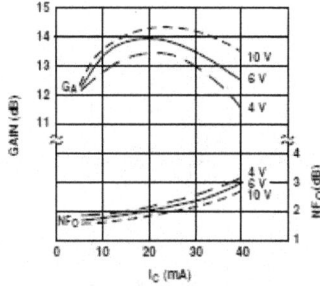

Figure 2. Optimum Noise Figure and Associated Gain vs. Collector Current and Collector Voltage. f = 2.0 GHz.

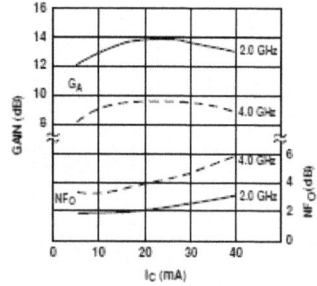

Figure 3. Optimum Noise Figure and Associated Gain vs. Collector Current and Frequency. $V_{CE} = 8$ V.

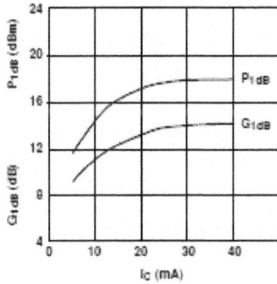

Figure 4. Output Power and 1 dB Compressed Gain vs. Collector Current and Frequency. $V_{CE} = 8$ V, f = 2.0 GHz.

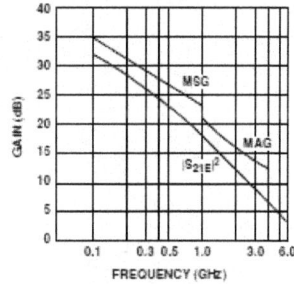

Figure 5. Insertion Power Gain, Maximum Available Gain and Maximum Stable Gain vs. Frequency. $V_{CE} = 8$ V, $I_C = 25$ mA.

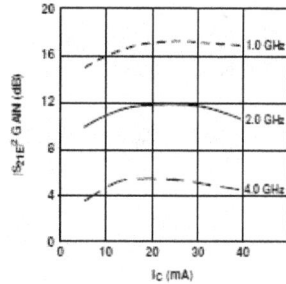

Figure 6. Insertion Power Gain vs. Collector Current and Frequency. $V_{CE} = 8$ V.

4

AT-41486 Typical Scattering Parameters, Common Emitter,

$Z_O = 50\ \Omega$, $T_A = 25°C$, $V_{CE} = 8\ V$, $I_C = 10\ mA$

| Freq. | S_{11} | | S_{21} | | | S_{12} | | | S_{22} | |
GHz	Mag.	Ang.	dB	Mag.	Ang.	dB	Mag.	Ang.	Mag.	Ang.
0.1	.74	-38	28.1	25.46	157	-39.6	.011	68	.94	-12
0.5	.59	-127	22.0	12.63	107	-30.2	.031	47	.60	-29
1.0	.56	-168	16.8	6.92	84	-27.7	.041	46	.49	-29
1.5	.57	169	13.5	4.72	69	-26.2	.049	49	.45	-32
2.0	.62	152	11.1	3.61	56	-24.8	.058	43	.42	-39
2.5	.63	142	9.3	2.91	47	-23.4	.068	52	.40	-42
3.0	.64	130	7.6	2.41	37	-22.2	.078	52	.39	-50
3.5	.68	122	6.3	2.06	26	-20.6	.093	51	.37	-60
4.0	.71	113	5.1	1.80	16	-19.5	.106	48	.35	-70
4.5	.74	105	4.0	1.59	7	-18.0	.125	48	.35	-84
5.0	.77	99	3.1	1.42	-4	-17.2	.139	43	.35	-98
5.5	.79	93	2.0	1.27	-13	-16.3	.153	38	.35	-114
6.0	.81	87	1.1	1.13	-22	-15.4	.170	34	.35	-131

AT-41486 Typical Scattering Parameters,

Common Emitter, $Z_O = 50\ \Omega$, $T_A = 25°C$, $V_{CE} = 8\ V$, $I_C = 25\ mA$

| Freq. | S_{11} | | S_{21} | | | S_{12} | | | S_{22} | |
GHz	Mag.	Ang.	dB	Mag.	Ang.	dB	Mag.	Ang.	Mag.	Ang.
0.1	.50	-75	32.0	40.01	142	-41.3	.009	54	.85	-17
0.5	.55	-158	23.2	14.38	97	-34.1	.020	48	.51	-24
1.0	.57	177	17.5	7.50	78	-29.9	.032	61	.46	-24
1.5	.57	161	14.1	5.07	65	-27.3	.043	62	.44	-28
2.0	.59	148	11.5	3.75	53	-24.8	.058	59	.43	-35
2.5	.61	139	9.6	3.02	45	-22.9	.072	58	.40	-41
3.0	.65	128	8.0	2.52	34	-21.6	.083	57	.38	-49
3.5	.70	121	6.7	2.17	24	-20.1	.099	56	.36	-59
4.0	.74	113	5.7	1.92	14	-18.8	.115	52	.34	-72
4.5	.78	107	4.7	1.72	3	-17.6	.132	47	.32	-87
5.0	.78	102	3.7	1.53	-8	-16.6	.149	42	.31	-106
5.5	.78	96	2.7	1.36	-19	-15.4	.169	36	.31	-125
6.0	.76	91	1.6	1.21	-29	-14.5	.188	31	.33	-144

A model for this device is available in the DEVICE MODELS section.

AT-41486 Noise Parameters: $V_{CE} = 8\ V$, $I_C = 10\ mA$

| Freq. | NF_O | Γ_{opt} | | $R_N/50$ |
GHz	dB	Mag	Ang	
0.1	1.3	.12	3	0.17
0.5	1.3	.10	16	0.17
1.0	1.4	.04	43	0.16
2.0	1.7	.12	-145	0.16
4.0	3.0	.44	-99	0.40

86 Plastic Package Dimensions

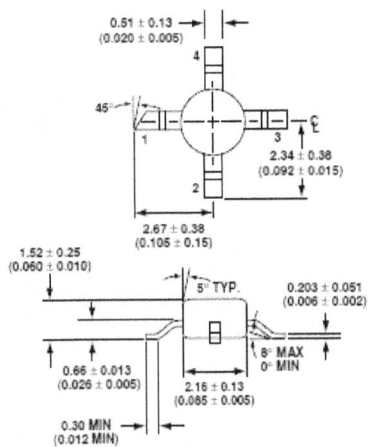

DIMENSIONS ARE IN MILLIMETERS (INCHES)

For product information and a complete list of distributors, please go to our web site:
www.avagotech.com

Avago, Avago Technologies, and the A logo are trademarks of Avago Technologies, Pte.
in the United States and other countries.
Data subject to change. Copyright © 2006 Avago Technologies Pte. All rights reserved.
Obsoletes 5968-2031EN
5989-2648EN August 22, 2006

Avago
TECHNOLOGIES

Index

About the Author

Manou Ghanevati received the Ph.D. degree from Drexel University and the B.S. degree from University of California at Irvine, both in electrical engineering. He has well over twenty years of professional experience as a RF engineer and an educator in academia. Dr. Ghanevati is a RF Microwave design engineer and has worked in several industries including Telecommunications, Semiconductor, and Aerospace / Satellite industries. His interests include design of low power RF front-end electronics, active antennas, RFIC design in CMOS technologies, ultra-wideband RF and millimeter wave circuits, linearized low noise amplifiers and power amplifiers, group III-V MMIC circuit, RF Photonic transceivers, and generation of highly stable RF / Photonic signal sources. Dr. Ghanevati is a senior member of IEEE and an occasional reviewer for various IEEE publications.

www.ingramcontent.com/pod-product-compliance
Lightning Source LLC
Chambersburg PA
CBHW082130210326
41599CB00031B/5931

* 9 7 8 0 5 7 8 5 7 5 3 0 8 *